01689

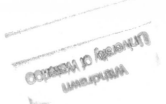

Optical Properties of
Highly Transparent Solids

OPTICAL PHYSICS AND ENGINEERING

Series Editor: **William L. Wolfe**
Optical Sciences Center, University of Arizona, Tucson, Arizona

M. A. Bramson
Infrared Radiation: A Handbook for Applications

Sol Nudelman and S. S. Mitra, Editors
Optical Properties of Solids

S. S. Mitra and Sol Nudelman, Editors
Far-Infrared Properties of Solids

Lucien M. Biberman and Sol Nudelman, Editors
Photoelectronic Imaging Devices
  Volume 1: Physical Processes and Methods of Analysis
  Volume 2: Devices and Their Evaluation

A. M. Ratner
Spectral, Spatial, and Temporal Properties of Lasers

Lucien M. Biberman, Editor
Perception of Displayed Information

W. B. Allan
Fibre Optics: Theory and Practice

Albert Rose
Vision: Human and Electronic

J. M. Lloyd
Thermal Imaging Systems

Winston E. Kock
Engineering Applications of Lasers and Holography

Shashanka S. Mitra and Bernard Bendow, Editors
Optical Properties of Highly Transparent Solids

A Continuation Order Plan is available for this series. A continuation order will bring delivery of each new volume immediately upon publication. Volumes are billed only upon actual shipment. For further information please contact the publisher.

Optical Properties of Highly Transparent Solids

Edited by

Shashanka S. Mitra

University of Rhode Island
Kingston, Rhode Island

and

Bernard Bendow

Air Force Cambridge Research Laboratories
Bedford, Massachusetts

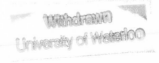
PLENUM PRESS · NEW YORK AND LONDON

Library of Congress Cataloging in Publication Data

International Conference on Optical Properties of Highly Transparent Solids, Waterville Valley, N. H., 1975.
Optical properties of highly transparent solids.

(Optical physics and engineering)
Proceedings of the conference held at Waterville Valley, N. H., Feb. 3-5, 1975, sponsored by Division of Engineering Research and Development, University of Rhode Island.
Includes bibliographical references and index.
1. Transparent solids–Optical properties–Congresses. I. Mitra, Shashanka Shekhar, 1933-   II. Bendow, Bernard, 1942-   III. Rhode Island University. Division of Engineering Research and Development. IV. Title.
QC176.8.06I53   1975   530.4'1   75-22006
ISBN 0-306-30861-4

Proceedings of the International Conference on
Optical Properties of Highly Transparent Solids
held at Waterville Valley, New Hampshire, February 3-5, 1975

© 1975 Plenum Press, New York
A Division of Plenum Publishing Corporation
227 West 17th Street, New York, N.Y. 10011

United Kingdom edition published by Plenum Press, London
A Division of Plenum Publishing Company, Ltd.
Davis House (4th Floor), 8 Scrubs Lane, Harlesden, London, NW10 6SE, England

Printed in the United States of America

ORGANIZING COMMITTEE

S. S. Mitra *(University of Rhode Island)*, Conference Chairman
E. Burstein *(University of Pennsylvania)*
M. Cardona *(Max Planck Institute)*
M. Hass *(U.S. Naval Research Laboratory)*
D. W. Pohl *(IBM, Zurich)*
S. H. Wemple *(Bell Labs, Murray Hill)*
P. D. Gianino *(AF Cambridge Research Laboratories)*, Conference Secretary

PROGRAM COMMITTEE

B. Bendow *(AF Cambridge Research Laboratories)*
J. L. Birman *(City University, New York)*
A. A. Maradudin *(University of California, Irvine)*
J. Tauc *(Brown University)*

Advisory: C. A. Pitha, AFCRL (Local Arrangements)
            A. Hordvik, AFCRL (Program)

Sponsored by: Division of Engineering Research and Development
                University of Rhode Island

Preface

Although much work has been performed on measurements and interpretation of light absorption by opaque or nearly opaque solids, it is surprising to note that until recently relatively little reliable experimental data, and much less theoretical work was available on the nature of transparent solids. This, in spite of the fact that a vast majority of engineering and device applications of a solid depend on its optical transparency. Needless to say, all solids are both transparent and opaque depending on the spectral region of consideration. The absorption processes that limit the transparency of a solid are either due to lattice vibrations, as in ionic or partially ionic solids, or due to electronic transitions, both intrinsic and impurity-induced. For most materials, a sufficiently wide spectral window exists between these two limits, where the material is transparent. In general, the absorption coefficient, in the long wavelength side of, but sufficiently away from, the fundamental absorption edge, is relatively structureless and has an exponential dependence on frequency. Recent evidence suggests that in the short wavelength side of the one-phonon region, but beyond two- or three-phonon singularities, the absorption coefficient of both polar and nonpolar solids is also relatively structureless and depends exponentially on frequency. The main contributions to the residual absorption in the transparent regime are (i) multiphonon processes, (ii) various defects and impurities, (iii) phonon-assisted electronic transitions in the long-wavelength tail of the fundamental absorption edge and (iv) multiphoton electronic transitions in the case of high photon flux. The principal purpose of this conference was to discuss fundamental optical processes and properties of solids in this regime. The emphasis was on highly transparent solids (insulators and semiconductors) suitable as substrates and/or coatings for applications such as laser windows, fiber optics and integrated optics. Bulk, surface and thin film properties of both crystalline and amorphous materials were considered.

A topical conference of this nature was thought to be best organized if the number of contributed papers and the number of participants were limited. As a result the program was planned without any overlapping sessions with ample time for discussions after each presentation. In all, forty-four papers were accepted of which seven were twenty-five minute invited talks and the rest were twelve or fifteen minutes each contributed papers. The papers were divided into six sessions (topics) each starting with an invited review talk. We present below the highlights of the papers presented in each of the sessions.

Multiphonon Absorption

Due to the advent of high power gas lasers in the near- and mid-infrared regions, the problem of suitable windows which are highly transparent in this region has attracted considerable attention in the last few years. Accurate measurements of the low absorption coefficient in highly transparent solids in the infrared have been undertaken, and several theories have been attempted to explain the residual intrinsic absorption in this region, in an otherwise pure crystal. T. C. McGill (Cal. Tech.) summarized recent theoretical results on multiphonon absorption in insulators. He discussed the mechanisms leading to multiphonon absorption, anharmonic potential and non-linearities in the dipole moment. The large number of proposed theoretical models were classified with an attempt to point out their similarities and differences. A. Nedoluha (Naval Elec. Lab.) calculated the multiphonon absorption in KCℓ to test the effect of form of the potential used on the agreement between calculated and observed results. For this purpose he used the perturbation theory with vertex corrections. Of the potentials examined, viz., inverse power, exponential, exponential with pre-exponential power factor and the Morse potential, the first one was found to give the best agreement. B. Bendow (AFCRL) and S. P. Yukon (Brandeis) demonstrated that the dielectric response in covalent solids may be formulated in terms of a cumulant series involving lattice displacement correlators, which avoids the usual perturbative expansions of the moment and anharmonic potential in powers of displacements and thus accounts for various classes of phonon processes to infinite order.

Experimental results and their interpertations according to one or the other existing theories were presented. D. W. Pohl (IBM,Zurich) in an invited talk

presented results on NaF and their interpretation via
the model of Namjoshi and Mitra based on joint phonon
density of states, and mechanical anharmonicity estimated from a Born-Mayer potential. The analysis of the
experimental data in terms of phonon occupation numbers
allows a decomposition of the absorption wing into N-
phonon spectra $(N = 2....5)$ with considerable structure.
The latter was interpreted in terms of phonon combination
of critical point phonons of high density. A. J. Barker
and G. R. Wilkinson (King's Coll.,London) and N. E. Massa
and S. S. Mitra (URI) showed from an examination of absorption spectra in the transparent regime of a large
number of simple crystals that the onset frequency of
transparency of pure crystals may be defined as $3\nu_{LO}$.
They also discussed the temperature dependence of the absorption coefficient of NaCℓ, KBr and LiF upto their melting points in the pre-transparent regime $(10>\alpha>0.1$ cm$^{-1})$.
Higher order multiphonon spectra of a number of alkali
halides were interpreted by Boyer, et.al.(Naval Res. Labs.).
Rowę and Harrington (U.Alabama) obtained low temperature
measurements of absorption in NaCℓ and KCℓ, and interpreted their findings in light of existing theories.
Multiphonon measurement on alkaline earth fluorides were
reported by Lipson, et.al. (AFCRL) and Chen, et.al. (Cal.
Tech). Koteles and Datars (McMaster Univ.) reported on
two-phonon absorption spectra of III-V semiconductors.

Electronic Processes

In an invited paper, J. D. Dow (Univ. of Illinois)
discussed Urbach's rule, which states that the fundamental optical (electronic) absorption edge of a variety of
nonconducting materials exhibits an exponential frequency
dependence. For semiconductors the Urbach rule is attributed to ionization of the exciton by internal electric
microfields. The extent to which the field-ionization
mechanism is responsible for all Urbach edges was also
discussed.

Magnetic circular dichroism of the Urbach edge in
KI, CdTe and TℓCℓ was reported by R. T. Williams and S.
E. Schnatterly (Princeton). Their conclusion is that
characterization of the processes giving rise to Urbach
edges as either fully "electrostatic" or fully "elastic"
is not supported as a generally applicable rule. M. J.
Frankel and J. L. Birman (CCNY) reported on a theory of
new transients and optical phenomena in bounded dispersive media. K. Vedam and E. D. Schmidt (Penn State)

reported on the variation of the refractive index of alkali halides with pressure (to 14 kbars). C. W. Litton
and D. C. Reynolds (ARL) and M. M. Kreitman and S. P.
Faile (U.Dayton) described progress in improving optical
transmission in α - HgS.

Impurity Effects

A. A. Maradudin (U.C.,Irvine) surveyed the effects
of impurities on infrared lattice vibration absorption
in transparent crystals. Absorption associated with localized gap and resonance modes due to substitutional
impurities were defined and illustrations given. Defect
activated absorption in the frequency range allowed the
phonons of the pure host crystal, and absorption by mixed
crystals were also covered. D. A. Abramsohn, et al.
(Purdue U.) presented several techniques, viz., Raman
scattering, photoconductivity and acoustoelectric probes,
to test and characterize GaAs for residual deep trap densities and weak optical absorption at 1.06 μm. Identification of Fe^{4+} and Fe^{5+} charge transfer photochromic
absorption bands in $SrTiO_3$ was reported on by O. F.
Schirmer, et al. (IBM, Zurich). W. L. Faust, et al.(NRL)
described investigations of defects in KCℓ by picosecond
spectroscopy. Using two-phonon absorption of the Nd:YAG
fourth harmonic to produce electron-hole pairs in KCℓ,
they monitored the time evolution of F-band absorption
by means of the (delayed) second harmonic pulse.

Glasses

The limiting loss mechanisms in highly transparent
glasses for applications such as fiber optics were reviewed by J. Tauc (Brown U.). Three papers, two experimental [D. Treacy (USNA) and P. C. Taylor (NRL) and R.E.
Howard, et al.(Catholic U.)] and one theoretical, discussed multiphonon absorption in amorphous solids. A well-
defined set of multiphonon peaks were observed to dominate the spectra of chalcogenide glasses, and were interpreted, in general, as overtones and combinations of the
fundamental peaks. Theoretical work by Tsay et al.(AFCRL)
on amorphous semiconducting films predicted substantially
enhanced multiphonon absorption and suppressed temperature
dependence of absorption in certain of these solids relative to their crystalline counterparts. Other papers
described light scattering experiments from composition
fluctuations in phase separating oxide glasses; two-photon
absorption in glass wave guides detected by monitoring

the absorption of a weak CW probe during and after a
strong probe at a different wavelength, which enables
sensitive detection of low concentrations of impurities;
and optical enhancement of photoluminescence in calcho-
genide glasses which provide evidence of broad bands of
localized levels in the "forbidden" gap.

Multiphoton Processes and Nonlinear Effects

In an invited paper N. Bloembergen, et al. (Harvard
U), reviewed three-wave light mixing techniques, which
permit the study of strongly dispersive features in the
ultraviolet and infrared regions, while all observations
are restricted to light waves in the transparent non-dis-
persive region of the material. They obtained a complete
and accurate determination of all elements of the third
order nonlinear susceptability tensor in cubic and isotropic materials. J. A. Weiss (NRL) reported on the
measurements of the nonlinear susceptibility for second
harmonic generation of several zinc blende type crystals
in the mid-infrared region. S.S.Mitra, et al.(Redstone
Arsenal)presented calculations of multiphoton ionization
probability as functions of wavelength and field intensity
for a number of semiconductors and insulators using three
different models, viz., those of Keldysh, Braunstein and
Basov after appropriate corrections. The calculated two-
photon absorption coefficients showed orders of magnitude
differences among the three models with Braunstein and
Keldysh underestimating, while Basov overestimating. Inspite of the large amount of disagreement among available
experimental data, Keldysh predictions seem to be best.
Another surprising feature of the Keldysh results however,
the latter predict both the absolute value and the proper
exponential for the frequency dependence of the one-photon
absorption coefficient of direct gap semiconductors. S.
D. Kramer and N. Bloembergen (Harvard U.) reported on the
measurement of the intrinsic two-photon absorption parameter in cuprous chloride, using a three-wave light mixing technique discussed by Bloembergen earlier. N. Tzoar
and J. I. Gersten (CCNY) discussed self-phase modulation
in transparent media, which results in broadening of output spectra. Effects of electronic absorption lines on
the broadened spectrum was also considered. W. L. Smith
et al. (Harvard U.) reported a study of picosecond laser-
induced damage in transparent media. Single amplified
picosecond pulses from a mode-locked Nd:YAG laser system
were used to measure the dielectric breakdown threshold
field and the nonlinear refractive index of several optical

materials, including NaCℓ, NaF, LiF and fused quartz.
The intrinsic damage field and the nonlinear refractive
index were determined with an accuracy of twenty percent.

Measurement Techniques

Small absorption measurement techniques were summarized by L. H. Skolnik (AFCRL). The techniques were
categorized as either thermal (or calorimetric), and
direct loss. Thermocouple laser calorimetry sensitivity
is being pushed to its limit; acoustic calorimetry approaches appear more promising for measuring losses in
the 10^{-5} - 10^{-6} cm^{-1} range. Hass et al. (NRL) presented
a novel calorimetric technique for separating bulk and
surface absorption by using long rodlike geometries.
Nurmikko et al. (MIT) presented a new optical calorimetric approach for measuring losses in semiconductors by
monitoring shifts in electronic edges with temperature,
while Bennett and Forman (NBS) provided calculations on
a steady state photoacoustic approach to measuring small
losses.

A surface acoustic wave method for profiling surface
losses was described by Parks et al. (USC). Index matching calorimetric techniques for measuring bulk losses
in fibers were described by White and Midwinter (British
P.O.).

Two direct loss techniques were presented. Merz
et al. (Bell Labs.) discussed the use of right angle Raman scattering for measuring direct attenuation as a
function of length along a dielectric waveguide which
averts requirements for multiple length samples. Krause
(Schott) described an accurate multiple pass spectroscopic
method which almost eliminates the index dependence of
the transmission in highly transparent solids.

Three papers dealing with measurement of refractive
properties were presented. Allen et al. (Hughes) talked
about a versatile ellipsometer for measuring dielectric
properties of thin films. Usable at 10.6 μm, the instrument incorporates a unique CO_2 waveguide laser and ZnSe
polarizers and modulators. Stress optic measurements on
various infrared transparent materials obtained by applying Michels method in the IR as well as in the visible
were presented by Friedman and Pitha (AFCRL), while Feldman et al. (NBS) reported photoelastic constants of CVD
ZnSe obtained by modified Twyman-Green interferometry.

We thank Dr. L. Skolnik for valuable discussions, and Mrs. Joan Lamoureux of URI and C. Guarente of AFCRL for their competent and conscientious secretarial assistance. Our thanks are also due to A. Corey and J. Goltman for their assistance in preparing the index.

S. S. Mitra

B. Bendow

Contents

SECTION VI
MEASUREMENT TECHNIQUES

Section I
Multiphonon Absorption

THEORY OF MULTIPHONON ABSORPTION: A REVIEW[+]

T. C. McGill[++]

California Institute of Technology

Pasadena, California   91125

ABSTRACT

The theories of multiphonon absorption in ionic insulators for frequencies several times the reststrahl frequency are reviewed. The physical properties of the solid, anharmonic potential and non-linear dipole moment, which are responsible for multiphonon absorption are discussed. The rather large number of theoretical calculations are classified with an attempt to highlight simularities and basic differences between the theoretical techniques used and the physics contained in or implied by the calculations. Finally, a result connecting the analyticity of the potential and dipole moment operator, and the behavior of $\varepsilon_2(\omega)$ at high frequencies is presented.

I.   INTRODUCTION

Recent compilation of experimental values of the absorption of insulators for frequencies immediately above the reststrahl frequency[1] has led to the realization that the absorption has an almost universal frequency dependence. The absorption β in a wide class of materials which includes alkali halides, alkaline earth

[+]Research supported in part by the Air Force Office of Scientific Research under Grant No. (73-2490).

[++]Alfred P. Sloan Foundation Fellow.

fluorides, oxides, and semiconductors is found to have a frequency
dependence of the form

$$\beta(\omega) \simeq Ae^{-\gamma\omega} \qquad\qquad , \quad (I-1)$$

where A and γ are material dependent parameters. Experimentally,
Eq. (I-1) is valid for frequencies between two to three times the
transverse optical phonon frequency and the highest frequencies at
which bulk absorption can be measured, five to six times the
transverse optical phonon frequency.

The temperature dependence of β at the above stated frequencies
has been studied in a number of the materials. Generally the absorption varies as some power of the temperature at high temperatures
and then levels off becoming temperature independent at low
temperatures.[2,3]

In this paper I will review what is known theoretically about
the absorption. Since most of the work so far has been aimed at
the more ionic crystals such as the alkali halides, and alkaline
earth fluorides, this paper will deal primarily with the results
which are appropriate to these materials.

Section II contains some general remarks. In Section III a
simple theory is presented. Section IV contains a more elaborate
discussion based on the so called molecular model using a linear
dipole. Section V presents the results of a simple calculation
of the absorption which includes at least approximately the effect
of a nonlinear dipole moment. Section VI gives the results of a
moment argument which rules out certain transitions of $\beta(\omega)$ for
very high frequencies. Finally, Section VII contains a brief
summary of the results and some concluding remarks.

II.   GENERAL REMARKS

The absorption due to multiphonon processes can in general be
derived from the equation[4]

$$\beta(\omega) = \frac{4\pi^2\omega}{cn\Omega Z}\left(1 - e^{-\hbar\omega/k_B T}\right)$$

$$\sum_{\substack{\text{Final States} \\ \text{Initial States}}} e^{-E_i/k_B T} \left|\langle \text{Final State}|M|\text{Initial State}\rangle\right|^2$$

$$\delta(\hbar\omega + E_i - E_f) \qquad\qquad , \quad (II-1)$$

where n is the index of refraction; Ω is the volume of the solid; Z is the partition function; and E_i and E_f are the initial and final state energies, respectively. M is the dipole moment operator for the solid. This expression clearly indicates the physical properties of the solid which are required to understand the absorption: the dipole moment as a function of the position of all the atoms in the solid, the potential as a function of all atoms in the solid, and the quantum mechanical initial and final eigenstates and the associated eigenenergies. These quantities are almost impossible to obtain for a realistic solid. The standard and highly useful harmonic approximation is not valid in the frequency range that we are interested in. Hence, we are forced to examine various simple models.

III. SIMPLE THEORY

The simplest theory one might think of is to expand the potential and dipole moment operator in a Taylor's series in the deviation of the atoms from their equilibrium positions. Schematically, this would produce expressions like

$$V(Q) = \sum_{n=2}^{\infty} \frac{V_n}{n!} Q^n \qquad , \quad (III-1)$$

and

$$M(Q) = \sum_{n=1}^{\infty} \frac{M_n}{n!} Q^n \qquad , \quad (III-2)$$

where Q is the dimensionless deviation of the atom from its equilibrium position produced by dividing the deviation by the amplitude of the zero point motion. Further, we might allow that each succeeding term in these series would be smaller by some characteristic number λ.[5] That is, we have that

$$\frac{V_{n+1}}{V_n} \cong \lambda \qquad , \text{ for each n} \qquad , \quad (III-3)$$

and

$$\frac{M_{n+1}}{M_n} \cong \lambda \qquad , \text{ for each n} \qquad . \quad (III-4)$$

In this case, we should expect that the matrix element to go from some initial state to a final state with n additional phonons (the process shown in Fig. 1) to go as λ^{n-1}. That is

$$\beta(\omega) \sim \sum_{n=1}^{\infty} \lambda^{2n-2} \sum_{\vec{Q}_1 P_1} \cdots \sum_{\vec{Q}_n P_n} \delta(\omega - \omega_{\vec{Q}_1 P_1} \cdots - \omega_{\vec{Q}_n P_n}) \quad , \quad \text{(III-5)}$$

where the subscripts $\vec{Q}$ and P indicate the wavevector and polarization, respectively, of the phonons in the final state.

Sham and Sparks[6] have shown how Eq. (III-5) leads to an expression for β which is of the form

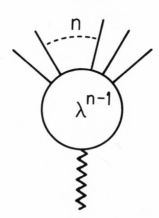

Fig. 1.   Schematic diagram illustrating a photon (〜) being
          absorbed by a solid with the creation of n phonons (—).

$$\beta(\omega) = Ae^{-\gamma\omega} \qquad\qquad\qquad (III-6)$$

However, for our purposes here, we can simply note that the n-phonon absorption β_n goes as

$$\beta_n \sim \lambda^{2n-2} \qquad\qquad , \quad (III-7)$$

which can be written

$$\beta_n \sim e^{\ln\lambda^2(n-1)} \qquad\qquad . \quad (III-8)$$

If we replace n by the ratio of the frequency to a single phonon frequency ω_0, then

$$\beta(\omega) \sim e^{\ln\lambda^2(\omega/\omega_0)} \qquad\qquad ; \quad (III-9)$$

which is of the desired exponential form.

While this theory is in agreement with experiment it does suffer from a number of serious difficulties. Some of these difficulties are: What is the value of λ? Does the matrix element go simply as λ^{n-1} or does it have some other dependence on n? What changes do dispersion, selection rules, more realistic potentials and dipole moment operators make?

IV. MOLECULAR MODELS

To attempt to answer some of the questions posed at the end of Section II, a number of authors have examined so called molecular models.[7-14] As illustrated in Fig. 2, these models consist of approximating the solid by a series of disconnected diatomic molecules. The interatomic potential is assumed to take on a number of different forms. Three of the most common ones, the inverse power law, Born-Mayer, and Morse potentials are given in Fig. 2. Further, the molecular dipole moment is assumed to be either linear[7-12] or nonlinear.[7,13,14]

A number of different calculational techniques have been applied to this problem including finite order perturbation theory,[7,8,9] infinite order perturbation theory,[12,14] and exact solutions for the Morse potential:[7,9-11,13] Once one has decided on the use of a molecular model, there are at least two ways of estimating β for a real solid from the model. One approach is to just calculate the absorption for the molecular model.[7,9,13] That is, one takes the matrix elements and energy eigenvalues for the molecules in the quantum case; or takes the transfer of energy from the electromagnetic

field to the diatomic molecule in the classical case. This approximation can be improved on somewhat by using the correct final and
initial state energies for the solid.[10-12,14] In this approximation,
one uses the molecular models to calculate the matrix element and
then replaces

$$\sum_{\text{Final States}} \delta(\hbar\omega + E_i - E_f)$$

by the n-phonon density of states $\rho_n(\omega)$. This n-phonon density of
states can be obtained from the expression

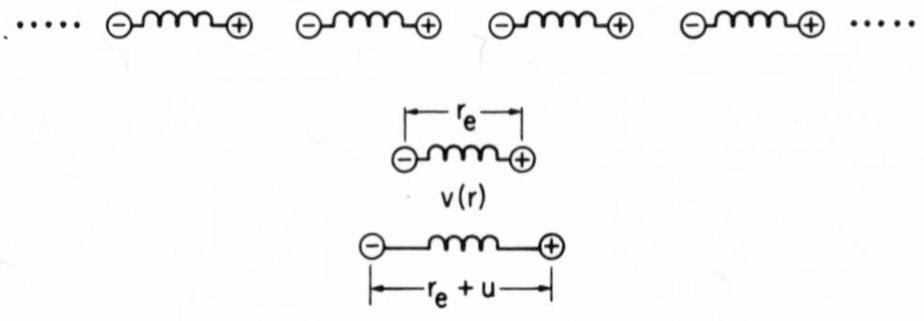

Potentials Used To Model Anharmonic Potential

$$v(r) \sim \begin{cases} A/r^n & \text{(Power Law)} \\ Ae^{-r/\rho} & \text{(Born Mayer)} \\ A(1-e^{-b(r-r_e)})^2 & \text{(Morse)} \end{cases}$$

Dipole Moment

$$M(u) \sim \begin{cases} e^* u \\ e^* u + M_{\text{nonlinear}}(u) \end{cases}$$

Fig. 2. Molecular model of a solid.

$$\rho_n(\omega) = \int \rho_{n-1}(\omega-\omega') \, \rho_1(\omega')d\omega' \qquad\qquad (IV\text{-}1)$$

The results of a calculation[11] for KCl using the latter procedure for a Morse potential and linear dipole moment operator is
shown in Fig. 3. The agreement between theory and experiment is
quite good.

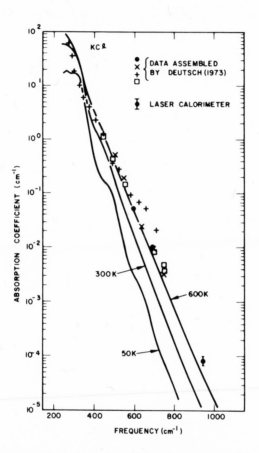

Fig. 3.   The absorption β versus the frequency for KCl (after Ref.
          11). The solid line gives the results of theoretical
          calculations performed using the Morse potential. See
          Ref. 11 for the source of the experimental data.

The model is also subject to a number of criticisms. It
neglects the variation of the matrix element with the polarization
and wavevector of the phonons. The assumed forms of the molecular
potential are not particularly good for diatomic molecules.[15]
Should we expect them to give an accurate estimate of the

anharmonic potential in a real solid? Sparks and Sham[16] suggest that
these simple models overestimate the contribution of various
processes in the absorption.

To explore Sparks and Sham's point further, we consider the
so called vertex correction. The vertex correction can be viewed as
the factor which we must multiply the simple summation process, see
Fig. 4 the diagrams at the left hand side, to take account of all
other processes of the type shown in the diagrams at the right hand
side of Fig. 4.

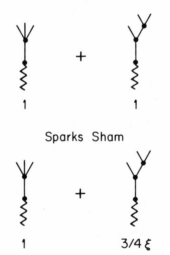

Fig. 4. The vertex correction due to the lowest order processes in
         perturbation theory which contribute to three phonon
         absorption. The numbers under the upper set of diagrams
         give the relative contributions for the two processes shown
         for a molecular model with a Born-Mayer potential as in
         Ref. 7. The expressions under the lower diagrams gives
         the results obtained by Sparks and Sham in Ref. 16. They
         find that ξ should be about 0.18 for NaCl.

The molecular model based on the Born-Mayer potential gives that the two processes in Fig. 4 contribute equally to the third order process. Sparks and Sham[16] content that for a more realistic model, the process on the right is down by a factor of 3/4 ξ from that on the left where ξ is about 0.18 for NaCl. While this correction is rather unimportant for a three phonon process the difference between the two results becomes larger and larger as the order of the process is increased. This result is shown in Fig. 5. Note that for a seven phonon process the difference between the Born-Mayer and that used by Sparks and Sham is some three orders of magnitude.

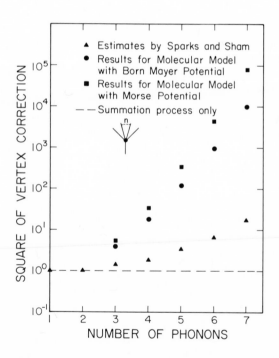

Fig. 5.   The square of the vertex correction versus the number of phonons absorbed. For one and two phonon processes all the theories give the same result since there is only one process in lowest order. In each case the contribution of the summation process shown in the inset is taken to be one.

Sparks and Sham[15] have used their procedure to calculate the β for NaCl. They assume a linear dipole moment using their approximation to the matrix operator and an approximate multiphonon density of states. Their results are shown in Fig. 6. Inspite of the change of emphasis on the importance of various processes in these calculations, they also obtain satisfactory agreement between theory and experiment. The rather good agreement with two types of theories which place very different weights on the mulitphonon processes suggest that we cannot use the experimental values of β to choose between the two approaches.

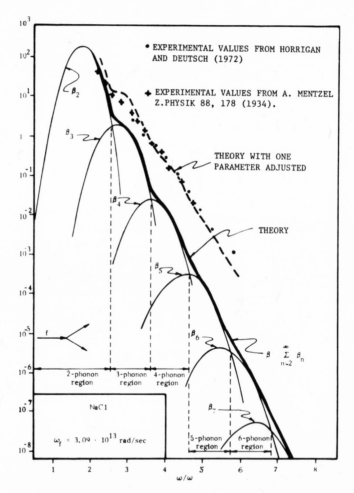

Fig. 6. The theoretical and experimental values of the absorption versus frequency for NaCl (after Ref. 16). See Ref. 16 for the sources of the experimental data and the exact procedure used to calculate β.

The temperature dependence of the absorption has also been treated by a number of authors.[8],[17-20] Reasonable agreement is found between theory and experiment using temperature dependent parameters in the potentials or by attempting to make an accurate calculation of the temperature dependence based solely on the molecular model.

V.   NONLINEAR DIPOLE MOMENT

Up until this point we have only discussed theories which have used a linear dipole moment operator. Recently a number of authors[7],[13],[14],[21],[22] have addressed themselves to the role of non-linear moment in the absorption. All of these calculations are in agreement on the points I want to illustrate here. Hence, we take a simple, but convenient, model consisting of a group of uncoupled diatomic molecules with an interatomic potential which is a Morse potential,

$$v(u) = A\left[1 - e^{-b(u/r_o)^2}\right] \qquad , \text{(V-1)}$$

where A and b are the parameters in the potential and

$$r_o = \left(\frac{\hbar}{\mu\omega_o}\right)^{1/2} \qquad , \text{(V-2)}$$

with ω_o the frequency of the oscillator and μ the reduced mass. Further we take the dipole moment to be of the form

$$M(u) = e^*u\left[1 + \frac{Sb}{2}\left(\frac{u}{r_o}\right)\right] \qquad , \text{(V-3)}$$

where e^* is the effective charge and S is a dimensionless parameter which gives the relative strength of the nonlinear term in the dipole moment. Estimates by Hellwarth and Mangir[22] for LiF give

$$S_{LiF} \approx 1.05 \qquad , \text{(V-4)}$$

other estimates for alkali halides give[13],[14],[21]

$$S \sim 1.0 \qquad . \text{(V-5)}$$

Calculations based on this simple model leads to a number of interesting results. First, if the sign S is positive, then the processes arising from the nonlinear term in the absorption destructively interfere with terms which are due to the linear moment as illustrated in Fig. 7. Second, for the values of S as large as those given in Eqs. (IV-4) and (IV-5), the terms in absorption due to the nonlinear moment dominate the terms due to the linear moment

for processes involving more than one phonon. These points are
shown graphically in Fig. 8 where we have plotted the ratio of the

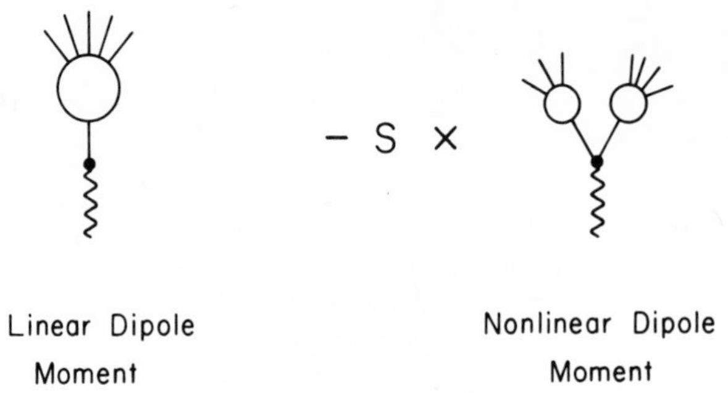

Linear Dipole
Moment

$- S \times$

Nonlinear Dipole
Moment

Fig. 7. Schematic illustration of the processes contributing to
the absorption. The diagram on the left occurs with a
linear dipole moment, that on the right occurs with a
quadratic term in the moment. The minus S is to illustrate the relative sign and magnitude of the two processes
as a function of S.

squared matrix element for finite S to that for S equal to zero.
The destructive interference shows up markedly when the ratio
goes to zero at two phonon processes for S equal to one. This graph
shows that the nonlinear moment can make very substantial corrections
to the absorption calculated using linear moment only.

While the addition of a nonlinear dipole moment can make
large modifications in β at frequencies where there is almost
perfect destructive interference, the variation of β with

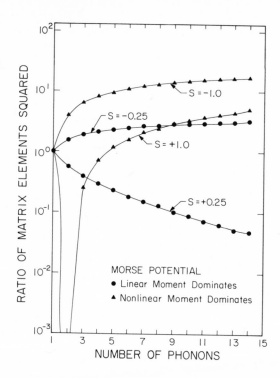

Fig. 8. The ratio of the square of the matrix element with
         S nonzero (nonlinear moment) to that with S zero
         (linear moment only) as a function of the number of
         phonons absorbed.

frequency away from this range is found to be almost exponential
by a number of authors.[7,13,20] Here again we find theories with
rather different internal physics can produce results in agreement
with the experimentally observed variation of β.

     Preliminary results on ZnSe[23] suggest that the nonlinear term
in the dipole moment operator is even more important in semiconductors.

VI.   EXACT RESULTS USING MOMENTS

Given the situation described above where there are a large number of theories each with very different internal physics and yet all giving that

$$\beta(\omega) \sim Ae^{-\gamma\omega} \qquad\qquad , \quad \text{(VI-1)}$$

one is led to wonder if there is not some underlying argument for an exponential which is independent of the details of the theory.

To examine this possibility we are led to ask what do we know about the high frequency behavior of $\varepsilon_2(\omega)$.   One way of investigating this high frequency behavior is to investigate the moments μ of $\varepsilon_2(\omega)$.[24],[25]   If we define

$$\mu_n \equiv \frac{2}{\pi} \int_0^\infty \omega^n \varepsilon_2(\omega) d\omega \qquad\qquad , \quad \text{(VI-2)}$$

then we find that

$$\mu_n = \frac{4\pi}{\hbar\Omega} <MA_n> \qquad\qquad , \quad \text{(VI-3)}$$

for n odd.[23-25]   In this expression, A_n is given by

$$A_n = \Big[H, A_{n-1} \Big] /\hbar \qquad\qquad , \quad \text{(VI-4)}$$

with

$$A_0 = M \qquad\qquad , \quad \text{(VI-5)}$$

where H is the Hamiltonian of the solid.

Further, we have that if H and M are analytic functions of then the right hand side of Eq. (VI-2) is finite.[24-26]   That is

$$\mu_n = \text{finite number} \qquad\qquad , \quad \text{(VI-6)}$$

for all odd n.   But we also have that

$$\varepsilon_2(\omega) \geq 0 \qquad\qquad . \quad \text{(VI-7)}$$

Hence for the integral in Eq. (VI-2) to satisfy Eq. (VI-6), we must have

$$\varepsilon_2(\omega) < A\omega^{-n} \quad \text{as} \quad \omega \to \infty \qquad\qquad , \quad \text{(VI-8)}$$

for all n. This result rules out an inverse power law for H and M analytic. However, it does not guarantee that $\varepsilon_2(\omega)$ will vary exponentially for large ω.

To illustrate the types of variations one might expect, we have evaluated the high frequency behavior of $\varepsilon_2(\omega)$ for two different molecular models: a square well, and a Morse potential. For a square well where the potential is not analytic we have that[26]

$$\varepsilon_2(\omega) \sim \omega^{-3} \qquad , \quad (V\text{-}9)$$

as expected. For a Morse potential,[26]

$$\varepsilon_2(\omega) \sim \varepsilon^{-\gamma\sqrt{\omega}} \qquad . \quad (V\text{-}10)$$

This variation for the Morse potential is not simply exponential. This leaves us three alternatives. First, the Morse potential is pathological. Second, the experimental results are not in the asymptotic or high frequency range and hence we cannot use asymptotic approximations to $\varepsilon_2(\omega)$. There is some hint that this is correct in the sense most theories do predict a noticeable deviation away from exponential a high phonon number. Finally, the molecular model may be too ideal a case. In the real crystal, we know that potentials and dipole moment operators are much more complicated and the potential is almost assuredly not analytic. Hence, it is rather difficult to attach a lot of importance to the asymptotic behavior obtained for the Morse potential.

VII. CONCLUSION

The theoretical situation seems somewhat confused. A number of very different theories all give an exponential variation of β with frequency. Many of the theories give results which are in satisfactory quantitative agreement with the experimental results. The few exact results that we can obtain from moment arguments rule out inverse power law behavior for wide classes of theories; yet, they do not guarantee an exponential variation with frequency. Given this very unsatisfactory theoretical situation, it becomes imperative to look for experiments which can distinguish between various contributions to the absorption and distinguish between the various theories.

ACKNOWLEDGEMENT

The author would like to acknowledge helpful conversations with R. W. Hellwarth, R. V. Winston, M. Hass, T. F. Deutsch, and R. Silver during the course of this work.

REFERENCES

1. T. F. Deutsch, J. Phys. Chem. Solids $\underline{34}$, 2091 (1973); and the references contained therein.

2. J. A. Harrington, and M. Hass, Phys. Rev. Letters $\underline{31}$, 710 (1973).

3. D. W. Pohl, and P. F. Meier, Phys. Rev. Letters $\underline{32}$, 58 (1974).

4. H. Bilz, Phonons in Perfect Lattice and Lattices with Point Imperfections (edited by R. W. H. Stevenson, Oliver and Boyd, London, 1966) pp 211-220.

5. L. Van Hove, N. M. Hugenholtz, and L. P. Howland, Quantum Theory of Many Particle Systems (W. A. Benjamin, New York, 1961) pp 5-101.

6. L. J. Sham and M. Sparks, Phys. Rev. $\underline{B9}$, 827 (1974).

7. T. C. McGill, R. W. Hellwarth, M. Mangir, and H. V. Winston, J. Phys. Chem. Solids $\underline{34}$, 2105 (1973); and T. C. McGill, unpublished.

8. K. V. Namjoshi and S. S. Mitra, Phys. Rev. $\underline{B9}$, 815 (1974); and Solid State Communications $\underline{15}$, 317 (1974).

9. D. L. Mills and A. A. Maradudin, Phys. Rev. $\underline{B8}$, 1617 (1973).

10. H. B. Rosenstock, Phys. Rev. $\underline{B9}$, 1963 (1974).

11. L. L. Boyer, J. A. Harrington, M. Hass, and H. B. Rosenstock, Phys. Rev. B (to be published).

12. B. Bendow, S. C. Ying, and S. P. Yukon, Phys. Rev. $\underline{B8}$, 1679 (1973); and B. Bendow, Phys. Rev. $\underline{B8}$, 5821 (1973).

13. D. L. Mills, and A. A. Maradudin, Phys. Rev. $\underline{B10}$, 1713 (1974).

14. B. Bendow, S. P. Yukon, and S. C. Ying, Phys. Rev. $\underline{B10}$, 2286 (1974).

15. R. L. Matcha, J. Chem. Phys. $\underline{47}$, 4595 (1967); $\underline{47}$, 5295 (1967); $\underline{48}$, 335 (1968), $\underline{49}$, 1264 (1968); and T. C. McGill, unpublished.

16. M. Sparks and L. J. Sham, Phys. Rev. $\underline{B8}$, 3037 (1973), and M. Sparks, Xonics, Inc. Final Report 1972, under Contract No. DAH 615-72-C-0129 (unpublished).

17. T. C. McGill and H. V. Winston, Solid State Communications $\underline{13}$, 1459 (1973).

18. B. Bendow, Appl. Phys. Letters $\underline{23}$, 133 (1973).

19. M. Sparks and L. J. Sham, Phys. Rev. Letters $\underline{31}$, 714 (1973).

20. A. A. Maradudin and D. L. Mills, Phys. Rev. Letters $\underline{31}$, 718 (1973).

21. M. Sparks, Phys. Rev. $\underline{B10}$, 2581 (1974).

22. R. Hellwarth and M. Mangir, Phys. Rev. $\underline{B10}$, 1635 (1974).

23. R. Hellwarth and M. Mangir, private communication.

24. P. C. Martin, in Probleme A N Corps; Many Body Physics (edited by C. DeWitt and R. Balian; Gordon and Breach, London, 1968) pp. 37ff.

25. G. Birnbaum and E. R. Cohen, J. Chem Phys. $\underline{53}$, 2885 (1970).

26. T. C. McGill and M. Chen, unpublished.

MULTIPHONON ABSORPTION FOR VARIOUS FORMS OF THE ANHARMONIC POTENTIAL

A. Nedoluha

Naval Electronics Laboratory Center

San Diego, California 92152

Multiphonon absorption in KCl has been calculated by perturbation theory (including vertex corrections) for four forms of the anharmonic potential: (1) exponential Born potential, (2) exponential with preexponential power factor, (3) inverse power, (4) Morse potential. The one-phonon density of states is approximated by a Debye model; an exponential approximation allows analytical calculation of the n-phonon density (including occupation probabilities) without resort to the central limit theorem. Measured values of the temperature dependent lattice constant and phonon frequencies have been employed. Depending on the chosen form of the anharmonic potential, the calculated $10.6\mu m$ absorption values differ from each other by several orders of magnitude. Without parameter adjustment, the inverse power potential gives good agreement with emittance data over the whole observed frequency range of intrinsic multiphonon absorption at 77K, 273K, and 373K, while potentials (2) to (4) give absorption values which are too low. Above 400K, theoretical absorption increases more rapidly with temperature than observed by calorimetric measurements at $10.6\mu m$. The same theory has been applied to predict multiphonon absorption in KI

Anharmonic multiphonon absorption depends on the form of the anharmonicity. We have calculated infrared multiphonon absorption in KCl for various forms of the anharmonic potential and compared theory with experimental data.[1-4] The two models which give the

21

best agreement with experiment have been employed to predict multi-phonon absorption in KI. Perturbation theory with vertex correc-tions has been used, following the approach by Sparks and Sham.[5] An alternate absorption mechanism, that due to nonlinear dipole moments, has recently found renewed interest,[6,7] but that type of process will be ignored here.

Four different forms of the anharmonic potential are investi-gated:

Model (1), the exponential Huggins-Mayer potential,

$$\phi_B(r) = C \exp (-r/\rho) \quad , \tag{1}$$

is the most commonly used form and has had good success in correla-ting crystal structure data.[8,9]

Model (2) assumes

$$\phi_B(r) = C (r/\rho)^k \exp (-r/\rho) \quad . \tag{2}$$

This form of ϕ_B is suggested by the asymptotic form of the hydrogen molecule potential,[9] in which case k=3. That value of k will be used in all our computations involving model (2).

Model (3), the inverse power potential,

$$\phi_B(r) = C (R_o/r)^P \quad , \tag{3}$$

has been employed by Pauling[8,10] for crystal structure considera-tions. Here R_o is the nearest neighbor distance in the lattice at absolute zero temperature.

Model (4), the Morse potential,

$$\phi(r) = C \{1-\exp[(R_o-r)/\rho]\}^2 \quad , \tag{4}$$

has found application [4,11-13] because it allows rigorous solution of the Schrödinger equation.

Equations (1) to (3) represent Born potentials which have to be supplemented by an attractive Coulomb potential

$$\phi_C(R) = -\alpha_M e^2/6R, \tag{5}$$

$$\phi = \phi_B+\phi_C , \tag{6}$$

where $\alpha_M = 1.7476$ is the Madelung constant for the KCl structure, e the electron charge, and R the nearest neighbor distance.

THEORY

Following Sparks and Sham[5] we find for the infrared absorption coefficient the expression

$$\beta = \frac{K/\omega^3}{n(\omega)+1} \sum_{m=2}^{\infty} \frac{\eta^m}{m!} [\Lambda_m \phi^{(m+1)}(R)]^2 S_m , \qquad (7)$$

where

$$K = \frac{4\pi^2 e_B^2}{\hbar c m_r^2 n_r R^3} , \qquad (8)$$

$$n(\omega) = [\exp (\hbar\omega/kT) - 1]^{-1} , \qquad (9)$$

$$\eta = 2\hbar/3m_r , \qquad (10)$$

$$S_m = \int \cdots \int \delta(\omega - \sum_{k=1}^{m} \omega_k) \prod_{j=1}^{m} d\omega_j\, g(\omega_j) \cdot [n(\omega_j)+1]/\omega_j , \qquad (11)$$

and the Λ_m are the vertex correction factors. Here e_B is the Born charge, $m_r = (m_<^{-1} + m_>^{-1})^{-1}$ the reduced mass with $m_<$ and $m_>$ the smaller and larger of the ionic masses, respectively, and $\omega = 2\pi c \bar{\nu}$ the photon frequency with $\bar{\nu}$ the wavenumber. The multiple integrations in Eq. (11) are to be performed from a lower cutoff frequency ω_C to the maximum phonon frequency ω_D. Assuming a Debye model, the normalized one-phonon density of states is given by

$$g(\omega_j) = 3\omega_j^2/\omega_D^3 . \qquad (12)$$

If ω_C is not too small (for our computations we take $\omega_C = \frac{1}{2}\omega_D$), then at any given temperature T parameters a and b may be chosen such that for $\omega_C \leq \omega_j \leq \omega_D$ in good approximation

$$\xi/(1-e^{-\xi}) \simeq a\, e^{b\xi} , \qquad (13)$$

where $\omega = \hbar\omega_j/kT$. The parameters a and b are assumed to depend on temperature, but not on frequency. Substitution of Eqs. (12) and

(13) in (11) allows to perform the integrations without resort to the central limit theorem.

The vertex corrections are given by

$$\Lambda_n = \sum_{m=o}^{n-2} \Lambda_n^{(m)} \quad , \tag{14}$$

with

$$\Lambda_n^{(m)} = n! \, A_n^{(m)} \, \zeta^m / \phi^{(n+1)}(R) \quad , \tag{15}$$

$$\zeta = 3/(5m_> \, \omega_D^2) \quad . \tag{16}$$

We have calculated coefficients $A_n^{(m)}$ up to $m=4$, where m is the number of internal phonon lines. For instance, there are nine types of diagrams with four internal phonon lines. An example, diagram type (4g), is shown in Fig. 1 and gives the contribution

$$A_n^{(4g)} = \sum_{i=2}^{n-5} \sum_{k=1}^{n-i-4} \sum_{\ell=2}^{n-i-k-2} \sum_{m=2}^{n-i-k-\ell}$$

$$\tag{17}$$

$$\phi^{(i+1)} \, \phi^{(j+4)} \, \phi^{(k+2)} \, \phi^{(\ell+1)} \, \phi^{(m+1)} / [2! i! j! k! \ell! m! i^2 m^2 (k+\ell)^2 \ell^2].$$

where $j=n-i-k-\ell-m$.

PARAMETER VALUES

The basic parameter values used in the computation of multi-phonon absorption are compiled in Table I.

The temperature dependence of R is obtained from the thermal expansion data given in the AIP Handbook.[14]

For KCl we calculate the Debye frequency $\omega_D = 2\pi c \bar{\nu}_D$ at room temperature as [17]

$$\omega_D = [2BR^3/(m_< + m_>)]^{\frac{1}{2}} k_D \quad , \tag{18}$$

where

$$k_D = (3\pi^2)^{\frac{1}{3}} /R \quad . \tag{19}$$

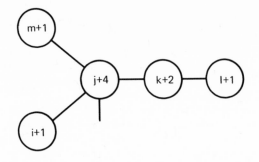

Fig. 1:   Diagram type (4g) for vertex corrections.

The resulting frequency value, corresponding to $\bar{\nu}_D = 156 \mathrm{cm}^{-1}$, agrees with that of the dominant peak in the KCl density of states.[18]   The temperature dependence of ω_D is assumed to be of the form

$$\omega_D = \begin{cases} \omega_{Do} & \text{for } T \leq T_c \ , \\ \omega_{Do} \ [1 - \alpha(T - T_c)] & \text{for } T \geq T_c \ . \end{cases} \tag{20}$$

With $T_c = 150°K$ and a temperature coefficient $\alpha = 4 \times 10^{-4} K^{-1}$ this approximates the temperature dependence of feature E of Harley and Walker.[19]

TABLE I. BASIC PARAMETER VALUES

Symbol	KCl	KI	Units	Ref.
$m_<$	5.89×10^{-23}	6.49×10^{-23}	g	(14)
$m_>$	6.49×10^{-23}	21.1×10^{-23}	g	(14)
Room temperature values:				
R	3.15×10^{-8}	3.53×10^{-8}	cm	(14)
B	1.78×10^{-11}	1.20×10^{11}	dyn/cm^2	(15)
n_r	1.4	1.6	–	(14)
e_B/e	1.12	1.11	–	(16)
Low temperature values:				
B_o	1.97×10^{11}	1.27×10^{11}	dyn/cm^2	(15)

In KI, because of the distinctly separated optical and acoustical peaks of the one-phonon density of states[20], we suspect Eq. (18) to be misleading and rather choose ω_D such as to give the same asymptotic n-phonon density of states peak as obtained from the actual density of states. Comparison of our Debye model with the 8-phonon density of states peak computed by Boyer et al.[4], gives $\bar{\nu}_D = 106 cm^{-1}$, which is 16% higher than the value calculated from Eq. (18). At 77K we assume ν_D to be increased by a factor 108/101, corresponding to the ratio of the fundamental phonon frequencies at low temperature and at room temperature.[16]

The temperature dependence of the Born charges is negligible.

For models (1) to (3) we obtain the values of the parameters characterizing ϕ_B from the equilibrium condition

$$\phi'(R_o) = 0 \quad , \tag{21}$$

and from

$$\phi''(R_o) = 3R_o B_o \quad , \tag{22}$$

where B is the bulk modulus and the subscript o refers to absolute zero temperature.

For model (4), Eqs. (21) and (22) are insufficient to determine C and ρ, because condition (21) is satisfied for arbitrary values of these parameters. According to the method of determining C and ρ, we distinguish several subcases. In model (4A) we take C and ρ as determined from thermal expansion data.[4] In model (4B) we use Eq. (21) which gives

$$C/\rho^2 = \frac{3}{2} R_o B_o \quad . \tag{23}$$

In model (4C) we put

$$C/\rho^2 = \tfrac{1}{2} m_r \omega_{fo}^2 \quad , \tag{24}$$

where ω_{fo} is the fundamental phonon frequency at absolute zero temperature. Eqs. (23) and (24) we supplement by C from thermal expansion data as in model (4A).

DISCUSSION

The factors of primary interest to us in Eq. (7) are the derivatives of the anharmonic potential and the vertex correction factors. Taking model (1) as the reference, we plot in Fig. 2 the ratios $(^X\phi(n+1)/^1\phi(n+1))^2$ vs. n, where left superscripts designate the model. The values of the second derivatives of $\phi(R_0)$ have to agree for all models which obey Eq. (22), i.e., models (1), (2), (3), and (4B). For higher derivatives, different models yield widely different values. Model (4B), not shown in Fig. 2, gives values between those for models (4A) and (4C).

In Fig. (3a) we plot for model (1) the relative contributions $\Lambda_n^{(m)}/\Lambda_n$ of diagrams with m internal phonon lines, where Λ_n has been approximated by Eq. (14) with m=4 as upper limit. Similar calculations have been performed for all other models and diagrams with up to four internal phonon lines were found sufficient for n at least up to 12.

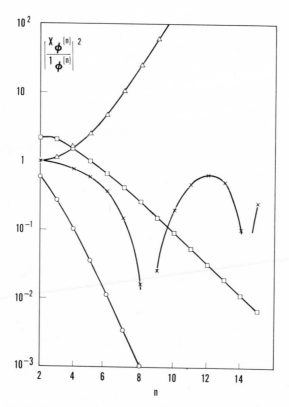

Fig. 2: Square ratio of n-th derivative of anharmonic potential in model (X) to that in model (1) for KCl at 373K. The symbols X , Δ , O , and □ indicate models (2), (3), (4A), and (4C) respectively.

Fig. (3b) shows the vertex correction factors Λ_n^2 for models
(1), (2), (3), and (4C). The vertex corrections for model (4A), not
shown in Fig. (3b), are very close to those for model (1).

In all our models, except model (4A), the absorption at 10.6μm
and 373K is dominated by 8-phonon processes; in model (4A), the contributions by the 7 and 8-phonon processes are about equal. In Table
II we show the 8-phonon absorption values at this wavelength and temperature for our various models, relative to the value for model (1);
also shown are the relative contributions of the factors $(\phi^{(9)})^2$ and
Λ_8^2. Notice that, with the exception of model (4A), the deviations
from model (1) due to the factors $\phi^{(n+1)}$ and Λ_n partially cancel each
other. For instance, while the magnitudes of the higher derivatives
of the anharmonic potential in model (3) are larger than those in
model (1), the vertex corrections are smaller.

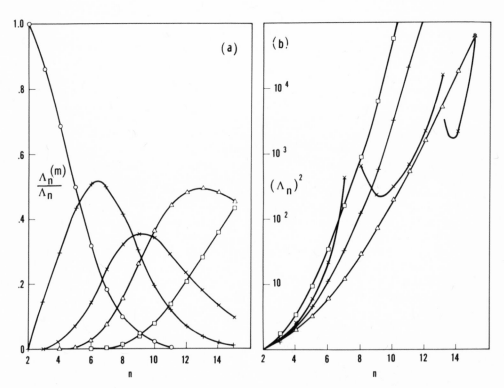

Fig. 3: Vertex corrections for KCl at 373K. (a) Relative contributions of diagrams with m internal phonon lines for model (1);
the symbols O , + , X , Δ , ☐ indicate m=0,1,2,3,4,
respectively. (b) Vertex correction factors Λ_n^2; the symbols + , X ,
Δ , and ☐ indicate models (1), (2), (3), and (4C), respectively.

TABLE II

8-phonon contributions to absorption at 10.6μm and 373K for models
(2), (3), (4A), and (4C), relative to contribution by model (1).

(X) (model)	$\left(\dfrac{X_\phi(9)}{1_\phi(9)}\right)^2$	$\left(\dfrac{X_{\Lambda_8}}{1_{\Lambda_8}}\right)^2$	$\dfrac{X_{\beta_8}}{1_{\beta_8}}$
(2)	0.026	5.4	0.15
(3)	68	0.24	17
(4A)	3.2×10^{-4}	0.87	2.8×10^{-4}
(4C)	0.15	7.6	1.2

In Fig. (4a) the absorption coefficients as functions of the
frequency are compared. Pauling's inverse power potential gives the
largest values, followed by the exponential potential, the hydrogen
molecule type potential, and the Morse potential (4A). Morse potential (4C), not shown in Fig. (4a), and a nonperturbative calculation
by Boyer et al.,[4] using the Morse potential, give results similar to
model (1). Also shown are emission data at 273K by Stierwalt,[1] supplemented by longer wavelength room temperature data from the AIP
Handbook, reported by Deutsch.[2] Data obtained by Deutsch at shorter
wavelength (up to 760 cm^{-1}) are not shown but are in good agreement
with Stierwalt's data.

Overall, the best agreement with experimental data in Fig. (4a)
is achieved by the inverse power potential, model (3). In Fig. (4b)
we extend the comparison for this model to data by Stierwalt[1] at
373K and 77K. The data points at 77K are adjusted values, obtained
by subtracting from the measured absorption values the measured 14μm
value $\beta = 2.5 \times 10^{-3}$cm^{-1}; at this wavelength Stierwalt's original 77K
data[1] exhibit a local minimum, indicating the presence of an additional absorption process. Considering the fact that all parameter
values entering the theory have been determined independently, the
agreement of theory with experiment is quite good.

In Fig. (5) we compare model (3) with experiment for absorption
as a function of temperature. For Stierwalt's data at 17, 16, and
15μm, below 400K, the agreement is satisfactory, but compared with
the 10.6μm data by Hass and Harrington[3] the theoretical absorption
curve becomes too steep at higher temperatures. This steep slope
which is not limited to model (3) may indicate a deficiency of our

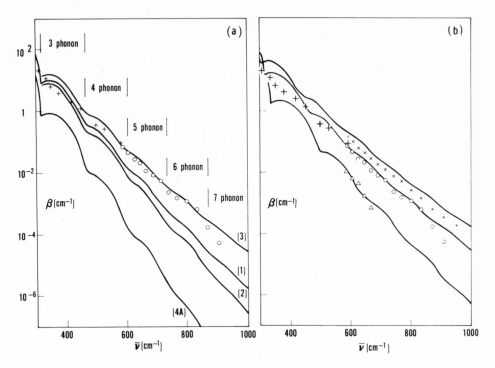

Fig. 4: Absorption coefficient β vs. wavenumber ν̄ for KCl.
(a) Solid lines: theory at 273K for models as labeled; experimental
values indicated by O (273K, Ref. 1) and + (room temperature,
Ref. 2).   (b) Solid lines: theory for model (3) at 373, 273, and 77K,
respectively (top to bottom); data points X , O , and △ : experimental values at 373, 273, and 77K, respectively (Ref. 1); data
points + as in (a).

models or a breakdown of the perturbation method. A reduction in
absorption at high temperatures may occur due to Debye-Waller-like
factors.[21] If one would assume ω_D as independent of T, much better
agreement with experiment would be achieved, as shown by the dashed
curve in Fig. 5, where ω_D has been taken as constant at its room
temperature value; however, we see no physical justification for
such an assumption.

Finally, because of renewed interest in KI, we show in Fig. 6
the absorption in that material at 273 and 77K, calculated according
to models (1) and (3). As for KCl, the absorption at 273K in the
7-phonon regime (around 600 wavenumbers for KI) calculated from

model (3) is about one order of magnitude greater than that from
model (1). Again as for KCl, the absorption values for KI calculated from model (1) are similar to those obtained by Boyer et al.[4]
from a nonperturbative calculation using the Morse potential, except
that we get less structure because of our use of the Debye density.

In conclusion, different choices of the form of the anharmonic potential may lead to strong differences of the absorption
coefficients. Pauling's inverse power potential gives the largest
values and, for temperatures below 400K, gives good agreement with
absorption data for KCl. However, models which yield much lower
absorption values are not necessarily to be abandoned: depending
on the outcome of the present controversy on the role of higher-
order-dipole-moments, an anharmonic potential which - by itself -
gives too low an absorption, may eventually lead to the experimentally observed absorption values when combined with the nonlinear-
dipole-moment absorption mechanism.

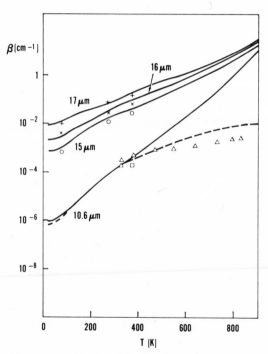

Fig. 5: Absorption coefficient β vs. absolute temperature T for
KCl. Solid lines: theory according to model (3) at wavelengths
(from top to bottom) 17, 16, 15, and 10.6μm, respectively. Dashed
line: theory according to model (3) at 10.6μm, but with Debye frequency independent of temperature. The data points +, X, O, ☐, and
Δ, represent experimental values at 17, 16, 15, 10.5 (all Ref. 1),
and 10.6μm (Ref. 2), respectively.

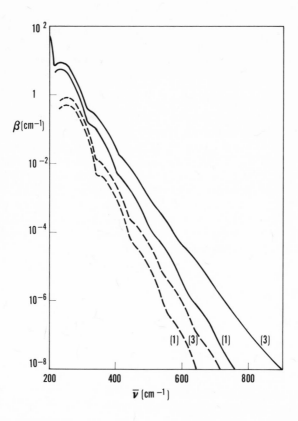

Fig. 6: Absorption coefficient β vs. wavenumber ν̄ for KI according to models (1) and (3) as labeled. Solid lines 273K; dashed lines 77K.

REFERENCES

1.   D.L. Stierwalt in High Energy Laser Windows, Semi-Annual Report No. 3, p.50 (1973), Naval Research Laboratory.

2.   T.F. Deutsch, J. Phys. Chem. Solids 34, 2091 (1973); Appl. Phys. Letters 25, 109 (1974).

3.   J.A. Harrington and M. Hass, Phys. Rev. Letters 31, 710 (1973); M. Hass, J.W. Davisson, P.H. Klein and L.L. Boyer, J. Appl. Phys. 45, 3959 (1974).

4.   L.L. Boyer, J.A. Harrington, M. Hass, and H.B. Rosenstock, to be published.

5.   M. Sparks and L.J. Sham, Phys'. Rev. B $\underline{8}$, 3037 (1973).

6.   B. Bendow, S.P. Yukon, and S.C. Ying, Phys. Rev. B $\underline{10}$, 2286 (1974).

7.   M. Sparks, Phys. Rev. B $\underline{10}$, 2581 (1974).

8.   F.G. Fumi and M.P. Tosi, J. Phys. Chem. Solids $\underline{25}$, 31 (1964).

9.   M.P. Tosi, Solid State Physics $\underline{16}$, 1 (1964).

10. L. Pauling, The Nature of the Chemical Bond (Cornell University Press, 1960).

11. T.C. McGill, R.W. Hellwarth, M. Mangir, and H.V. Winston, J. Phys. Chem. Solids $\underline{34}$, 2105 (1973).

12. D.L. Mills and A.A. Maradudin, Phys. Rev. B $\underline{8}$, 1617 (1973); A.A. Maradudin and D.L. Mills, Phys. Rev. Letters $\underline{31}$, 718 (1973).

13. H.B. Rosenstock, Phys. Rev. B $\underline{9}$, 1963 (1974).

14. American Institute of Physics Handbook, 3rd ed., Ed. D.E. Gray (McGraw Hill, New York, 1972).

15. M.H. Norwood and C.V. Briscoe, Phys. Rev. $\underline{112}$, $\underline{45}$, (1958).

16. G.O. Jones, D.H. Martin, P.A. Mawer, and C.H. Perry, Proc. Roy. Soc. $\underline{A261}$, 10 (1961).

17. K.V. Namjoshi and S.S. Mitra, Phys. Rev. B $\underline{9}$, 815 (1974).

18. J.R.D. Copley, R.W. Macpherson, and T. Timusk, Phys. Rev. $\underline{182}$, 965 (1969).

19. R.T. Harley and C.T. Walker, Phys. Rev. B $\underline{2}$, 2030 (1970).

20. A.N. Basu and S. Sengupta, Phys. Rev. B $\underline{8}$, 2982 (1973).

21. B. Bendow, S.C. Ying, and S.P. Yukon, Phys. Rev. B $\underline{8}$, 1679 (1973).

CUMULANT METHODS IN THE THEORY

OF MULTIPHONON ABSORPTION

Bernard Bendow

Solid State Sciences Laboratory, Air Force Cambridge
Research Laboratories (AFSC), Hanscom AFB, MA 01731

Stanford P. Yukon*

Parke Mathematical Labs., Carlisle, MA 01741, and
Dept of Physics, Brandeis University, Waltham, MA 02154

    Recent studies suggest that multiphonon absorp-
tion in covalent solids is substantially influenced
by contributions due to nonlinear electric moments.
We demonstrate that the dielectric response in the
latter case may be formulated in terms of a cumulant
series involving lattice displacement correlators,
which avoids the usual perturbative expansions of the
moment and anharmonic potential in powers of displacements, and thus accounts for various classes of phonon
processes to infinite order. To assess the utility
of the method we compare results obtained by truncating the cumulant series with exact ones, calculated
for a single-particle model. For parameters characteristic of typical binary semiconductors, the harmonic approximation to the cumulant yields poor results.
On the other hand, the use of anharmonic correlators
is shown to provide good agreement with exact behavior after several terms in the cumulant have been included. Overall, the results suggest that cumulant
methods are highly promising for calculating multiphonon absorption coefficients.

*Supported by Air Force Cambridge Research Labs (AFSC),
 Contract No. F19628-71-C-0142.

INTRODUCTION

Recent investigations have suggested that nonlinear moments can contribute significantly to the multiphonon absorption at frequencies far above the reststrahl of solids.[1] While the influence of nonlinear moments appears to be greatest for the more covalent solids, their effect in ionic solids remains a matter of some controversy.[1-4] An exact calculation of the multiphonon absorption in real crystals is a formidable task, especially when the nonlinear moments are accounted for. Recently, a variety of approaches[1-10] have been developed in order to calculate the multiphonon absorption coefficient α. To avoid the complexities of calculations for real crystals, various simplified models have been introduced, among them the single-particle model.[2,7-9] In the latter, the crystal is replaced by a collection of non-interacting particles, each subject to an anharmonic potential v, and interacting with light through a (nonlinear) moment m. For appropriate choices of v, one can obtain results for α which are exact in principle. Although the possibility of an exact solution is an attractive one, the adequacy of such an oversimplified model for predicting actual crystalline properties is questionable. On the other hand, straightforward application of perturbation theory to general crystals becomes very tedious, even when modified in the manner of Sparks and Sham,[6] say. In this paper we describe and assess a cumulant approach to the calculation of α, which is capable of accounting simultaneously for both anharmonicity and nonlinear moments, and of incorporating various real-crystal properties, yet is simpler mathematically than standard perturbation formulations. To aid in inferring the utility of cumulant methods for general crystal calculations, we compare cumulant results to exact ones which are available for the single-particle model. We will conclude that cumulant methods offer a variety of advantages for computations of multiphonon absorption.

MULTIPHONON ABSORPTION COEFFICIENT

In the present work we restrict attention to simplified forms of the electric moment which are most appropriate for purposes of illustration. Following Flytzanis,[11] the crystalline moment M may be expressed as a sum of individual bond moments $\tilde{m}_{ij}$ between atoms i and j as

$$\underset{\sim}{M} = \sum_{ij} \underset{\sim}{m}(\underset{\sim}{r}_i - \underset{\sim}{r}_j) \equiv \sum_{ij} \underset{\sim}{m}_{ij} \tag{1}$$

where $\underset{\sim}{r}_i = \underset{\sim}{r}_i^0 + \underset{\sim}{u}_i$, where $\underset{\sim}{u}_i$ is the displacement from equilibrium of atom i. For an optically isotropic crystal the absorption coefficient α is given by[12,13]

$$\alpha = \frac{2\pi\omega}{3nc} \mathrm{Im}\chi_\omega \; ; \; \chi_\omega = \langle \underset{\sim}{M}(t); \underset{\sim}{M}(0) \rangle_\omega \tag{2}$$

where $<A(t);B(0)>_\omega$ is the retarded Green's function in frequency space for the operators (A,B). Since $\text{Im}<A(t);B(0)>_\omega$ is proportional to the correlation function $<A(t)B(0)>_\omega$, calculation of α boils down to evaluation of moment correlators of the form

$$<\underset{\sim}{m}_{ij}(t)\underset{\sim}{m}_{k\ell}(0)>_\omega \tag{3}$$

For simplicity, we suppress the vector character of $\underset{\sim}{m}$, and consider a linear vibrational model for the crystal. Choosing

$$m(r) \propto e^{-Ar}, \tag{4}$$

we then require the function

$$Q_\omega = <e^{-u_1}e^{-u_2}>_\omega \tag{5}$$

where $u_1 = Au_{ij}(t)$ and $u_2 = Au_{k\ell}(0)$. Employing the standard cumulant expansion[13] for Q yields the form

$$Q = \exp(U_1+U_2+U_3+\cdots\cdots) \tag{6}$$

where the ith cumulant U_i is the sum of all terms involving a product of i u's. For crystals where all atoms are centers of inversion, $U_1 = 0$ and

$$U_2 = <u_1u_2> , \tag{7}$$

$$U_3 = \frac{1}{2}(<u_1^2u_2> + <u_1u_2^2>) , \text{ etc}$$

where we have displayed just the t-dependent terms which determine the spectral shape.

With the above formulation, truncation of the cumulant series after any number of terms yields an absorption coefficient accounting for various classes of phonon processes to infinite order. Also, the anharmonic and nonlinear moment effects may be treated essentially independently (compare, for example, with the cumbersome double perturbation series of Szigeti[14]). Another advantage of the present formulation is that the problem is cast in terms of the physically meaningful correlation functions[12] of the form $<u_1u_2\cdots>$, which already incorporate various aspects of the lattice dynamics of the crystal and, moreover, possess functional properties which often may be inferred on general physical grounds.[15]

Although the above results are especially simple because m was chosen as exponential, a similar treatment may be pursued for any m for which a Fourier transform exists (see Ref. 1 for details).

Effects due to the vector character of $\underset{\sim}{m}$, and due to sums over crystalline site indices, may be accounted for by methods described in Refs. 1 and 11.

As an example of the application of the above formulation, we consider the non-interacting cell approximation described in Refs. 1 and 5, in which just moments within the same cell are correlated. Then one finds, for the model of Eq. 4,

$$\mathrm{Im}\chi_\omega \sim m^2(r_o) \sum_{\underset{\sim}{r},n} \frac{1}{n!} \, \rho_n(\underset{\sim}{r},\omega) \tag{8}$$

where r_o is the interatomic equilibrium distance, and ρ_n is the nth convolution of the frequency Fourier transform of the cumulant U; for example, omitting Debye-Waller terms,

$$U_2 = A^2[C_{11}(\underset{\sim}{r},t) + C_{22}(\underset{\sim}{r},t) - 2C_{12}(\underset{\sim}{r},t)] \tag{9}$$

where C_{ij} is the displacement-displacement correlator for atoms i and j, $C_{ij} = <u_i(t)u_j(0)>$. For more general forms of the moment, the results are not so simple. However, truncation at the quadratic term U_2 leads to the following expression for the non-interacting cell case[1]:

$$\mathrm{Im}\chi_\omega \sim \sum_n \frac{1}{n!} [\frac{\partial m^n(r)}{\partial r^n}]^2_{r=r_o} \rho_n(\underset{\sim}{r},\omega) \tag{10}$$

where, for a diatomic crystal, $\rho_1 = U_2/A^2$ of Eq. 9. Due to their simplicity, these expressions are extremely convenient for actual computations. Yet, they are capable of realistically incorporating various crystalline features, since no assumptions have been made regarding the crystalline spectrum or the anharmonicity. As discussed in Ref. 1, they easily yield the dependence of α on the size of the anharmonic potential, the functional form of m, and the phonon spectrum. Calculations based on Eq. 10 utilizing simple models for m and C are described in Ref. 1.

CUMULANT COMPUTATIONS FOR SINGLE-PARTICLE MODEL

In the single-particle (sp) model the system eigenstates $|n>$ are obtained from

$$[p^2/2m + v(r)]|n> = \omega_n|n> . \tag{11}$$

At zero temperature,

$$\mathrm{Im}\chi_\omega = \sum_n |<0|m|n>|^2 \delta(\omega_{no}-\omega) \tag{12}$$

where $\omega_{no} = \omega_n - \omega_o$; results for non-zero temperature are given in Ref. 9, for example. Eq. 12 represents a line spectrum, which is generally related to crystalline behavior by averaging over a frequency distribution[9] to provide a continuous α. It is reasonable to employ the sp model to assess the utility of cumulant methods for the calculation of α in actual crystals, because the m and v employed are characteristic of the individual m's and v's in the sums which constitute, respectively, the crystalline moment and the crystalline potential.

In our calculations for zincblende semiconductors, we employ a Morse potential for v,

$$u = v_o(e^{-2\xi r} - 2e^{-\xi r}) \tag{13}$$

and the form

$$m = em_o r_o e^{-2\xi_1 r} \tag{14}$$

for the moment. For this case one can show that

$$<\ell|m(r)|o> \propto (-1)^\ell a(a+1)\cdots(a+\ell-1)$$

$$x \left\{\frac{\Gamma(2d-\ell+a-1)[(2d-2\ell-1)(2d-1)]^{1/2}}{\Gamma^{1/2}(2d-\ell)\Gamma^{1/2}(2d)\sqrt{\ell!}(2d)^a}\right\} \tag{15}$$

where $a = 2\xi/\xi_1$, $d = 4\pi v_o/\omega_o$, with ω_o the harmonic frequency. The determination of the parameters v_o, ξ, m_o and ξ_1 from various thermodynamic relations and phonon pressure dependence data is described in Refs. 1 and 16. We here employ parameters characteristic of GaAs (denoted by bars)

$$\overline{v}_o = 4.1 \times 10^{-12} \text{erg}; \quad \overline{\xi} = 1.25\text{\AA}^{-1}$$

$$\overline{m}_o = 2.44; \quad \overline{\xi}_1 = 0.54\text{\AA}^{-1} \tag{16}$$

After Eq. 12 has been calculated to yield the sp spectrum, it is integrated over a Debye frequency distribution with ω_{TO} as the upper limit. The same calculation is repeated utilizing the appropriate generalization of the cumulant expansion in Eqs. 6 and 7 for the case $v(x) \neq v(-x)$, and truncating at the quadratic or cubic term. Also, the harmonic approximation to U (for which just U_2 survives) was utilized to calculate α. The results are indicated in Fig. 1 for GaAs parameters. Clearly, the harmonic approximation for U yields extremely poor results in this case. Use of the anharmonic cumulant substantially improves the computation, with a

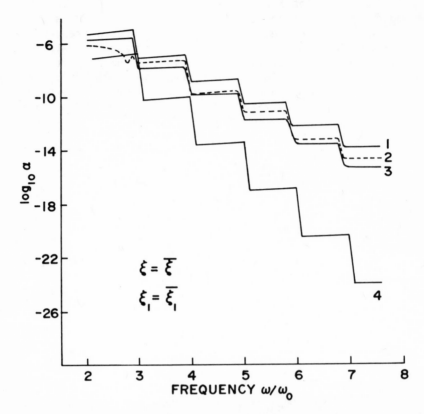

FIGURE 1. $\log_{10}$ absorption coefficient vs frequency ω/ω_0, for GaAs parameters. Curve 1 = exact; Curve 2 = cumulant truncated at cubic term; Curve 3 = cumulant truncated at quadratic term; Curve 4 = harmonic approximation to cumulant.

discrepancy of about an order of magnitude from the exact results for the cubic cumulant ($U=U_2+U_3$). Two observations are in order: First, the degree of discrepancy is associated with the rapid decrease in α predicted for this case; at finite temperature α decreases much more slowly, in which case an improved agreement would be expected. Second, the cumulant calculations do appear to be converging reasonably rapidly toward the exact results.

Since other solids may be characterized by substantially different values of parameters, it is instructive to consider the effect on α when they are varied. Since the cumulant series is a series in ξ_1, convergence should be most rapid for small values of ξ_1; this is evidenced by the results in Fig. 2, where $\xi_1 = 0.1\overline{\xi}_1$ is

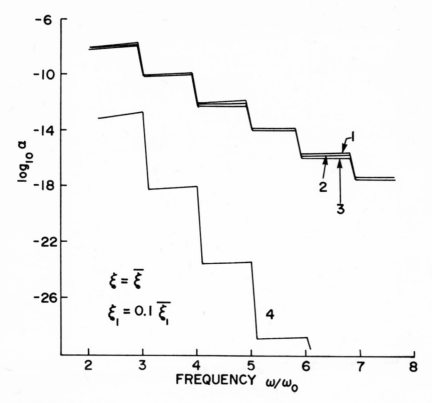

FIGURE 2.   Same as Fig. 1, but for parameters indicated.

employed.   Regarding ξ, it can be shown that U_n $(n>2)\rightarrow0$ for vanishing anharmonicity (which corresponds to $\xi\rightarrow0$, but $v_0\xi^2$ = constant).   Thus, use of just the quadratic term in U or, moreover, its harmonic value, will be an increasingly good approximation for small ξ.   This is indicated by Fig. 3, where results are displayed for $\xi = 0.01\overline{\xi}$.   Fig. 2 also indicates that the harmonic approximation becomes increasingly poor for small ξ_1.   This is because the lower order terms in m (which, for the harmonic limit, possess a vanishing contribution in the many-phonon regime) dominate α in this instance.

SUMMARY AND CONCLUSIONS

In the present treatment we have demonstrated the application of cumulant methods to the calculation of multiphonon absorption in solids.   Such methods are most advantageous for a

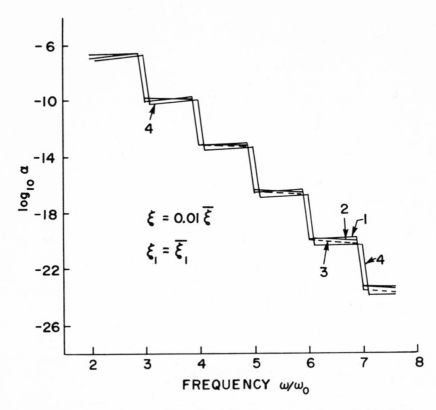

FIGURE 3.   Same as Fig. 1, but for parameters indicated.

simultaneous treatment of anharmonic and nonlinear-moment effects
for more realistic models of crystals. They are considerably less
cumbersome than standard perturbative techniques; in fact, they
are sufficiently attractive to dissuade the use of unphysical and/
or oversimplified models. Moreover, the cumulant approach provides
a physically motivated formulation in terms of lattice displacement correlation functions.

The computations carried out here indicate that truncation
of the cumulant series after the first few terms provides the best
results for α in cases where the moments $m(r_{ij})$ fall off slowly as
functions of r_{ij}, and in which the anharmonicity is small. For
parameters characteristic of III-V semiconductors, the cumulant
results appear to converge reasonably rapidly toward the exact
ones. Except for cases in which the crystal anharmonicity is
unrealistically small, the harmonic approximation to the cumulant
yields poor results for α.

Overall, the present observations suggest that cumulant methods are promising for calculations of α in actual crystals. To do this meaningfully, however, it will be necessary to account for selection-rule effects in the correlation functions, which are responsible for structure in the multiphonon spectrum, and to determine the crystalline moment accurately. Hopefully, such calculations will enable a detailed interpretation of the spectral properties observed in the multiphonon regime of various solids.

REFERENCES

1. S. C. Ying, B. Bendow and S. P. Yukon, in "Physics of Semi-conductors," M. Pilkuhn, ed. (Tuebner, Stuttgart, 1974); B. Bendow, S. P. Yukon and S. C. Ying, Phys. Rev. B10, 2286 (1974).

2. D. L. Mills and A. A. Maradudin, Phys. Rev. B10, 1713 (1974).

3. M. Sparks, Phys. Rev. B10, 2581 (1974).

4. R. Hellwarth and M. Mangir, Phys. Rev. B10, 1635 (1974).

5. B. Bendow, S. C. Ying and S. P. Yukon, Phys. Rev. B8, 1679 (1973); B. Bendow, Phys. Rev. B8, 5821 (1973).

6. M. Sparks and L. J. Sham, Phys. Rev. B8, 3037 (1973).

7. T. C. McGill, R. W. Hellwarth, M. Mangir and H. V. Winston, J. Phys. Chem. Sol. 34, 2105 (1973).

8. D. L. Mills and A. A. Maradudin, Phys. Rev. B8, 1617 (1973).

9. H. B. Rosenstock, J. Appl. Phys. 44, 4473 (1973); Phys. Rev. B9, 1973 (1974).

10. K. V. Namjoshi and S. S. Mitra, Phys. Rev. B9, 815 (1974).

11. C. Flytzanis, Phys. Rev. Lett. 29, 772 (1973); Phys. Rev. B6, 1264 (1972).

12. A. A. Maradudin et al, "Theory of Lattice Dynamics in the Harmonic Approximation", (Academic, NY, 1971).

13. A. A. Maradudin, in "Solid State Physics", F. Seitz and D. Turnbull, eds. (Academic, NY, 1966), Vols. 18 and 19.

14. B. Szigeti, Proc. Roy. Soc. (London) A258, 377 (1960).

15. P. C. Kwok, in "Solid State Physics", op. cit., Ref. 13, Vol. 20.

16. B. Bendow, P. D. Gianino, Y. F. Tsay and S. S. Mitra, Applied Optics 13, 2382 (1974).

THE HIGH FREQUENCY TAIL OF THE LATTICE ABSORPTION

SPECTRA OF SIMPLE CRYSTALS[*]

A. J. Barker and G. R. Wilkinson
Department of Physics, King's College,
University of London, London, U. K.
and
N. E. Massa[**] and S. S. Mitra[+]
Department of Electrical Engineering
University of Rhode Island, Kingston, R.I. 02881

ABSTRACT

The absorption coefficient in the pre-transparent regime ($100 > \alpha > 0.1$ cm^{-1}) has been measured for several alkali halides as functions of temperature extending up to their melting points. The frequency dependence of the absorption coefficient is calculated by means of a breathing shell model using multi-phonon density of states. The temperature dependence of the absorption coefficient is explained in terms of a temperature dependent effective phonon frequency.

I. INTRODUCTION

Ionic solids such as the alkali halides resonantly absorb far infrared radiation in the region of optical phonon frequencies. On the other hand, homopolar crystals

*Supported by a NATO Research Grant (No. 775).
**On leave from Physics and Astronomy Department, Vanderbilt University, Nashville, Tenn. 37235.
+Supported in part by Air Force Cambridge Research Laboratories, (AFSC), Contract F19628-72-C-0286.

45

such as diamond and silicon do not display such absorption. The strength of this absorption depends on the degree of the ionic character of the solid. However, at
frequencies two to three times the optical phonon frequencies all crystals, ionic, partially ionic and homopolar, absorb radiation due to multiphonon processes,
the absorption coefficient steadily decreasing with increasing frequency. In two- and three- phonon processes
the spectra occasionally show some structure due to singularities in the phonon density of states. Beyond the
three-phonon region the spectrum is essentially structureless for most solids. The absorption coefficient in
this transparent regime has nearly exponential frequency
dependence.

An examination of the absorption spectra of a simple
solid in the transparent regime reveals an abrupt decrease in its absorption coefficient. The frequency at
which this occurs may be defined as the onset frequency
of transparency. An operational definition of a transparent solid is that it transmits at least 99°/o radiation flux for a thickness of 1 cm. In Table I we record
the onset frequency (ω_{onset}) of transparency $(\alpha=0.01 \text{ cm}^{-1})$
for a number of simple solids. In this table we also
list the maximum one-phonon frequencies, which are the
long-wavelength longitudinal optical mode frequencies (ω_{LO})
It is interesting to note that irrespective of the struc-
ture or binding type, ω_{onset} almost coincides with $3\omega_{LO}$.
Since beyond $3\omega_{LO}$ no three-phonon processes are possible
it is usually termed the three-phonon cut off frequency.
Thus it is safe to say that most crystals are transparent
at photon frequencies larger than the three-phonon cut
off frequency.

In this paper we have examined the pre-transparent
regime $(10.0 > \alpha > 0.1 \text{ cm}^{-1})$ of three alkali halides. The
absorption coefficient, and its temperature dependence
are measured. The results are discussed in terms of current theories of higher order multiphonon interactions.

II. EXPERIMENTAL RESULTS

Infrared absorption by KBr, NaCℓ and LiF were measured by the obscured-mirror technique[1]. The apparatus
consisted of a small water-cooled furnace surrounding a
horizontal platinum crucible with associated external optics to focus the measuring beam into the sample. The
unit was used in conjunction with a Perkin-Elmer Model

457 spectrophotometer[2,3]. The experimental arrangement is shown in Figure 1. All samples (1 - 2 mm thick) were prepared from high quality single crystals obtained from Scientific Supplies Co. Ltd., Vine Hill, London. For each compound studied, measurements were taken at three elevated temperatures in the solid state and at one temperature slightly above the melting point. The reflectivity of each material was derived from measurements using a shotblasted platinum disc in place of the platinum mirror. Absorption coefficients were evaluated from an approximation of Genzel formula[4].

Figure 2, shows the infrared absorption coefficients for LiF, NaCℓ and KBr in the pre-transparent regime. Each curve shows mean data points derived from three or four independent measurements using different samples. The random error by successive determination of α varied from 0.03 cm^{-1} ($\alpha \sim 0.1$) to 0.3 cm^{-1} ($\alpha \sim 10.0$ cm^{-1}).

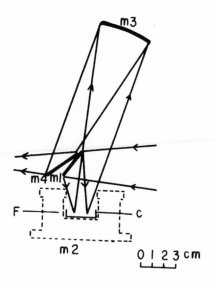

Fig. 1 Schematic diagram of reflectance furnace: m1,m4 aluminized plane mirrors, m2 solid platinum plane mirror, m3 aluminized concave spherical mirror, C platinum crucible, F furnace.

TABLE I. Frequency of Onset of Transparency Compared
with the Three-Phonon Cut Off Frequency

Material	$\omega_{max}=\omega_{LO}$ (cm^{-1})	$3\omega_{LO}(cm^{-1})$	ω_{onset}^{*}
LiF	665	1995	2000
NaF	422	1226	1350
NaCℓ	265	795	700
KCℓ	205	615	580
KBr	163	489	430
CsBr	114	342	330
CsI	90	270	240
MgO	728	2184	1750
AgCℓ	189	567	500
CuCℓ	216	648	620
CuBr	169	507	500
ZnS	347	1041	1000
AℓSb	340	1020	1110
InSb	197	591	550
GaP	403	1209	1200
GaAs	286	858	810
CaF$_2$	479	1437	1470
SrF$_2$	389	1167	1250
BaF$_2$	338	1014	1070
Si	508	1524	1520
Ge	300	900	880
C(Diamond)	1332	3996	3950

*For α = 0.01 cm^{-1} (see text)

III. DISCUSSION

It is well known that the absorption coefficient in
the transparent regime varies exponentially with frequency. Present data when plotted on a semilog scale
also yield straight lines as is shown for LiF in Fig. 2.

A number of attempts have been made to explain[5,6]
experimentally observed exponential dependence of the absorption coefficient in the high frequency side of the
reststrahlen band based on multi-phonon interactions.
Here, we attempt to obtain the frequency and temperature

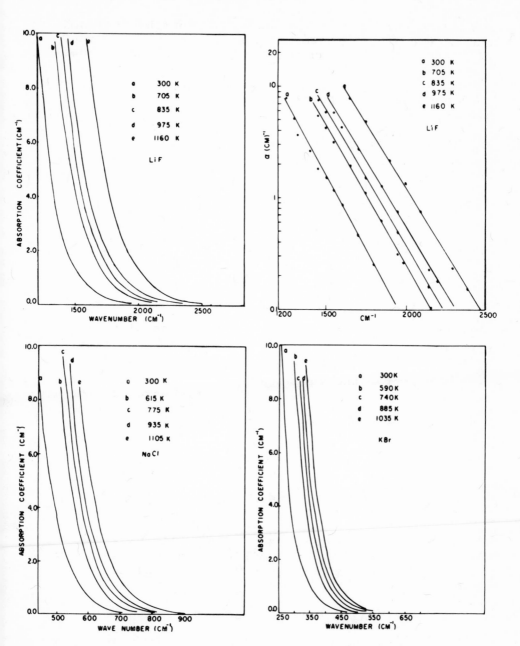

Fig. 2   Infrared absorption of LiF at several temperatures; same data on a semilog plot; infrared absorption
of NaCℓ at several temperatures; infrared absorption of
KBr at several temperatures.

dependence of the absorption coefficient by the Namjoshi and Mitra method[6] in terms of multiphonon density of states distributions using the breathing shell model[7].

The absorption coefficient, α, is given in terms of real and imaginary parts of the dielectric constant; ε' and ε''

$$\alpha = \varepsilon''\omega/c\varepsilon'^{1/2} \qquad (1)$$

where

$$\varepsilon = \varepsilon' + i\varepsilon''$$

$$= \varepsilon_\infty + \frac{A_o}{[1 - (\frac{\omega}{\omega_o})^2] + i(\frac{\gamma_o\omega}{\omega_o})} + \sum_i \frac{A_i}{[1 - (\frac{\omega}{\omega_i})^2] + i(\frac{\gamma_i\omega}{\omega_i2})} \qquad (2)$$

γ_i's are the damping factors, ω_i's the oscillator frequencies, ε_∞ the high frequency dielectric constant, and A_i's the oscillator strength parameters. The subscript zero stands for the fundamental process and the subscript i stands for a multi-phonon process. In the limit when γ_i's go to zero[6], the imaginary part of the dielectric constant reduces to

$$\varepsilon'' = \frac{-A_o\gamma_o\omega}{[1 - (\frac{\omega}{\omega_o})^2]^2 + (\frac{\gamma_o\omega}{\omega_o2})^2} - \sum_j A_j(\omega)\, \rho_j(\omega)\omega\pi/2 \qquad (3)$$

$\rho_j(\omega)$ corresponds to the j-phonon density of states. Contribution to α from the multi-phonon processes of the order higher than two is shown[6] to be equal to

$$\alpha = \sum_{j \geqslant 2} A_j\, \rho_j(\omega)\pi\, \omega^2/2c\varepsilon^{1/2} \qquad (4)$$

To obtain the absorption coefficient from multi-phonon processes the density of states functions were calculated from the relation

$$\rho_n(\omega) = \int_{\omega_1=0}^{\omega} \int_{\omega_2=0}^{\omega-\omega_1} \int_{\omega_{n-1}=0}^{\omega-\omega_1-\omega_2\cdots\omega_{n-2}} \rho(\omega_1)\rho(\omega_2)\cdots\rho(\omega_{n-1})$$

$$\rho(\omega_{n-1})\rho(\omega-\omega_1-\omega_2\cdots\omega_{n-1})d\omega_1 d\omega_2\cdots d\omega_{n-1}$$

$$(5)$$

where $\rho(\omega)$ is the one-phonon density of states function. Figure 3 shows the result of these iterations for the one- to six-phonon processes for KBr. It may be noted that higher order phonon densities of states gradually approach a normal distribution.

The oscillator strengths were assumed to depend on the corresponding derivatives of the crystal potential energy, that in this work was assumed to be a Morse type potential:

$$\Phi(r) = b(1 - e^{-\beta(r-r_o)})^2 \qquad (6)$$

The higher order derivatives used in the calculation of the oscillator strength only involved the potential constant β, as evaluated by

$$\beta^2 = 4\sqrt{3}\ \frac{B_T r_o}{U} = \omega_h^2\left(\frac{\mu}{2U}\right) \qquad (7)$$

where B_T is the isothermal bulk modulus, r_o the nearest neighbor distance, U the cohesive energy, μ the reduced ionpair mass, and ω_h the harmonic Morse potential frequency. Thus, explicity, the oscillator strength of the nth order process is given by

$$A_n(r) \sim (\omega_h(\tfrac{m}{2U})^{1/2})^{2n-4}(2^{n-1}-1)^2(\langle r_1^2\rangle)^{n-1} \qquad (8)$$

with $\langle r_1^2\rangle$ the mean square ionic displacement given by

$$\langle r_1^2\rangle = \frac{1}{N}\int_{\omega=0}^{\omega_{max}} \rho(\omega)\ d\omega\left[\frac{\hbar\omega}{e^{\hbar\omega/kT}-1} + \frac{\hbar\omega}{2}\right]\frac{1}{m\omega^2} \qquad (9)$$

Here we calculate the absolute values of the absorption coefficient for NaCℓ, KBr and LiF using the method outlined above. The results of such calculation at room temperature are shown in Figure 4. The present experi-

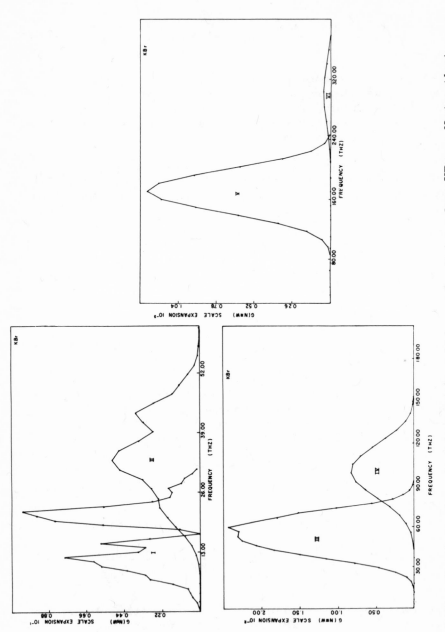

Fig. 3 One- to six-phonon densities of states for KBr. Note that the vertical scale has been expanded to show the normal distribution behavior of higher order processes. One-phonon density of states was normalized to one.

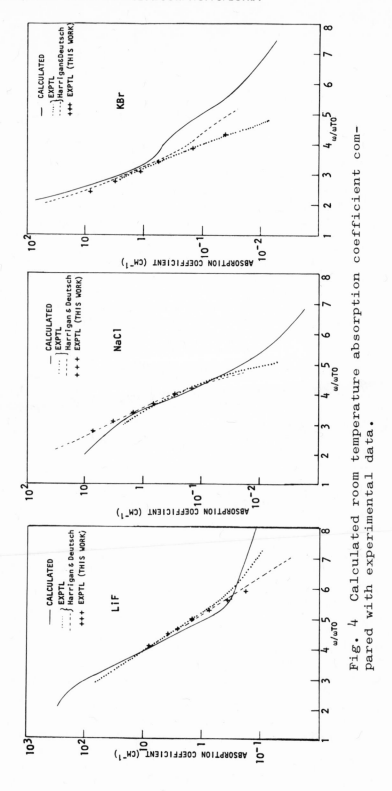

Fig. 4 Calculated room temperature absorption coefficient compared with experimental data.

mental data $(300K)$, and those compiled by Deutsch[8] are
also displayed on this figure. The agreement between the
calculated and experimental data is good.

From the expression for α, it is possible to see
that at a fixed frequency the absorption coefficient de-
pends on temperature dependence of $\rho_j(\omega)$, $A_j(\omega)$ and ε'.
At this frequency $\varepsilon' \approx \varepsilon_\infty$ and therefore has a relatively
small temperature dependence. Thus temperature depen-
dence of α is mainly due to the temperature dependence
of A_j and $\rho_j(\omega)$. A_j is related to ω_h and $<r_1{}^2>$ and its
temperature dependence is essentially manifested through
the phonon occupation number. $\rho(\omega)$ will depend on temperature since phonon frequencies, in general, decrease
with increasing temperature, with a different temperature
dependence for each phonon mode. Hence, one-phonon density of state function will shift to lower frequencies,
in addition to undergoing deformation in its shape. Accordingly, the higher order phonon density of states functions also shift toward lower frequencies, thus somewhat
suppressing higher order processes at higher temperatures.
This problem can be explicitly handled by obtaining j-
phonon densities of states as functions of temperature,
provided each of the 6N phonon frequencies were known as
functions of temperature, which is a formidable, if not
impossible, task. However, a reasonable approximation
is possible, by representing all the normal mode frequencies by a suitable "average" or "effective" phonon frequency and an effective temperature dependence of this
representative phonon. In such an approximation, one may
readily use the formula given by Bendow[9]

$$\alpha(\omega) = \alpha_o \frac{[n(\omega_o)+1]^{\omega/\omega_o}}{[n(\omega)+1]} \exp\left(-\frac{A\omega}{\omega_o}\right) \qquad (10)$$

where ω_o is a representative phonon frequency and it is
temperature dependent and the phonon occupation number
n is given by

$$n(\omega) = \left[\exp\left(\frac{\hbar\omega}{kT}\right) - 1\right]^{-1} \qquad (11)$$

Temperature dependence of α_o and A are neglected.

There are two obvious choices for ω_o. If it is assumed that all phonons contribute to the higher order processes, the choice should be the Debye frequency, ω_D. On
the other hand, since the contributions due to acoustic
phonons are relatively suppressed by energy conservation,

ω_0 may be an averaged optical phonon frequency, e.g., the Brout frequency, ω_B

$$\omega_B = \left(\frac{\omega_{LO}^2 + 2\omega_{TO}^2}{3}\right)^{1/2} \tag{12}$$

where ω_{TO} is the long-wavelength transverse optic frequency. It turns out that ω_D and ω_B are almost equal (see Table II) due to the fact that in general high frequency phonons have higher densities. The temperature dependence of a mode frequency ω_i results mainly from thermal expansion, and is given by[10]

$$\omega_i(T) = \omega_i(0) \exp\left(-3\gamma_i \int_0^T \alpha \, dT\right) \tag{13}$$

where α is the linear coefficient of thermal expansion and γ_i is the mode-Grüneisen parameter. A small contribution arises from anharmonicity of the crystal potential and may be neglected in many cases[10]. But as we are using a representative phonon frequency, ω_0 instead of the N individual ω_i's, the T-dependence of ω_0 may be represented by

$$\omega_0(T) = \omega_0(0) \exp\left(-3\gamma \int_0^T \alpha \, dT\right) \tag{14}$$

where

$$\gamma = \frac{\sum_{i=1}^{3N} C_i \gamma_i}{\sum_{i=1}^{3N} C_i} \tag{15}$$

is the thermally averaged Grüneisen constant and C_i's are the Einstein heat functions. γ used in this calculation is that given by the Grüneisen relation[11],

$$\gamma = \frac{3V\alpha B_T}{C_V} \tag{16}$$

In our calculation the constants α_0 and A of eq. (10) were obtained by fitting the equation to room temperature data. When, ω_D or ω_B (Table II) was used for ω_0 and its temperature dependence represented by eq. (14), reasonable but not perfect agreement was obtained with the experimental data on the temperature dependence of the absorption coefficient. To improve the agreement, ω_0 was next used as a fitting parameter, the values of which are also given in Table II. As may be noticed,

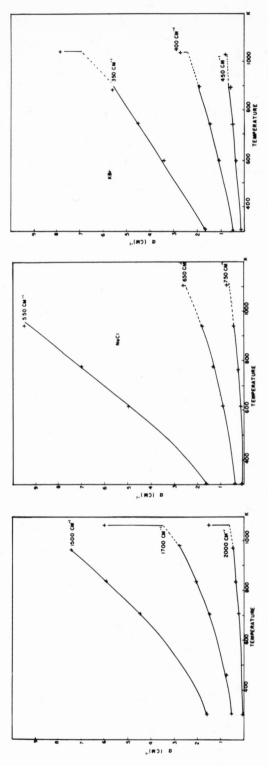

Fig. 5 Temperature dependence of absorption coefficient of LiF, NaCl
and KBr. Lines: calculated; +: present experimental data. Vertical
lines in the right hand side of the Fig. correspond to temperatures
above the melting points.

TABLE II. Characteristic Phonon Frequencies $\left(\text{cm}^{-1}\right)$ and the Grüneisen Constant of LiF, NaCℓ and KBr

Crystal	ω_{TO} [a]	ω_{LO} [a]	ω_B	ω_D [b]	ω_o	γ [c]
LiF	306	659	454	451	414.3	1.99
NaCℓ	164	264	202	195	184.8- 211.0 [d]	1.43
KBr	113	165	132	125	131.9	1.43

a. from reference 10, p. 413.
b. from M. Blackman in "Handbuch der Physik", Vol. VII/1 Springer-Verlag, Berlin, 1955, pp. 325-382.
c. from reference 11, p. 52.
d. The range of ω_o correspond to photon frequency range of 580 to 750 cm^{-1}.

these values are in the vicinity of ω_B and ω_D but not identical with either. For NaCℓ a range of ω_o was needed for fitting the temperature dependence of α at various frequencies. The experimental and calculated results are shown in Figure 5.

It may be observed that the experimental points for the molten salts are higher than those predicted by extraplation of the calculated results for the solids. Nevertheless, the relatively small change in the spectra which accompanies the melting transitions implies that the respective vibrational spectra, close to the melting point, are not too different in the solid and liquid phases.

ACKNOWLEDGMENTS

The authors are gratefull to Dr. K. V. Namjoshi for helpful discussions. Thanks are also due to the University of Rhode Island Computer Laboratory where all the calculations were performed.

REFERENCES

1.   A. J. Barker, J. Phys. C 5, 2276 (1972).
2.   A. J. Barker, J. Phys. E, to be published.
3.   A. J. Barker, Ph.D. Thesis, University of London, London, U.K., 1972.

4.   L. Genzel, Glastech Ber <u>24</u>, 5563 (1951).

5.   M. Sparks and L. J. Sham, Solid State Commun. <u>11</u>,
      1451 (1972); D. L. Mills and A. A. Maradudin, Phys.
      Rev. B<u>8</u>, 1617 (1973); B. Bendow, S. C. Ying and S.
      P. Yukon, Phys. Rev. B<u>8</u>, 679 (1973); H. B. Rosen-
      stock, Phys. Rev. B<u>9</u>, 1963 (1974).

6.   K. V. Namjoshi and S. S. Mitra, Phys. Rev. B<u>9</u>, 815
      (1974); Solid State Commun. <u>15</u>, 317 (1974).

7.   K. V. Namjoshi, S. S. Mitra and J. F. Vetelino, in
      <u>Phonons</u>, edited by M. A. Nusimovici, Flammarion,
      Paris, 1971, pp. 79-83.

8.   T. F. Deutsch, J. Phys. Chem. Solids <u>34</u>, 2091
      (1973).

9.   B. Bendow, Appl. Phys. Letters <u>23</u>, 133 (1973).

10.   S. S. Mitra, in <u>Optical Properties of Solids</u>, edi-
      ted by S. Nudelman and S. S. Mitra, Plenum Press,
      New York, 1969, p. 389.

11.   M. Born and K. Huang, <u>Dynamical Theory of Crystal
      Lattices</u>, Oxford, 1954, p. 52.

OPTICAL ABSORPTION BY ALKALI HALIDES: POSSIBLE STRUCTURE IN THE

MULTIPHONON REGION*

L.L. Boyer, James A. Harrington**, Marvin Hass, and
Herbert B. Rosenstock

Naval Research Laboratory
Washington, DC   20375

Earlier theories of the multiphonon absorption in insulators
have either treated anharmonicity as a perturbation rather than
exactly, or have ignored the lattice in favor of a single-oscil-
lator model. Here we attempt to incorporate the important features of both models into a single theory applicable to many substances. For the heavier alkali halides, the calculation predicts
some structure in the absorption spectrum in the "transparent"
region well above the reststrahl line. The nature of this structure is examined in detail by focusing specifically upon results
for the potassium halides. The amount of structure is shown to
be related to the amount of overlap between the optic and acoustic
branches of the phonon dispersion curves. Experimental possibilities for observing the multiphonon bumps are discussed.

I.   INTRODUCTION

The absorption of infrared radiation by ionic crystals in the
"transparent" frequency region well above the reststrahl line has
been of recent interest because of their application in windows
for high power infrared lasers. At these frequencies the absorption can be attributed to "intrinsic" processes involving several
phonons or to defect modes involving impurities, vacancies or
surfaces. A number of theoretical approaches[1-7] have been developed to calculate the frequency and temperature dependence of the
intrinsic multiphonon absorption.

The "intrinsic" absorption may be due to two possible mechanisms: anharmonic coupling of phonons to the reststrahl mode which
is possible even in a crystal of rigid ions, though not in a wholly
harmonic one ("anharmonicity"); and a displacement-induced electric

moment of the ions themselves, which can couple directly to the radiation ("higher order moments"). Strictly speaking, these two mechanisms are not wholly distinct, since they both result physically
from "charge overlap". Traditionally, however, theories have made
this distinction. The bulk of past work on ionic crystals deals with
anharmonically induced absorption with the higher order moments mechanism only receiving attention very recently [2c,3,4b,4c,6c]. The
effect of higher-order moments is expected to be less important for
the highly ionic compounds since the constituent ions are generally
less polarizable.

Two fundamentally different approaches have been taken to
calculate the anharmonically induced absorption in ionic crystals.
The most common approach[2-5] requires the solution of the harmonic
lattice problem followed by a perturbation treatment of the anharmonicity. Such calculations using these methods become increasingly
complex for higher order processes and consequently simplifying
assumptions and approximations are made.

Other workers[6-7] stress the importance of treating the anharmonicity exactly but in order to do so the lattice model is abandoned
in favor of a single oscillator picture. The Morse potential is
especially useful in this regard since it has an exact quantum
mechanical solution.

In our recent work[8] we attempted to incorporate the important
features of both approaches into a single theory. The Morse potential was retained as a device for evaluating matrix elements
while the lattice properties were included through the use of a
realistic n-phonon (rather than 1-phonon) frequency distribution.
An extensive series of calculations of the frequency and temperature dependence of multiphonon absorption for a number of alkali
halides and alkaline earth fluorides was carried out and compared
with existing experimental data. The overall agreement was quite
good except for the lighter compounds at high temperatures, where
the model is inadequate. An interesting feature of these results
is the prediction of structure in the multiphonon ($n \geq 3$) absorption
spectrum for some heavier alkali halides which has not been observed.

Here we review the essential features of this theory and
analyze the nature of the multiphonon structure by focusing our
attention on results for the potassium halides. The experimental
possibilities of observing such structure are also discussed.

II. THEORY

The absorption coefficient of a gas of diatomic ionic molecules
at Temperature T is

$$\beta(\nu,T) = \frac{4\pi^2 \sigma e^2 \Omega}{3\hbar c n_o Z} \sum_m \sum_{n>o} \left| \langle m+n | r | m \rangle \right|^2$$

$$\times (e^{-E_m/kT} - e^{-E_{m+n}/kT}) \, \delta \, (\Omega -E_{m+n}/\hbar + E_m/\hbar) \tag{1}$$

where $\nu = \Omega/2\pi$ is the frequency of the absorbed radiation, σ is the number of molecules per unit volume, e is the magnitude of the charge of the ions, $2\pi\hbar$ is Planck's constant, c/n_o is the velocity of the light in the medium, k is Boltzmann's constant,

$$Z = \sum_m e^{-E_m/kT} \tag{2}$$

is the partition function, and E_m and $|m\rangle$ are the eigenvalues and eigenvectors of the Hamiltonian for the vibrational motion of a single molecule. The Morse potential,[9]

$$V(r) = D(1-e^{-a(r-r_0)})^2 , \tag{3}$$

will be used to determine $|m\rangle$ and E_m; it has energy levels

$$E_m = \hbar\omega (m+(1/2)) [1 - \tfrac{1}{j} (m+(1/2))], \tag{4}$$

$$m = 0, 1, 2,\dots m_{max}$$

where m_{max} is the largest integer below $(j - 1)/2$, with

$$j = \frac{4D}{\hbar\omega} \tag{5}$$

D is the dissociation energy, $\omega/2\pi$ is the classical frequency of small amplitude oscillations of a particle of mass μ in the potential $V(r)$, and is related to the other parameters by

$$a = \omega \left(\frac{\mu}{2D}\right)^{\frac{1}{2}} \tag{6}$$

Two parameters are required to define the Morse potential: D and a, related respectively to the depth and width of the potential well, or D and ω. Strictly speaking r_0 is also an adjustable parameter, but the quantum mechanics of the Morse potential is independent of it.

The dipole matrix elements $r_{mm'} = \left| \langle m |r| m' \rangle \right|$ in Eq. (1) are given by[10]

$$r_{m,m'}^2 = \frac{(j-1-2m) (j-1-2m')}{[a(j-1-m-m')(m'-m)]^2} \frac{\binom{m'}{m}}{\binom{j-m-1}{m'-m}} \tag{7}$$

where $\binom{m'}{m}$ denotes the usual binomial coefficient, and may be approximated as

$$r^2_{m+n,m} = \frac{(m+1)(m+2)\ldots(m+n)}{n!} \quad \frac{\hbar}{2\mu\omega}\left(\frac{\hbar\omega}{4D}\right)^{n-1} \quad (8)$$

for low lying levels (small m). Eq. (8) is useful for physical insight, but the exact expression (7) will be used in our calculations.

As $D \to \infty$, keeping ω fixed, $V(r)$ becomes a parabola, the second term in Eq. (4) goes to zero, and, from Eq. (8), only transitions to adjacent states are allowed (in agreement with the well-known[11] result for harmonic oscillators). Thus equations (4) and (5) show that the parameter $\hbar\omega/4D$ is a measure of the amount of anharmonicity present. As usual, we shall refer to a transition involving states m and $m+n$ as an "nth order" or "n-phonon" transition.

The main virtue of the Morse potential is that is provides a convenient formalism for including the anharmonicity. Once the anharmonicity parameter is determined, the deviation of energy levels from their harmonic values is included implicitly through Eq. (4). The difficulty with the Morse potential treatment, or any independent oscillator approach for that matter, is adapting the formalism to quantitatively describe β for the solid in a way that is physically convincing.

This is accomplished by integrating first-order transitions over a one-phonon frequency distribution, ρ_1, second-order transitions over a two-phonon frequency distribution, ρ_2, etc. Explicitly, Eq. (1) becomes

$$\beta = \frac{4\pi^2 \sigma e^2 \Omega}{3\hbar c \, n_0} \sum_m \sum_{n>0} \int_0^\infty d\omega \frac{1}{Z(\omega)} r^2_{m+n,n}(\omega)$$

$$x(e^{-E_m(\omega)/kT} - e^{-E_{m+n}(\omega)/kT}) \, \rho_n(n\omega) \quad (9)$$

$$x \, \delta \, (\Omega - E_{m+n}(\omega)/\hbar + E_m(\omega)/\hbar)$$

where σ is now the oscillator density of the solid and ρ_1 is normalized to unity;

$$\int_0^\infty \rho_1(\omega) \, d\omega = 1. \quad (10)$$

Given ρ_1, we approximate the higher order distributions by

$$\rho_n(\omega'') = \int_0^\infty d\omega \int_0^\infty d\omega' \ \rho_{n-1}(\omega') \rho_1(\omega) \delta(\omega''-\omega'-\omega) \quad (11)$$

using an iteration procedure beginning with $n=2$. We recognize ρ_n in Eq. (11) as the probability of finding n phonons in the lattice whose energies add up to ω''. Thus, in the present approach, the fact that transitions from levels m to $m+n$ involve physically the creation of n lattice phonons is properly weighted. The Morse potential model now serves the sole purpose of providing a convenient tool for computing the strength of the anharmonic interaction in a physically reasonable way. To be sure, this still neglects wave vector conservation and "difference processes" in which some phonons are destroyed as well as created. Proper account of the former would replace Eq. (11) by complicated multiple integrals over wave vector space; we can argue that detailed effects would probably average out for higher order transitions. Difference processes must, on the average, be of higher order than the "summation processes" we consider, and will therefore make a relatively small contribution to β.

Recall that two parameters, the harmonic frequency and the dissociation energy, are required to define the Morse potential. Integrating over the appropriate frequency distribution accounts for the parameter ω. The quantity $\rho_1(\omega)$ is obtained from published density of states calculations. The published results are adjusted (thermal and anharmonic correction) so that the "measured" frequency is the $T=0$ value of $(E_1-E_0)/h$. The dissociation energy is determined by fitting to expansion coefficient data as described in Ref. 8.

III. MULTIPHONON ABSORPTION IN THE POTASSIUM HALIDES

Values for the oscillator density, index of refraction, reduced mass, and dissociation energy appropriate to the potassium halides are listed in the table below. We use Buhrer's[12] density of states calculations for KF and Basu and Sengupta's[13] results for KCl, KBr, and KI.

	KF	KCl	KBr	KI
$\sigma(\text{\AA}^{-3})$	0.158	0.097	0.084	0.068
n_0	1.36	1.45	1.50	1.60
$\mu(\text{a.m.u.})$	12.8	18.4	26.2	29.8
$D(\text{cm}^{-1})$	4660	3950	4350	3930

Results for KF and KI (Figs. 1 and 2) illustrate the depend-
ence of $\rho_n(\omega)$ on ω and n. We see that detailed structure tends
to disappear as n increases, with ρ_n becoming more and more
Gaussian in shape, but the gap in the density of states for KI
causes this to happen much more slowly than for KF, which has no
gap. This shows that the central limit theorem is not appropriate
for all materials, especially those which have a gap in their
density of states, as noted by Sparks.[2b]

The gap in the density of states of KI merely reflects the
fact that the optic and acoustic branches of the phonon dispersion
curves do not overlap. Thus, the three peaks in ρ_2, for example,
indicate the creation of two acoustic modes, one optic and one
acoustic mode, and finally two optic modes. Similarily the bumps
in ρ_3 correspond to 3 acoustic, 2 acoustic and 1 optic, 2 optic
and 1 acoustic, and finally 3 optic modes. The qualitative features
of similar curves for KCl and KBr lie between the two extremes
shown in Figs. 1 and 2. This has a marked effect on the multiphonon
absorption spectrum, as we shall see.

The calculated frequency dependence of the absorption coefficients of the potassium halides is shown in Fig. 3 for three
temperatures (50°K, 300°K, and 600°K). Experimental data are also
included where available. The most striking feature of these results
is the appearance of pronounced structure in the absorption spectrum
of the heavier compounds.

It is important to analyze in some detail the nature of this
multiphonon structure in order to assess the reliability of the
prediction. Structure in the multiphonon spectrum can result from
1) abrupt changes in absorption upon going from an n to a n+1
dominated region, or 2) directly from structure in the density of
states itself. The first type of structure can be rather subtle as
it depends upon the analytic behavior of the high frequency tail of
the distribution function as well as the dependence of the matrix
elements on the order of the transition. For example, if the
frequency distribution goes discontinuously to zero, as in a Debye
distribution, the absorption coefficient will show this type of
structure as discontinuities between n and n+1 dominated regions.
At the same time, the magnitude of the discontinuity will depend
upon the ratio of the nth to the (n+1)th matrix elements
($\sim \hbar\omega/4D$ for the Morse Potential). Structure of this sort should
not be taken very seriously because it is sensitive to details of
the model which may not be reliable. By contrast, the structure we
see for the heavier potassium halides is due to structure in
the density of states itself. We see that for KI (Fig. 2), a

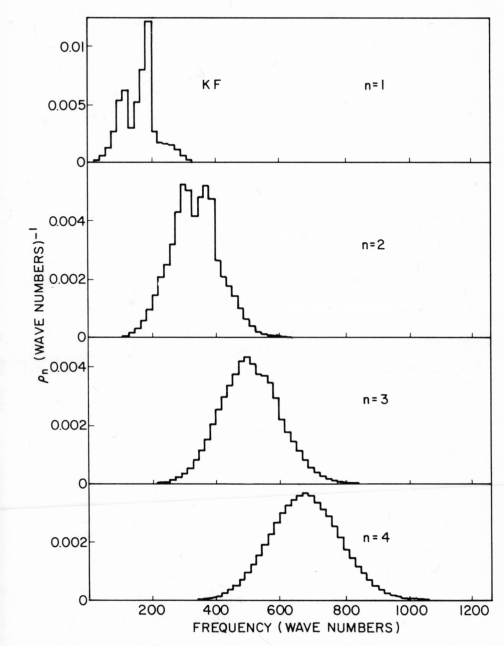

Fig. 1. Normalized n-phonon frequency distribution, $\rho_n(\nu)$, for KF.

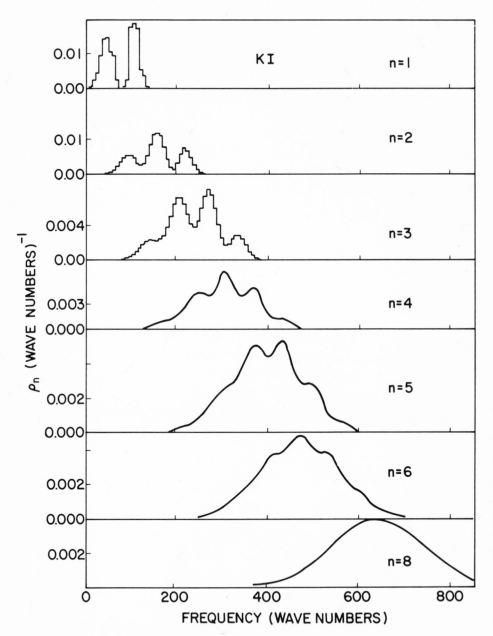

Fig. 2. Normalized n-phonon frequency distribution, $\rho_n(\nu)$, for KI.

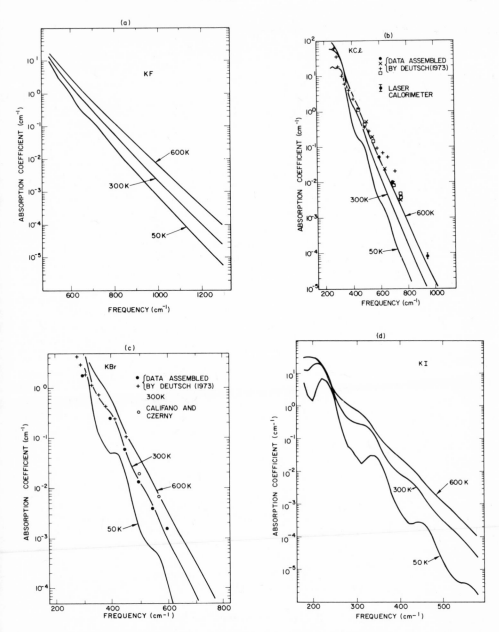

Fig. 3. Theoretical and experimental results for the frequency dependence in the multiphonon region of the absorption coefficients of the potassiumhalides at T = 50K, 300K, and 600K. (a) KF; (b) KCl; (c) KBr; (d) KI. T.F. Deutsch, J.Phys.Chem.Solids 34, 2091 (1973); S. Califano and M. Czerny, Z.Physik 150, 1 (1958).

significant amount of structure remains in ρ_{n+1} at frequencies for which ρ_n is already zero for all n up to 5 or 6. For KF, this is true only for n up to 2. The effect of this is clearly evident in the absorption spectrum (see Fig. 3d). For example, the slight shoulder at ~ 280 cm^{-1} is due to the third peak in ρ_3 while the large band at ~ 340 cm^{-1} is due to the adjacent highest frequency peak in ρ_3. For lower frequencies second order processes dominate and therefore the effect of structure in the low frequency end of ρ_3 is not visible. The higher order bands are each due to the outermost peak in ρ_n.

In order to further examine the nature of this structure we have calculated the absorption in KF using an artificially high value of the dissociation energy. In principle the structure will be enhanced by increasing D because the anharmonicity parameter, $\hbar\omega/4D$, will be smaller, causing a more abrupt transition from n to n+1 dominated regions. However the effect was only slight, with the absorption spectrum for KF remaining essentially structure free.

In conclusion, the structure in the absorption spectrum of the heavier potassium halides results directly from structure in the n-phonon frequency distributions and this is ultimately due to the gap or near-gap between the optic and acoustic branches of the phonon dispersion curves. As this is probably the most reliable ingredient of our model we expect the proclivity for multiphonon structure in the heavier alkali halides to be a genuine effect rather than some artifact such as a peculiar property of the Morse potential.

In searching for this experimentally, the room temperatue absorption spectrum of KI should be studied in the multiphonon region between about 200 and 600 cm^{-1}. This is relatively easily carried out with existing transmission or emittance spectrometers. Here departures from an exponential absorption with bumps or shoulders at nearly integral multiples of an optical lattice frequency in KI are expected. If traces exist at room temperature they may be enhanced at low temperatures (liquid nitrogen).

At low temperatures and even at room temperature for that matter, there is a special problem with KI as it is extremely hygroscopic. As a result surface deterioration and resultant absorption can occur which could be confused with multiphonon absorption.

REFERENCES

*   Supported in part by the Defense Advanced Research Projects
    Agency under ARPA Order No. 2031.
** Present address: University of Alabama in Huntsville, Hunts-
    ville, Alabama 35807.

1.   J.R. Hardy and B.S. Agrawal, Appl. Phys. Lett. $\underline{22}$, 236 (1973).
2.   (a) L.J. Sham and M. Sparks, Phys. Rev. $\underline{B9}$, 827 (1974);(b) M.
     Sparks, Xonics Inc., Technical Progress Final Report, 1972,
     under contract no. DAHC 15-72-C-0129 (unpublished); (c) M.
     Sparks, Phys. Rev. B, to be published.
3.   T.C. McGill, R.W. Hellwarth, M. Mangir and H.V. Winston, J.
     Phys. Chem. Solids $\underline{34}$, 2105 (1973).
4.   (a) B. Bendow, S.C. Ying, and S.P. Yukon, Phys. Rev. $\underline{B8}$, 1679
     (1973); (b) B. Bendow and S.C. Ying, Phys. Lett. $\underline{42A}$, 359 (1973);
     (c) B. Bendow, S.P. Yukon, and S.C. Ying, Phys. Rev. B., to be
     published.
5.   K.V. Namjoshi and S.S. Mitra, Phys. Rev., $\underline{B9}$, 1617 (1974).
6.   (a) D.L. Mills and A.A. Maradudin, Phys. Rev. $\underline{B8}$, 1617 (1973);
     (b) A.A. Maradudin and D.L. Mills, Phys. Rev. Lett. $\underline{31}$, 718
     (1973); (c) D.L. Mills and A.A. Maradudin, Phys. Rev. B., to
     be published.
7.   (a) H.B. Rosenstock, Phys. Rev. $\underline{B9}$, 1963 (1974); (b) Proceedings
     Third Conf. on High Power Laser Window Materials, Nov. 1974 (Air
     Force Cambridge, Research Lab., TR-74-0085(1), C.A. Pitha and B.
     Bendow, editors) p. 205.
8.   L.L. Boyer, J.A. Harrington, M. Hass, and H.B. Rosenstock, Phys.
     Rev. B., to be published.
9.   P.M. Morse, Phys. Rev. $\underline{34}$, 57 (1929); (see also Ref. 7 for a
     review of the quantum mechanics of the Morse potential).
10. K. Scholz, Z. Phys. $\underline{78}$, 751 (1932).
11. L. Pauling and E.B. Wilson, Introduction to Quantum Mechanics
     (McGraw-Hill, 1935) p. 82.
12. W. Buhrer, Phys. Stat. Sol. $\underline{41}$, 789 (1970).
13. A.N. Basu and S. Sengupta, Phys. Rev. $\underline{B8}$, 2982 (1973).

EXPERIMENTAL STUDIES OF MULTIPHONON IR ABSORPTION

Dieter W. Pohl

IBM Zurich Research Laboratory

Säumerstrasse 4, 8803 Rüschlikon, Switzerland

ABSTRACT

The temperature dependence of multiphonon absorption in NaF was measured and analyzed in terms of phonon occupation numbers. This procedure allows to decompose the exponential reststrahl absorption wing into N-phonon spectra (N = 2 ... 5) with considerable structure. The latter coincides with phonon combinations with particularly large densities of states. Comparison with data on KBr, NaCl, and LiF demonstrates the influence of zero point motion on the magnitude of multiphonon absorption.

I.   INTRODUCTION

Our investigations on multiphonon absorption in alkalihalides were motivated by the problem of second sound* detection in materials such as NaF.

Second sound had been excited and detected by means of heat pulses so far. For more detailed investigations, it was highly desirable to employ light scattering techniques. It turned out, however, that the coupling of second sound to light gets extremely small at low temperatures. We evaluated that second sound scattering would become feasible, only, if a thermal wave would be excited

---

*Second sound is wavelike – rather than diffusive – transport of heat. This is possible if elastic phonon collisions in a material become more frequent than inelastic ones. NaF and some other good crystals satisfy this condition at low temperatures.

artificially, for example by a moving thermal grating. Such a
grating can be generated by means of interference and absorptive
heating of intersecting laser-beams. We hence became interested
in the ir absorption of NaF. Employing a CO_2 laser, thermal
gratings in NaF of several millidegree amplitude could be produced.[1]

In order to understand the formation of the thermal grating
correctly, it was necessary to get better insight into the absorption process. The question was which and how many phonons
were produced by the absorption of one 10.6 μ photon.

In the following Sections, we first (II) shall give an overview about the experimental work on multiphonon absorption which
came to our knowledge so far. Emphasis will be put on investigations of frequency *and* temperature dependence of the absorption.
A simple theoretical approach will also be discussed in this con-
text. The following Section (III) describes our investigation of
NaF. The experimental data will be analyzed with respect to temperature and frequency dependence. The final Section (IV) is devoted to the magnitude of the absorption at different orders and
in different alkalihalides. The resumé of the extensive analysis
of experimental data will strongly support the theoretical
approach and, as I hope, a stimulus for further work along
these lines.

II. OVERVIEW

A key to the understanding of multiphonon absorption is its
temperature dependence. The larger the number of phonons participating, the steeper the increase of absorption with temperature.
Unfortunately, up to now, few experimental investigations covered
both frequency and temperature dependence of the ir absorption
wing.

Detailed experimental data of absorption vs. frequency exist
for a large number of crystals listed (probably incomplete) in
Table I. All of these follow more or less precisely the exponential law; but the temperature dependence is known to some
detail only for the last group, the alkalihalides. An overview
over the zones of investigation is given in Figs. 1(a) to (c).

Figure 1(a): The parallelograms represent areas in the ϑ/T
plane for which the absorption is known. The deviation from rectangular shape reflects the increase of absorption with temperature which allows to push the low frequency limit to the left at
small temperatures, and the high frequency limit to the right at
high temperatures. The parallelograms represent the results of
Klier,[2] Barker[3] and Pohl and coworkers.[4]

Table I: Crystals with exponential ir absorption wing

SUBSTANCE	REF.	α vs. T ?
1. OXIDES		
MgO	a,b	
CaO	b	
AL_2O_3	c,d,e	
SiO_2	a	
TiO_2	a	
2. II-VI COMP.		
ZnSe	e,f	x
CdTe	e	
3. III-V COMP.		
GaAs	f	x
4. FLUORIDES		
MgF_2	a,e	
CaF_2	a,e,g	
SrF_2	e,g	
BaF_2	e,g	
5. PEROVSKITES		
$MgTiO_3$	h	
$CaTiO_3$	h	
$SrTiO_3$	a,h	
$BaTiO_3$	a,h	
$CdTiO_3$	a,h	
$CdTiO_3$	h	
$ZnTiO_3$	h	
$PbTiO_3$	h	
$KNbO_3$	h	
$NaNbO_3$	h	
6. ALKALIHALIDES		
"all"		
in particular		
LiF	i,j,k	x
NaF	j,l,m,n	x
NaCl	k,n,o	x
KCl	n,o	(x)
KBr	k	x

References

a) G. Rupprecht, *Phys. Rev. Lett.* 12: 580 (1964).

b) J.T. Gourley and W.A. Runeiman, *J. Phys. C* 6: 583 (1973).

c) Linde Bullet. F-917-D.

d) Oppenheim and Even, *J. Opt. Soc. Am.* 52: 1078 (1962).

e) T.F. Deutsch, *J. Phys. Chem. Sol.* 34: 2091 (1973).

f) L.H. Skolnik, H.G. Lipson, and B. Bendow, to be published.

g) W. Kaiser, W. Spitzer, and R. Kaiser, *Phys. Rev.* 127: 1950 (1962).

h) J.T. Last, *Phys. Rev.* 105: 1740 (1957).

i) H.W. Hohls, *Ann. Phys.* 29: 433 (1937).

j) M. Klier, *Z. Physik* 150: 49 (1958).

k) A.J. Barker, *J. Phys. C* 5: 2276 (1972).

l) D.W. Pohl and P.F. Meier, *Phys. Rev. Lett.* 32: 58 (1974).

m) T.F. McNelly and D.W. Pohl, *Phys. Rev. Lett.* 32: 1305 (1974).

n) J.A. Harrington and M. Hass, *Phys. Rev. Lett.* 31: 710 (1973).

o) L.L. Boyer, J.A. Harrington, M. Hass, and B. Rosenstock, to be published.

Barker's data are particularly interesting since they extend beyond the melting point. They demonstrate that multiphonon absorption is quite insensitive with respect to melting. It is a pity that they were measured at 5 temperatures only which prevents a detailed comparison with theory. The same is true for Klier's investigation which also is restricted to 5 temperatures.

In order to get more detailed information, Thomas McNelly and I measured the absorption of NaF between 600 and 1500 cm^{-1} at about 40 different temperatures between 90 and 830 K. This field hence may be considered the best-mapped area of the whole graph.

Actually, there is one exception: The large vertical bars at 940 cm^{-1} indicate the careful absorption vs. temperature measurements by means of CO_2 laser radiation. These investigations – important for laser design, and for our second sound experiment as well – will be discussed in a number of other papers and shall not be included here.

Figure 1(b): A very important parameter for the classification of multiphonon processes is the minimum number of phonons participating. The latter depends on the LO(Γ) frequency $\tilde{\nu}_{LO}$ which therefore is a useful scaling parameter for the frequency axis. We now find that the positions of the parallelograms almost coincide, being centered in the 3-phonon regime with extensions towards 2 phonons. Note that the numbers N on the abscissa only represent the *lowest* order allowed. Higher order processes may also exist, but they are generally less intense. Hence, the average phonon number $N(\omega)$ adjusts itself in such a way that

$$\tilde{\nu}/\tilde{\nu}_{LO} \times \bar{N} \simeq \tilde{\nu}/\tilde{\nu}_{LO} \times N \simeq 1 \quad . \tag{1}$$

The concentration of experimental data in the 3-phonon regime is an indication that the absorption is much the same for the different alkalihalides when scaled to $\tilde{\nu}_{LO}$. The 10.6 μ lines now have fallen apart being located in the centers of three, four, and five (!) phonon absorptions.

Figure 1(c): Our overview becomes more complete if the temperature scale is normalized, too. In the simplest expression[5a] for N-phonon absorption α_N, the temperature dependence is given by the occupation n of phonon states,

$$n_N = (e^{h\tilde{\nu}/ckTN} - 1)^{-1} \tag{2}$$

with

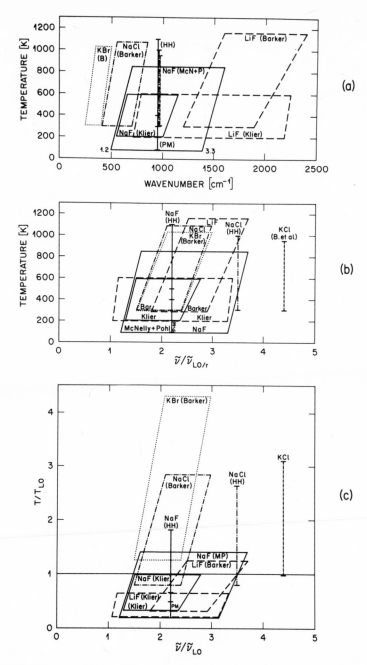

Fig. 1. Fields of experimental investigation, a) in the $\tilde{\nu}/T$ plane,
b) in the $(\tilde{\nu}/\tilde{\nu}_{LO})/T$ plane, c) in the $(\tilde{\nu}/\tilde{\nu}_{LO})/(T/T_{LO})$ plane.

$$\alpha_N(\tilde{\nu}) = (2\pi^3 c/n) <f_N> \tilde{\nu}^2 \rho_N(\tilde{\nu}) \times [(n_N + 1)^N - n_N^N] . \quad (3)$$

$<f_N>$ and ρ_N are average oscillator strength and N-phonon density of states, respectively. From Eq. (1), it is quite natural to normalize T to $T_{LO} = 1.439 \tilde{\nu}_{LO}$. Equation (1) implies that $n_N \approx (e^{T_{LO}/T} - 1)^{-1}$. Equation (3) then simplifies to

$$\ln \alpha_N \propto T_{LO}/(T \times N) \qquad \text{for} \qquad T < T_{LO} \qquad (3a)$$

$$\alpha_N \propto (T/T_{LO})^{N-1} \qquad \text{for} \qquad T > T_{LO} . \qquad (3b)$$

The variation of α_N with N is larger in the high-temperature regime than in the low one. At the first glance, it seems more favorable to determine N at $T > T_{LO}$. However, the quantities in front of the bracket in Eq. (3) are not completely independent from temperature as has been pointed out in various papers.[5] ρ_N, for example, is a function of temperature via thermal expansion and temperature dependent phonon frequencies. These effects are very small at low temperatures and completely negligible. However, they may have some influence at $T > T_L$ as indicated by some experimental results which we shall discuss later.

Figure 1(c) is the frequency and temperature normalized presentation of Fig. 1(a). The areas of validity of the approximations (3a) and (3b) are separated by the line $T/T_{LO} = 1$. Most of the experimental work extends to both sides of the line indicating that the full formalism of Eq. (3) is needed for analysis.

III. MULTIPHONON ABSORPTION OF NAF

a) Temperature Dependence

Let us now have a closer look to the NaF field investigated by Thomas McNelly, P.F. Meier and me.[4,6] Two NaF samples [(A) 54.98 mm,[7] and (B) 3.82 mm [8] thickness] of extreme purity were employed for the spectroscopy. There were no indications of any extrinsic absorption which is important in view of the very small absorption at large frequencies. The crystals were mounted either in a cryostat (90 to 400°K) or in a small oven (300 to 850°K). The absorption was measured by means of a Beckman Acculab-6 ir spectrophotometer. Values of the transmission between 1 and 98-99% could be well detected. The corresponding absorption ranges from about 10 cm^{-1} down to less than 0.002 cm^{-1}. At temperatures different from ambient, corrections to the measured nominal transmission were required. The instrument was not compensated for the thermal radiation from the sample with respect to the variable attenuator in the reference beam. Knowledge of the source characteristic,[9] i.e., its temperature (1240°K)

and emissivity ($\sim$ 0.85, NiCr wire), however, allowed the required
corrections with fair accuracy. Up to 500 K, the correction was
quite unimportant except for the smallest wave numbers investigated. At 940 cm^{-1}, the absorption was also detected by means of
a stabilized CO_2 laser between 4.2 and 400 K.

Entering new territory with respect to the experimental temperature dependence, our first aim was to check the validity of
Eq. (3). Using our own data and those of Harrington and Hass[10] at
10.6 μ, good agreement was obtained with N = 3 and an
appropriately adjusted value of

$$\alpha_N^{(0)} = (2\pi/nc) \ \langle f_N \rangle \ \omega^2 \rho_N(\hat{\nu}) \qquad . \qquad\qquad (4)$$

Figure 2 gives an impression of the quality of the fit which goes
from 4.2 to 1100 K, i.e., from T/T_{LO} = 0.007 to 1.8, covering
practically the whole solid phase. The good fit even at the
largest temperatures indicates that the $\langle f_N \rangle \ \rho_N(\hat{\nu})$ are fairly
constant in the 3-phonon regime. For comparison, we also intro-
duced the corresponding calculated curve for 4-phonon absorption.
It deviates drastically from experimental data.

At 10.6 μ, the situation is particular simple since the
absorption is completely dominated by three phonon processes. In
general, however, the actual absorption will be a sum over pro-

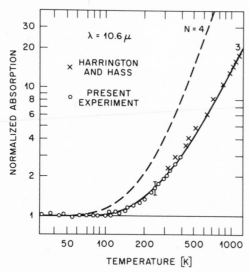

Fig. 2. Experimental and theoretical absorption of NaF at the
CO_2 laser frequency.

cesses with different order:

$$\alpha = \sum \alpha_N = \sum \alpha_N^{(0)} \times [(n_N + 1)^N - n_N^N] \quad . \tag{5}$$

The sum can be restricted to a few elements since the oscillator
strength $<f_N>$ decays exponentially with N (cf. Section IV).

Equation (5) lends itself readily to a least square fit of
the experimental α vs. T curves with respect to $\alpha_N^{(0)}$. It
turned out that the best fit at most frequencies was obtained by
putting all $\alpha_N^{(0)}$ equal to zero except for the two of lowest
orders allowed. At frequencies $700 < \tilde{\nu} < 1000$ cm^{-1} the quality
of the fit was comparable to that of Fig. 2. Below and above the
absorption became very large and very small, respectively, and
the experimental accuracy decreased correspondingly. Figure 3 de-
picts a set of experimental isochromates for $\tilde{\nu} = 960$ to 1420 cm^{-1}
(thick lines) which were obtained from the long sample A. The
fitted calculated curves have also been drawn (thin lines). The
deviations at low temperatures are overemphasized by the logarithmic presentation of the data. The high-temperature deviations
are within the experimental uncertainty of the correction for black
body radiation. At large $\tilde{\nu}$, they are in fact somewhat too large
to be explained by this instrumental effect alone and might indicate some influence of the above-mentioned variation of $<f_N> \times$
$\rho_N(\tilde{\nu})$ with T.[5] The fit to Eq. (5) therefore was restricted to
$T < 500 \ldots 600$ K corresponding to $30 \ldots 40$ data points for each
isochromate.

The resultant decomposed multiphonon spectrum $\alpha_N^{(0)}(\tilde{\nu})$ of
NaF is depicted in Fig. 4. Clearly resolved are the contributions
from 2,3,4 and 5-phonon processes. At the largest frequencies, a
small contribution from 6-phonon processes was found but not included since it scarcely exceeds the experimental uncertainty. As
mentioned above, only two orders of phonon processes were detected
almost at each frequency. It is quite clear from Fig. 4 that the
subsequent higher order processes are weaker by one or two orders
of magnitude. Therefore they cannot easily be detected in the
presence of lower order ones.

At the lowest frequencies, for instance, two phonon absorption is beginning to die out and three phonon absorption starts
taking over. Computation yields $2 \ldots 4\%$ of 4-phonon absorption
in addition - just the right amount to continue smoothly the
$\alpha_4^{(0)}$ curve towards smaller wavelengths. However, with 5 to 10%
experimental accuracy, this low frequency tail of $\alpha_4^{(0)}$ might as
well be a computational artifact and therefore should not be considered too seriously.

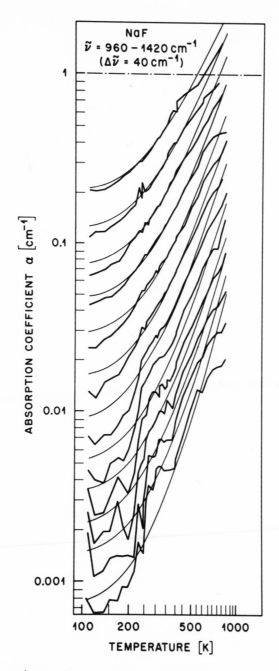

Fig. 3. Experimental and fitted isochromates between 960 and
1420 cm^{-1}. Note the logarithmic display which overemphasizes
deviations at small T.

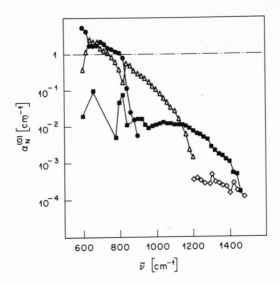

Fig. 4. Experimental 2 (●), 3 (△), 4 (■) and 5 (◇) phonon zero
temperature spectrum of NaF.

b) Frequency Dependence

The frequency dependence $\alpha_N^{(0)}(\tilde{\nu})$ is strongly influenced by
$\rho_N(\tilde{\nu})$, the N-phonon density of states. $\rho_N(\tilde{\nu})$ is an N-fold convolution of $\rho_1(\tilde{\nu})$ with the constraints of energy and momentum
conservation. $\rho_1(\tilde{\nu})$ is known for most alkalihalides from extended investigations on neutron scattering and lattice dynamics calculations.[11] Difficulties with the computation of $\rho_N(\tilde{\nu})$ arise
from the condition of momentum conservation which reduces the
large number of energetically allowed phonon combinations to
those with a resultant moment equal zero. Since the number of
combinational modes grows exponentially with increasing N, the
reduction due to momentum conservation gets more and more independent from specific phonon frequencies. As a consequence, it
can be expected that the N-fold convolution $\bar{\rho}_N(\tilde{\nu})$ without
momentum conservation will be a good approximation to $\rho_N(\tilde{\nu})$ at
larger N.

In the following figures we shall compare the experimental
$\alpha_N^{(0)}$ with prominent (mostly zone edge) combinational modes and
these $\bar{\rho}_N$.

2-Phonon Spectrum [Fig. 5(a)]: The experimental data start
at $\tilde{\nu} > 2\tilde{\nu}_{TO}$ only. Little structure is seen in agreement with
the fact that the major combinational modes are below $2\tilde{\nu}_{TO}$. An

interesting finding is the strong absorption around 800 cm^{-1}, i.e.
close to $\tilde{\nu}_{LO}$ (k = 0). It can be caused only by off-zone boundary
LO phonons which do not show up in the one-phonon spectrum. As

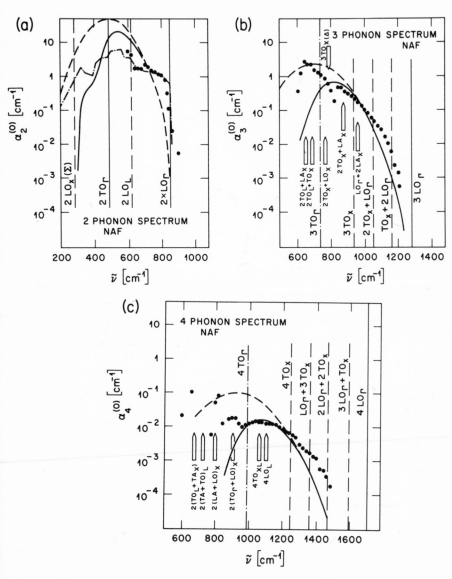

Fig. 5. Experimental multiphonon spectra, prominent combinational
modes, and multiphonon density of state curves. Solid curves: all
phonons included; dashed curves: optical phonons only; dash–dotted
curve in a): $\rho_1(2\tilde{\nu})$.

to be expected, there is poor agreement between $\alpha_2^{(0)}(\tilde{\nu})$ and
$\tilde{\nu}^2 \times \bar{\rho}_2(\tilde{\nu})$ due to the lack of momentum conservation. The latter,
in fact, is much better satisfied if with $\rho_2(\tilde{\nu})$ put equal to
$\rho_1(2\tilde{\nu})$ also shown in Fig. 5(a).

3-Phonon Spectrum [Fig. 5(b)]: $\alpha_3^{(0)}$ has been obtained over
a wide range of frequencies with good precision. Again, little
structure is seen but there are some rudimentary peaks coinciding
with prominent zone-boundary combination modes. It should be
noted that $\alpha_3^{(0)}$ near $3\tilde{\nu}_{TO}$ is quite smooth, i.e., $3\tilde{\nu}_{TO}$ is
not an important mode. The agreement of $\alpha_3^{(0)}(\tilde{\nu})$ and $\tilde{\nu}^2\bar{\rho}_3(\tilde{\nu})$
turns out to be quite satisfactory except for the minimum near
$3\tilde{\nu}_{TO}$ which may be caused by a particularly strong reduction in
allowed modes due to momentum conservation.

4-Phonon Spectrum [Fig. 5(c)]: $\alpha_4^{(0)}$ also was followed
through a wide range of frequencies. The tendency to flatten out
is obvious. The number of prominent combinational modes becomes
very large; a few only have been introduced in the figure in order
to give a feeling "where" the 4-phonon spectrum is sitting. It is
interesting to see that $\alpha_4^{(0)}$ distinctly extends beyond the
largest pure zone boundary combinational mode 4 TO(X) as well
as LO(Γ) + 3 TO(X) but starts to disappear at $2[LO(\Gamma) +$
TO(X)]. A surprisingly good agreement exists between $\alpha_4^{(0)}(\tilde{\nu})$
and $\tilde{\nu}^2\bar{\rho}_4(\tilde{\nu})$ supporting our previous assumption that $\rho_N \propto \bar{\rho}_N$
with increasing N.

5-Phonon Spectrum: $\alpha_5^{(0)}$ further supports the above find-
ings. Again, a good fit to the corresponding $\tilde{\nu} \times \bar{\rho}_5(\tilde{\nu})$ is
found. Irregular fluctuation in the experimental data indicate
that the limit for meaningful results has been approached and the
frequency analysis was finished at this point.

IV. MAGNITUDE OF MULTIPHONON ABSORPTION

a) Different Orders

So far, attention was focused on the temperature and fre-
quency dependence of $\alpha_N(\tilde{\nu}, T)$. The results supported the validity
of Eq. (3). Next the relation between the $<f_N>$'s of different
order shall be examined. $<f_N>$ determines the average magnitude
of $\alpha_N^{(0)}$.

$<f_N>$, for simplicity assumed to be frequency independent, is
proportional to the anharmonicity of the interionic potential φ
which is felt by the ions through their zero point motion $<\Delta r^2>$.
Following Namjoshi and Mitra,[5a] we write

$$<f_N> \propto <\Delta r^2>^{N-1} [\varphi^{(N+1)}/\varphi^{(2)}]^2 \quad , \tag{6}$$

where $\varphi^{(N)}$ is the N-th derivative of φ. Introduction of the Born-Mayer Potential $e^{-r/\rho}$ + Coulomb term yields

$$<f_N> \propto (<\Delta r^2>/\rho^2)^{N-1} \quad . \tag{7}$$

The zero point motion is

$$<\Delta r^2> = (\hbar/2m) \int \rho_1(\tilde{\nu}) \, d\tilde{\nu}/\tilde{\nu} \quad , \tag{8}$$

where m is the reduced mass of the alkali and halide ions. With the value of the repulsive range parameter $\rho_{NaF} = 0.29$ Å [12] we find

$$<f_N>_{NaF} \propto \begin{cases} 0.036^{N-1} \text{ for optical phonons alone} \\ 0.049^{N-1} \text{ for optical and accoustical phonons} \\ 0.054^{N-1} \text{ for the Debye approximation.} \end{cases} \tag{7a}$$

The absorption coefficients of subsequent order hence should differ by a factor of 20 to 30 for equivalent frequencies. A glance at Fig. 4 shows that this condition is quite well satisfied by the experimental results. Best fits of the $\tilde{\nu}^2 \rho_N(\tilde{\nu})$ curves to the experimental $\alpha_N^{(0)}$ spectra provide the values listed in Table II. The tendency of the experimental ratios to approach the "optical phonon" value 0.036 is obvious and supports the assumptions entering Eqs. (6) to (8).

b) Different Alkalihalides

The availability of experimental data $\alpha(\tilde{\nu},T)$ for LiF,[2,3] NaCl and KBr[3] allows a further check of the validity of the above

Table II: Experimental and theoretical decay of
oscillator strength $<f_{N+1}>/<f_N>$

N	Expt	Calculated (a)	(o)	"Debye"
2	0.020 ± 0.07			
3	0.023	0.049	0.036	0.054
4	0.037			
5	(0.04?)			

theory. An analysis of Barker's absorption curves[3] with the same technique as before shows up that most of his data are located in the 3-phonon regime. Figure 6 represents the resultant $\alpha_3^{(0)}$ together with the NaF curve. To allow for comparison, the frequencies are normalized to $\tilde{\nu}_{LO}$. The $\alpha_3^{(0)}$ curves go fairly parallel indicating that the respective $\rho_3(\tilde{\nu})$ cannot be very different. The NaF curve decays particularly smooth in agreement with a smooth decline of $\rho_1(\tilde{\nu})$ at $\tilde{\nu} \to \tilde{\nu}_{LO}$.[11]

The absolute magnitude of the $\alpha_3^{(0)}$ curves differ distinctly for the four alkalihalides. According to Eq. (7) they are expected to vary like

$$(<\Delta r^2>/\rho^2)^2 \propto 1/(m\rho^2)^2 \qquad . \qquad (9)$$

A comparison of theoretical and experimental ratios of absorption is presented in Table III. The experimental ratios, of course, are approximate averages. In view of the assumptions made, the agreement between theory and experiment is very satisfactory.

Reviewing the successful comparison of experimental and calculated data with respect to temperature, frequency, order, and mass, we feel tempted to conclude that the simple analytical apparatus expressed in Eqs. (1) to (9) is fully adequate to describe multiphonon absorption in alkalihalides. However, it would be desirable to have better methods for the computation of $\rho_N(\tilde{\nu})$

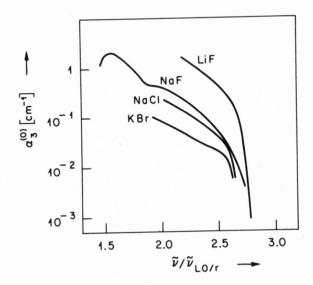

Fig. 6. Zero temperature three-phonon spectra for four alkalihalides.

Table III: 3-Phonon absorption of LiF, NaF, NaCl, KBr

SUBSTANCE (x)	REDUCED MASS	REPULSIVE RANGE PARM. ρ [Å]	$\dfrac{<\Delta r^2>_x}{<\Delta r^2>_{NaF}}$	$<\alpha_{3,x}^{(0)}/\alpha_{3,NaF}^{(0)}>$ (experim.)
LiF	10.2	0.29	4.4	5
NaF	21	0.29	–	–
NaCl	27.8	0.32	0.5	0.6
KBr	41.6	0.34	0.19	0.25

and more experimental data on other substances. It then might be possible to get detailed information on such interesting properties like the shape of the interionic potenial or the oscillator strength at particular combinational modes.

REFERENCES

1. D.W. Pohl, S.E. Schwarz, and V. Irniger, *Phys. Rev. Letters* 31: 32 (1973).
2. M. Klier, *Z. Physik* 150: 49 (1958).
3. A.J. Barker, *J. Phys. C* 5: 2276 (1972).
4. T.F. McNelly and D.W. Pohl, *Phys. Rev. Letters* 32: 1305 (1974).
5. D.L. Mills and A.A. Maradudin, *Phys. Rev.* 8: 1716 (1973); B. Bendow, S. Ying, and S.P. Yukon, *Phys. Rev.* B 8: 679 (1973); M. Sparks and L.J. Sham, *Phys. Rev.* B 8: 3037 (1973); B. Bendow, *Phys. Rev.* B 8: 5821 (1973); K.V. Namjoshi and S.S. Mitra, *Sol. State Commun.* 15: 317 (1974); a) see, for example, K.V. Namjoshi and S.S. Mitra, *Phys. Rev.* B 9: 815 (1974) and references cited therein.
6. D.W. Pohl and P.F. Meier, *Phys. Rev. Letters* 32: 58 (1974).
7. This crystal previously had been used for second-sound experiments; cf. T.F. McNelly *et al.*, *Phys. Rev. Letters* 24: 100 (1970).
8. Supplied by Sonderforschungsbereich 67 an der Universität Stuttgart, Stuttgart, W. Germany; cf. Ref. 6.
9. Obtained from the nominal transmission at $\tilde{\nu} > 600$ cm^{-1} where the samples are practically opaque; the values are in agreement with Beckman technical data.
10. J.A. Harrington and M. Hass, *Phys. Rev. Letters* 31: 710 (1974).
11. For NaF: A.M. Karo and J.R. Hardy, *Phys. Rev.* 181: 1272 (1972).
12. C. Kittel, *Introduction to Solid State Physics*, Wiley, New York (1967), p. 98.

TEMPERATURE DEPENDENCE OF MULTIPHONON ABSORPTION IN FLUORITE CRYSTALS

H. G. Lipson and B. Bendow
Air Force Cambridge Res. Labs., Hanscom AFB,
Mass. 01731
and
S. S. Mitra*
Department of Electrical Engineering
University of Rhode Island
Kingston, R. I. 02881

We report measurements of the absorption coefficients of CaF_2, SrF_2 and BaF_2 in the frequency range ~700 to ~1600 cm^{-1}, at temperatures ranging from 300 to 800°K. The frequency dependence is found to be nearly exponential-like in the temperature range investigated, in agreement with existing multiphonon theories. A detailed analysis of the measured temperature dependence, utilizing the theories of Bendow, and of Namjoshi and Mitra, shows that the observed behavior conforms overall to that predicted for intrinsic multiphonon processes. The role of the temperature dependence of the phonon spectrum in suppressing the temperature dependence of the multiphonon absorption is also investigated.

I. INTRODUCTION

Multiphonon absorption is generally believed to be responsible for the residual absorption in the transparent regime of pure non-metallic solids. In this regime the spectrum is essentially structureless and the

*Supported by AFCRL(AFSC), contract F19628-72-C-0286.

absorption coefficient usually has an exponential dependence on frequency.   Such observations have been made
on a variety of materials including the alkali halides[1]
and the alkaline earth fluorides[2,3].   Theoretical inter-
pretations based on multiphonon interactions have been
put forward within a variety of formalisms and models[4].
In this paper we report the measurement of the absorption coefficient of CaF_2, SrF_2 and BaF_2 in the spectral
region covering 700 to 1600 cm^{-1}.   The temperature dependence of the spectra has also been measured in the
temperature range of 300 to 800 K.   The results are discussed in terms of currently accepted theories of multiphonon spectra[4].

II. EXPERIMENTAL

The BaF_2, SrF_2 and CaF_2 single crystal samples used
for the higher temperature optical absorption measurements were prepared from high purity Optovac material.
The room temperature transmissions of long samples (CaF_2-
10.12 cm, SrF_2 - 7.50 cm, and BaF_2 - 5.40 cm) with polished ends were determined with a double beam Beckman
IR-7 spectrophotometer.   No impurity absorption bands
were found in the frequency range investigated.   For the
high temperature measurements, samples 3/4" in diameter
with the following polished thicknesses CaF_2 - 0.645 and
1.285 cm, SrF_2 - 1.640 cm, BaF_2 - 1.416 cm, were cut from
the long samples.   The transmissions of these samples were
also measured on the Beckman IR-7 spectrophotometer.

A schematic drawing of the optical set-up used for
the high temperature transmission measurements is shown
in Fig. 1.   The exit beam from a Perkin Elmer Model 99
spectrometer equipped with a NaCℓ prism is focused and
imaged at the center of the furnace with a spherical and
a plane mirror.   The sample is positioned at the focal
point inside the furnace.   The beam from the sample is
re-imaged by a spherical mirror at one of the focal points
of an elliptical mirror and focused on a Perkin Elmer
thermocouple detector.   A thermocouple was used since it
was less affected by fluctuations in background thermal
radiation than more sensitive photoconductive detectors.
The sample holder consisting of two concentric quartz
tubes separated by a layer of quartz wool is movable
throughout the length of a quartz liner tube placed in
the center of the furnace.   The wire wound tube furnace
was 12" long with a uniform temperature region longer
than the average sample size (<2 cm).   The furnace power

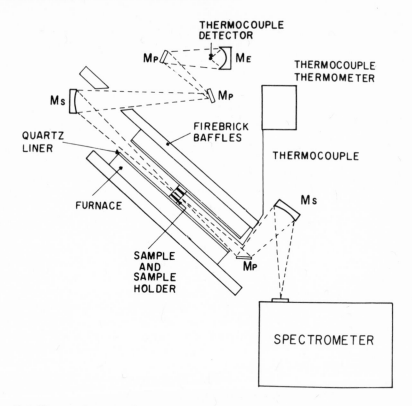

Fig. 1 Experimental arrangement for the high temperature absorption spectra measurements.

was supplied with a Variac which was also used to set the temperatures. Temperatures were measured with a Cu-constantan thermocouple placed at the edge of the sample and monitored with a Doric digital thermocouple meter. At temperatures above 400°C an extended calibration was used[5]. The furnace was allowed to stabilize at each temperature with a variation during a measurement of less than ±5°C at the highest temperatures (∼535°C). The temperature in the center of the sample was also determined by drilling a hole in a sample and inserting a Cu-constantan thermocouple. Before complete stabilization the difference in temperature between the center and edge of the

sample ranged from 14°C at 490°C to 3°C at 95°C. After
about one hour the temperature was nearly uniform between the center and edge of the sample. The entire furnace was baffled with fire brick to eliminate direct heat
radiation from reaching the detection thermocouple.
Opening in the baffle was just large enough to allow the
sample beam to pass through.

Optical transmission measurements were made by an
in-out technique. An I_0 level was first measured throughout the spectral range at a temperature without the sample in position. The sample was then moved to a pre-de-
termined focal position in the furnace and allowed to
stabilize in temperature. Transmission measurements were
made between 700 and 1600 cm^{-1} at temperatures ranging
from 295 to 800°K. For high temperature transmission
measurements at the CO_2 laser frequency the beam was di-
rected through the sample in the center of the furnace.
For these measurements where the temperature rise and decay was continuous rather than at a single stabilized
value, the sample was mounted in a platinum ring and the
measuring thermocouple clamped between this ring and the
edge of the sample. Transmission through the sample was
measured on a Coherent 201 power meter and temperature
determined from Cu-constantan. Thermocouple voltage was
recorded continuously as the sample heated from room temperature to about 800°K and during the cooling period.
I_0 measurements for laser beam intensity were checked previous to the run and at intervals during the heating and
cooling cycles.

Calorimetric measurements at the laser frequency
(943 cm^{-1}) indicated a rapid rise in temperature due to
laser heating, particularly at the higher absorption coefficients obtained at higher temperatures. Temperatures
at the edge of the sample may be considerably lower than
those in the sample center where the transmission is actually measured. This temperature uncertainty is larger
at the higher temperatures.

Absorption values were determined from transmission
measurements using a computer program of the expression
which takes multiple reflections into account

$$\alpha = \frac{1}{d} \ln\{ \frac{(1-R)^2}{2T} + [R^2 + (\frac{(1-R)^2}{2T})^2]^{1/2}\} \tag{1}$$

where R is the reflectivity, T the transmission and d the
sample thickness. The principal error in absorption values

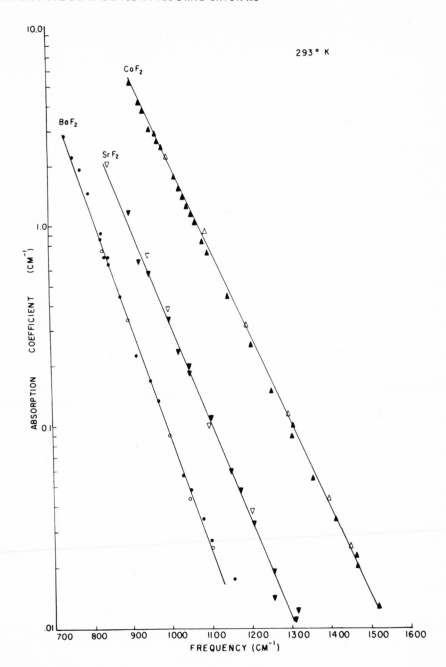

Fig. 2 The room temperature absorption coefficient as a function of frequency for the three alkaline earth fluorides. Closed symbols: present experimental data; open symbols: experimental data of Deutsch (Ref. 2); lines: results of theoretical calculation (Ref. 3).

results from the difficulty in estimating the maximum
transmission of the sample in a region of no absorption.
With the long optical path length involved small sample
positioning variations can contribute to this error, as
well as differences in furnace emission with and without
the sample in position at temperatures whose blackbody
peak is in the spectral region of interest. Repeat runs
were made when necessary and the effect of blackbody
emission on thermocouple signal was determined at the
higher temperatures. This transmission value determines
the reflectivity R used and leads to a background absorption when improperly estimated. Refractive index variation with frequency and temperature in the ranges of investigation were estimated to be small and were not taken
into account in the absorption calculations. Room temperature measurements in the furnace were also made on
each sample for comparison with the Beckman IR-7 spectrophotometer data. The reproducibility of room temperature
absorption values between the two spectrometers was used
as an error estimate for the measurements, as well as the
difference in absorption calculated for a spread in R
values determined from possible error in estimating the
maximum transmission.

III. RESULTS AND DISCUSSION

The measured absorption coefficient (α), at room
temperature for the three fluorides are shown as functions
of photon wave number in Fig. 2 on a semilog scale. For
comparison the room temperature data of Deutsch[2] are also
displayed in the same figure. The agreement between the
two sets of data are excellent and they both show the expected exponential dependence on frequency (ω). The absorption data for SrF_2 as functions of frequency at six
temperatures, ranging from the room temperature to 800°K
are shown in Fig. 3. It is to be noted that for each
temperature $\log \alpha$ is a linear function of ω with a temperature dependent slope. Similar results were also obtained for CaF_2 and BaF_2 and will be reported elsewhere.

To interpret the measured room temperature spectra,
we consider the multiphonon anharmonicity theory of
Namjoshi and Mitra[6], in which the absorption in the highly
transparent regime is given by

$$\alpha = \frac{\pi\omega^2}{2c\sqrt{\varepsilon(\omega)}} \sum_{n\leq 2} A_n \rho_n(\omega) \qquad (2)$$

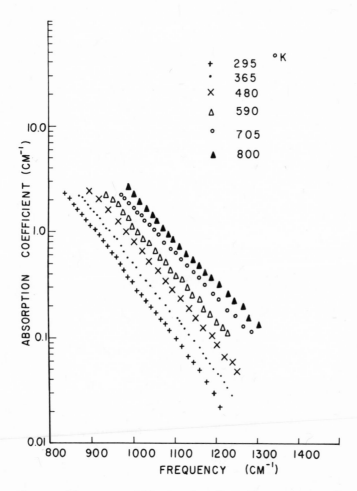

Fig. 3 Absorption coefficient of SrF_2 versus frequency
at various temperatures.

where ρ_n is the n-phonon density of states. For large n ρ_n is, to an excellent approximation, the nth convolution of the one-phonon density of states, which approaches a normal distribution with increasing n. The coefficient A_n is obtained from the Born-Mayer potential as describ-ed in Ref. 6. Such a calculation for the alkaline earth fluorides has recently been reported by Namjoshi, et al[3], and is indicated by the lines in Fig. 2.

In calculating α as a function of temperature T, one must consider both explicit and implicit T-dependences[6]. As explained before, α is a function of T through the T-dependence of A_j and $\rho_j(\omega)$. The T-dependence of A_j is primarily manifested in the temperature dependence of the phonon occupation number, and the T-dependence of $\rho_j(\omega)$ reflects the temperature dependence of the normal modes of the crystal. In the work by Namjoshi, et al[3], the latter was approximated, over a much shorter range of temperature, by the temperature dependence of the bulk modulus. Here, we employ a somewhat simpler approach, as given by Bendow[7]. He obtains the following approxi-mate expression for the temperature dependence of multi-phonon absorption applicable to linear electric moments and high frequencies, where α versus ω is a smooth curve.

$$\alpha(\omega) = \alpha_o \frac{[n(\omega_o)+1]^{\omega/\omega_o}}{[n(\omega)+1]} \exp\left(-\frac{A\omega}{\omega_o}\right) \qquad (3)$$

where ω_o is a T-dependent representative lattice freque-ncy, and n is the phonon occupation number given by

$$n(\omega) = \left[\exp\left(\frac{h\omega}{kT}\right)-1\right]^{-1} \qquad (4)$$

Strictly speaking, α_o and A also depend on T; however, their variation is relatively weak compared to that a-rising from the T-dependence of ω_o and n.

Parametrization of α in terms of an appropriately averaged optical frequency $\omega_o(T)$ is motivated by the fact that selection rules break down in the highly transparent regime but, on the other hand, contributions due to acous-tic phonons are suppressed by energy conservation. The problem then boils down to the choice of a representative optical phonon and its temperature dependence. The long wavelength optical phonon frequencies of the three fluo-rides are listed in Table I. The temperature dependence of the $\vec{k} \simeq 0$ longitudinal optical (LO) and transverse

TABLE I

Characteristic Phonon Frequencies (cm^{-1})
of the Alkaline Earth Fluorides

	CaF_2	SrF_2	BaF_2
ω_{TO}	262	223	188
ω_{LO}	482	395	344
ω_R	322	283	243
ω_B	337	288	247
ω_D	329	288-299	215-240

optical (TO) phonon frequencies has been reported in the
literature[8]. It is also well known[9] that the TO phonons
have a much stronger temperature dependence than the LO
phonons. The absorption coefficient of SrF_2 was calcu-
lated by means of Eq. (3), and using ω_{TO} and ω_{LO}, respectively for ω_O along with their measured temperature dependence. At all frequencies, the absorption values obtained from $\omega_{TO}(T)$ were systematically higher and those
from $\omega_{LO}(T)$ systematically lower than the experimental
values, with the departures becoming more pronounced with
increasing temperatures. The $\vec{k} \simeq 0$ LO and TO phonons form
the two extrema of the optical phonon band; obviously,
better agreement with experiment is expected when an average optical phonon frequency is employed for ω_O. For this
purpose, the following two average phonon frequencies were
considered: (i) The Debye frequency, representing an average phonon for the entire phonon spectrum, and (ii) the
Brout sum phonon frequency, averaging the optical phonons
only. These are also given in the table. The Brout average frequency ω_B was calculated[10] from:

$$\omega_B = [\frac{2\omega_{TO}^2 + \omega_{LO}^2 + 3\omega_R^2}{6}]^{1/2} \tag{5}$$

where ω_R is the Raman frequency. It may be noted that
the long-wavelength Raman frequency, the Brout frequency
and the Debye frequency are not too far from each other.
Calculations of α with a temperature independent Debye

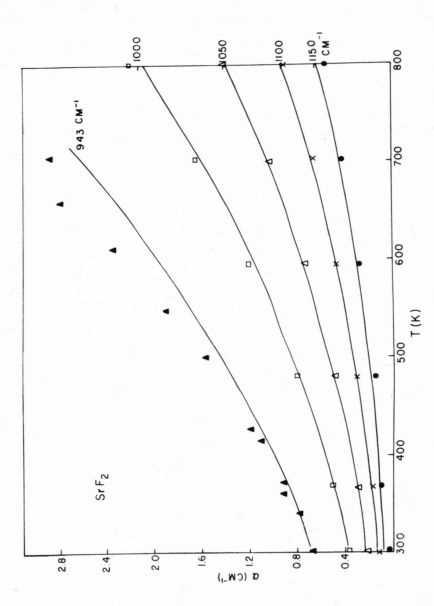

Fig. 4 Temperature dependence of absorption coefficient of SrF_2 at various frequencies. Lines are results of calculation (see text); various symbols are experimental data. The results for 943 cm-1 (Δ) are from CO_2 laser measurements.

frequency also produced higher values, which is not surprising because the effect of temperature is to shift the frequency towards lower values, and suppress[6,7] the temperature dependence of α. From the foregoing discussion it becomes evident that a temperature dependent frequency of the order of ω_B will provide the best agreement with experiment. We here, employ ω_R in calculating α, since $\omega_R \approx \omega_B$ and the T-dependence of ω_B is not readily available. The temperature dependence of ω_R is given by[11]

$$\omega_R(T) = \omega_R(0) \, \exp\left(-3\gamma_R \int_0^T \alpha \, dT\right) \tag{6}$$

where α is the linear coefficient of thermal expansion, and γ_R the mode Grüneisen parameter obtained from pressure dependent Raman measurements[12]. The results of calculation are shown in Fig. 4 for SrF_2. Similar results were obtained for other two fluorides also, and will be presented elsewhere. It may be mentioned that no fitting parameters were used in this calculation. The constants of Eq. (3) were obtained from the slope and the intercept of the calculated[3] straight line (300°K) of Fig. 2. It is noted that the agreement is excellent for the spectroscopic data on SrF_2. Some disagreement is evident in the case of measurements made with a CO_2 laser at 943 cm[-1], particularly at high temperatures. This discrepancy may be partly accounted for by greater local heating of the sample center by the laser beam which introduces a larger uncertainty in the measured temperature than would be expected for results obtained with a spectrometer.

ACKNOWLEDGMENTS

The authors are grateful to Mr. P. Ligor for technical assistance and Mr. N. E. Massa for helpful discussions.

REFERENCES

1. See, for example, Proceedings of the 1972 and 1973 Conference on Electronic Materials, published in the Feb. and May issues of J. Elec. Mats. 2 and 3 (1973, 1974); and the Proceedings of the First, Second and Third Conferences on Infrared Laser Window Materials (AFCRL, Bedford, Mass., 1972, 1973, 1974).

2. T. F. Deutsch, J. Phys. Chem. Solids 34, 2091 (1973).

3.   K. V. Namjoshi, S. S. Mitra, B. Bendow, J. A.
     Harrington, and D. L. Stierwalt, Appl. Phys. Letters,
     26, 41 (1975).

4.   M. Sparks and L. J. Sham, Solid State Comm., 11,
     1451 (1972); D. L. Mills and A. A. Maradudin, Phys.
     Rev. B8, 1617 (1973); B. Bendow, S. C. Ying and S.
     P. Yukon, Phys. Rev. B8, 1679 (1973); K. V. Namjoshi
     and S. S. Mitra, Phys. Rev. B9, 815 (1974); Solid
     State Commun. 15, 317 (1974).

5.   National Bureau of Standards RP1080 (1938).

6.   K. V. Namjoshi and S. S. Mitra, Phys. Rev. B9, 815
     (1974); Solid State Commun. 15, 317 (1974).

7.   B. Bendow, Appl. Phys. Letters, 23, 133 (1973).

8.   I. F. Chang, Ph.D. Thesis, University of Rhode Island
     (1968); R. P. Lowndes, J. Phys. C. (London) 4, 3083
     (1971).

9.   See for example, S. S. Mitra in "Optical Properties
     of Solids" ed. S. Nudelman and S. S. Mitra, Plenum
     Press, New York 1969, p. 385.

10. See Ref. 9, p. 421.

11. See Ref. 9, p. 389.

12. S. S. Mitra and O. Brafman, unpublished. The ex-
     perimental value of γ_R of SrF_2 used in this calcu-
     lation is 1.9.

MULTIPHONON ABSORPTION IN THE ALKALINE EARTH FLUORIDES[*]

M. Chen, M. Hass[+], and T. C. McGill[†]

California Institute of Technology

Pasadena, California 91125

ABSTRACT

The temperature dependence of the intrinsic multiphonon absorption in CaF_2, SrF_2 and BaF_2 was obtained at 10.6 μm from 300°K to 800°K. These results were compared with two different theoretical approaches, both of which have given reasonable agreement for the alkali halides. Best agreement with experiment was obtained with the approach involving a quantum-mechanical perturbative treatment of the potential whose parameters are determined from compressibility and lattice frequency at each temperature.

INTRODUCTION

A successful theoretical approach for multiphonon absorption would be capable of predicting both the frequency and temperature dependence of the absorption for a wide variety of compounds with no adjustable parameters. At the present time, a number of different theoretical approaches involving mechanical anharmonicity have succeeded in accounting for the frequency and temperature dependence of the absorption for most of the alkali halides in a semi-quantitative manner.[1-7] The alkaline earth fluorides are a different class of materials whose frequency dependence is known[8] and can be accounted for by present approaches.[7,9] The present work involves the determination of the temperature dependence of the absorption at 10.6 μm and comparison with various theoretical approaches. In the case of the alkali halides, temperature dependence studies[10] have been of significance in achieving a more accurate theoretical description.

Most present theoretical approaches are based upon Einstein models of the crystal lattice. These approaches fall into two main classes. In the first class, an interatomic potential is derived from experimental parameters such as the compressibility and fundamental lattice frequency.[1-4] This potential function is allowed to be temperature dependent in the same way as the experimental parameters from which it is derived. The second class of approaches also involves the deduction of an interatomic potential from different experimental parameters such as the thermal expansion coefficient. However, the potential constants are not assumed to be temperature dependent.[6,7] In the case of the heavier alkali halides, both approaches have given good agreement with experiment with few, if any, adjustable parameters. In the case of the alkaline earth fluorides, it will be seen that where calculations are available, the temperature dependences are quite different.

EXPERIMENTAL

The absorption measurement technique involves a simple laser transmission experiment at 10.6 μm. While use of a fixed frequency laser severely limits the wavelengths range of measurement, this technique does have the advantage of simplicity, freedom from emittance correction, and wide dynamic range.

The absorption coefficient β can be related to the transmittance Tr by the relation

$$Tr = (1 - R)^2 e^{-\beta L} \qquad , \qquad (1)$$

where L is the path length and R is the calculated single surface reflectivity. This relation is an approximation which holds over the range of interest. The laser power ranged from 2 to 4 watts. Single crystals of CaF_2, SrF_2, and BaF_2 obtained from Hughes Research Laboratory were employed. The sample thicknesses were selected to provide adequate transmission over the range of measurement and the absorption coefficients at room temperature agreed with values reported previously. The sample temperature was measured using a thermocouple cemented to the crystal. The crystals were placed at the center of a small cylindrical oven and the transmittance was obtained as a function of temperature. On recycling down to room temperature, the absorption level increased slightly for SrF_2, but not enough to affect the trend of the results. Such an affect has been noted previously by heating in air and associated with formation of hydroxides and carbonates as well as increased scattering.[11] Consequently, prolonged heating at high temperatures was avoided.

RESULTS

The experimentally observed absorption coefficients of CaF_2, SrF_2, and BaF_2 at 10.6 μm are shown as a function of temperature in Figs. 1-3 along with low temperature data[12] and high temperature data for BaF_2 reported elsewhere[13] and with the theoretical calculations to be discussed shortly. In the higher temperature region, the absorption coefficient can be described as varying as T^x, where the value of x are found to be about 1.7 for CaF_2, 1.9 for SrF_2, and 2.5 for BaF_2.

DISCUSSION

The analysis of these results will be compared with the two theoretical approaches mentioned in the Introduction. Before doing this, however, the temperature dependences can be analyzed using simple Bose-Einstein population factors in which the number of phonons participating in the absorption process and a mean frequency ν_0 is selected by allowing $\nu_L = n\nu_0$ where ν_L is the laser frequency. Here the matrix elements for the transition are assumed to be temperature independent and the temperature dependence is associated with variations in the Bose-Einstein population factors.[14] In this way the temperature dependence data for many compounds can be fitted. In the case of the alkali halides NaCl and KCl, the frequency ν_0 would be close to the longitudinal optical mode frequency at which the density-of-states is small and so this approach is not justified. In the case of NaF, the frequency ν_0 lies closer to a region where the density-of-states is higher and so might be regarded as more physically reasonable in this case.[15]

In the case of the alkaline earth fluorides, the temperature dependence data obtained here has been analyzed with this simple approach and the resulting mean frequencies ν_0 and number of phonons n participating in the process for integral n have been given in Table I along with typical lattice frequencies. It can be seen that the required frequencies ν_0 do lie within the range of the optical branch.

The calculation of the magnitude of the multiphonon absorption and its frequency and temperature dependence involves recourse to a model. Here as indicated in the Introduction, a theoretical approach involving an interatomic potential with temperature dependent parameters has been employed by McGill, et al.[1,2], Sparks and Sham,[3] and Namjoshi and Mitra.[4] While the details of the calculation and the way in which the lattice dispersion is introduced is quite different, the underlying physical concepts are similar. The potential constants of a Born-Mayer potential or a

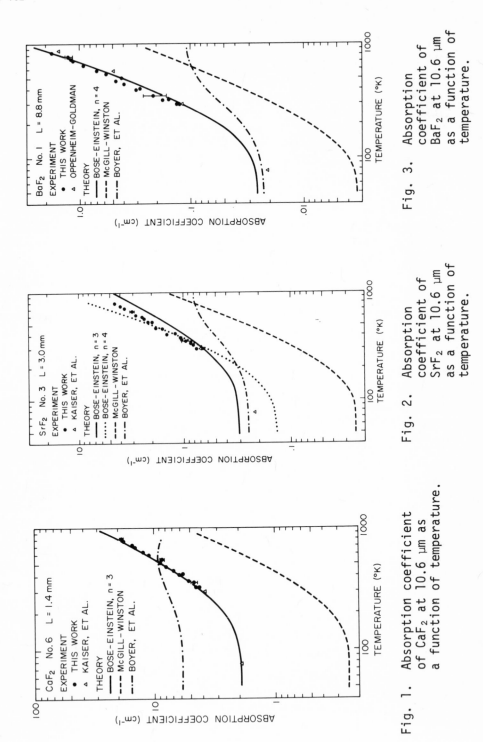

Fig. 3. Absorption coefficient of BaF₂ at 10.6 µm as a function of temperature.

Fig. 2. Absorption coefficient of SrF₂ at 10.6 µm as a function of temperature.

Fig. 1. Absorption coefficient of CaF₂ at 10.6 µm as a function of temperature.

similar potential are obtained from compressibility and lattice
frequency experimental data. These quantities are temperature
dependent and the corresponding potential constants are allowed
to vary in the same way.

Table I. Lattice frequencies using simple Bose-
Einstein temperature factors at 10.6 μm.

Material	No. of phonons n	Mean Phonon Freq ν_o (cm^{-1})	Trans.[a] Opt. Freq (cm^{-1})	Long.[a] Opt. Freq (cm^{-1})
CaF_2	3	314	266	474
SrF_2	3	314	219	382
BaF_2	4	236	189	330

[a]300 K values from Ref. 21.

In the present work, the experimental results have been
compared with the perturbative treatment of McGill and Winston[2] as
an example of this class of approaches. The extension of their
approach for diatomic lattices to fluorite lattices and the
parameters chosen is outlined in the next section. A comparison
of the theoretical results with experiment is shown in Figs. 1-3.
On the whole, the temperature dependence agrees fairly well
with experiment and might be improved slightly by a choice of a
frequency slightly different from the transverse optical mode.
The magnitude of the absorption cannot be predicted as well.
However, the discrepancy is within the range expected for various
kinds of crystal potentials. It should also be noted that the
temperature variation of the experimental parameters for the
alkaline earth fluorides is not as large as for the alkali halides.

A different approach was developed by Maradudin and Mills,[6]
Rosenstock,[7] and Boyer et al.[7] Here a potential with temperature
independent parameters is employed. For calculational simplicity,
a Morse potential has been employed. In the treatment by Boyer,
et al. for which the most extensive results are available, the
parameters of the potential were deduced from the thermal expansion
coefficient and its temperature dependence. Using this approach,
the multiphonon absorption and its temperature dependence can be
calculated in a systematic way with no adjustable parameters.
The results of such a calculation are also shown in Figs. 1-3. At
least for the alkaline earth fluorides, the predicted temperature

dependence is much less than that obtained experimentally. The exact reasons for this have not been explored. It should be pointed out that part of the problem may lie in the use of the Morse potential. For compounds containing lighter atoms such as fluorine, the fundamental frequencies are higher and the finite number of levels in a Morse potential results in lower absorption at higher temperatures. It may be possible to improve agreement with experiment by including transitions to unbound as well as bound states.

CALCULATIONS OF MULTIPHONON ABSORPTION

The calculations in the present work have been carried out according to the approach of McGill, et al.[1,2] As stated here the potential energy for the one-dimensional diatomic lattice can be written as

$$V = \frac{\hbar\omega_0}{2(\lambda g)^2} \left(e^{\lambda gQ} - 1 - \lambda gQ\right) . \tag{2}$$

which can be obtained by rewriting the Hamiltonian (2.9) in Ref. 1. Here Q is a normalized displacement coordinate, ω_0 is the fundamental frequency (here the transverse optical mode frequency was chosen) and λ is a parameter which can be directly related to the usual Born-Mayer distance ρ appearing in the exponential short-range repulsive term by

$$\lambda = \left(\frac{\hbar}{2\omega_0\mu\rho^2}\right)^{1/2} . \tag{3}$$

where μ is the reduced mass. The factor g is introduced as an adjustable parameter which is expected to be close to unity. Thus the potential function employed is quite similar to the usual Born-Mayer potential and the short range part, $V_{SR} \sim \exp(\lambda gQ)$, which is the primary origin of most of the anharmonicity associated with multiphonon absorption, is the same. The long range part, which is linear in the displacement coordinate, is of a different form than the usual Coulomb term, but this is not expected to affect the results to a first approximation.

In extending these calculations to the fluorite structure, the parameter ρ was computed from the compressibility and found to be close to 0.28 A,[16] rather than 0.345 Å as in the alkali halides.[17] The reduced mass $\mu = m_+m_-/(m_+ + 2m_-)$ was employed and an ionic charge of -e and +2e were assumed for the fluoride and metal ions respectively. Here linear moment operator (3.3) of Ref. 1 becomes

$$M = \sqrt{2}e \sqrt{\frac{\hbar}{2\omega_0 \mu}} \, Q \qquad , \qquad (4)$$

for this structure. The values of the parameters employed are shown in Table II. The absorption expression is given by Eq. (5) in Ref. 2.

Table II. Parameters in present calculations of multiphonon absorption

Material	ρ	λg	g
	$(10^{-8}$ cm$)$		
CaF_2	0.268	0.277	0.925
SrF_2	0.289	0.283	1.08
BaF_2	0.297	0.281	1.08

The compressibilities were computed from the elastic constants for CaF_2,[18] SrF_2,[19] and BaF_2.[20] The temperature dependence of the absorption frequencies,[21] elastic constants, and thermal expansion coefficients[22] were all extrapolated from low temperature data (up to 300 K) and this procedure is not as accurate as it might be. The parameter g was adjusted slightly so that the slope of calculated multiphonon absorption curve agreed with the observed experimental data. However, as can be seen from the values of g in Table II, the departure from unity is less than 10%.

On the whole, the agreement between theory and experiment for the temperature dependence is quite good in view of the approximations involved. Slightly better agreement might be obtained by using a slightly higher frequency than that of the transverse optical mode, but this is not significant. The magnitude of the absorption is several times lower than that expected. It is not known if this stems from uncertainty in the nature of the potential[23] or some other factor.

Other results on the temperature dependence of multiphonon absorption of the alkaline earth fluorides have also been obtained by Lipson, Bendow, and Mitra.[24]

CONCLUSION

The temperature dependence of the multiphonon absorption at 10.6 μm in the alkaline earth fluorides have been obtained and can be explained by the use of an Einstein-type model employing an anharmonic potential of the Born-Mayer type in which the potential constants are allowed to vary with temperature in the same way as the experimental parameters from which it is derived. The predicted temperature dependence using a different approach in which the potential constants are temperature independent and are derived from the thermal expansion coefficient is appreciably lower than observed. This may be due to the choice of the Morse function for the choice of potential coupled with the low reduced mass of fluorides.

REFERENCES

*Supported in part by AFOSR under Grant No. 73-2490.

+On sabbatical from Naval Research Laboratory.

†Alfred P. Sloan Foundation Fellow

1. T. C. McGill, R. W. Hellwarth, M. Mangir, and H. V. Winston, J. Phys. Chem. Solids $\underline{34}$, 2105 (1973).

2. T. C. McGill and H. V. Winston, Solid State Commun. $\underline{13}$, 1459 (1973).

3. M. Sparks and L. J. Sham, Phys. Rev. $\underline{B8}$, 3037 (1974); M. Sparks and L. J. Sham, Phys. Rev. Lett $\underline{31}$, 7$\overline{14}$ (1973).

4. K. V. Namjoshi and S. S. Mitra, Phys. Rev. $\underline{B9}$, 815 (1974); Solid State Commun. $\underline{15}$, 317 (1974).

5. B. Bendow, S. P. Yukon, and S. C. Ying, Phys. Rev. $\underline{B8}$, 1679 (1974); Phys. Rev. $\underline{B10}$, 2286 (1974); B. Bendow, Phys. Rev. $\underline{B8}$, 5821 (1973).

6. D. L. Mills and A. A. Maradudin, Phys. Rev. $\underline{B8}$, 1617 (1973); A. A. Maradudin and D. L. Mills, Phys. Rev. Lett. $\underline{31}$, 718 (1973).

7. L. L. Boyer, J. A. Harrington, M. Hass, and H. B. Rosenstock, Phys. Rev. B. March 15, 1975 and Proc. Conference Optical Properties of Highly Transparent Materials; H. B. Rosenstock, Phys. Rev. $\underline{B9}$, 1963 (1974).

8. T. F. Deutsch, J. Phys. Chem. Solids, 34, 2091 (1973).

9. K. V. Namjoshi, S. S. Mitra, B. Bendow, J. A. Harrington,
    and D. S. Stierwalt, Appl. Phys. Lett. 26, 41 (1975).

10. J. A. Harrington and M. Hass, Phys. Rev. Lett. 31, 710 (1973).

11. W. Bontinck, Physica 24, 650 (1958); K. A. Wickersheim and
    B. M. Hanking, Physica 25, 569 (1959).

12. W. Kaiser, W. G. Spitzer, R. H. Kaiser, and L. E. Howard,
    Phys. Rev. 127, 1950 (1962).

13. U. P. Oppenheim and A. Goldman, J. Opt. Soc. Amer. 54, 127
    (1964).

14. J. R. Hardy and B. S. Agrawal, Appl. Phys. Lett. 22, 236 (1973).

15. D. W. Pohl and P. F. Meier, Phys. Rev. Lett. 32, 58 (1974);
    T. F. McNelly and D. W. Pohl, Phys. Rev. Lett. 32, 1305 (1974).

16. J. R. Reitz, R. N. Seitz, and R. W. Genberg, J. Phys. Chem.
    Solids 19, 73 (1961).

17. M. Born and J. E. Mayer, Z. Phys. 75, 1 (1932); F. Seitz,
    Modern Theory of Solids (McGraw-Hill, New York) 1940, p. 82.

18. D. R. Huffman and M. H. Norwood, Phys. Rev. 117, 709 (1960).

19. D. Gerlich, Phys. Rev. 136, A1366 (1964).

20. D. Gerlich, Phys. Rev. 135, A1331 (1964).

21. P. Denham, G. R. Field, P. L. R. Morse, and G. R. Wilkinson,
    Proc. R. Soc. A317, 55 (1970).

22. A. C. Bailey and B. Yates, Proc. Phys. Soc. 91. 390 (1967).

23. A. Nedoluha, Proc. Conf. Optical Properties of Highly
    Transparent Materials.

24. H. G. Lipson, B. Bendow and S. S. Mitra, Proc. Conf. Optical
    Properties of Highly Transparent Materials.

MULTIPHONON ABSORPTION IN KCl, NaCl, AND ZnSe*

J. M. Rowe and J. A. Harrington

The University of Alabama in Huntsville

Huntsville, Alabama 35807

The frequency and temperature dependence of the absorption coefficient has been measured in NaCl, KCl, and ZnSe using dual beam infrared spectroscopy and tunable CO_2 laser calorimetry. The frequency dependent measurements indicate a difference in slope of the log of absorption versus frequency curves between single crystal and the same host hardened by the addition of a small amount of impurities or hot press forged. The temperature dependence of the absorption has been measured from room temperature to near 100K. These measurements extend our earlier high temperature results and thus make possible a more complete comparison between theory and experiment. The results indicate that both an elementary Bose-Einstein treatment and current multiphonon theories are in reasonable agreement with experimental data.

INTRODUCTION

The dependence of the absorption coefficient β on frequency and temperature has been of great interest of late in the study of optical absorption in highly transparent solids. At frequencies from two to eight times higher than the Reststrahl frequency intrinsic (multiphonon) and extrinsic (impurity) processes give rise to a small residual absorption. The frequency dependence of this absorption has been measured (1) in many solids in this regime and always found to vary exponentially, i.e. $\beta \sim e^{-A\omega}$. This exponential behavior has been explained in terms of a variety of multiphonon theories (2). The variation of β with temperature has also been studied as a means of further understanding multiphonon effects (3-5). In general, the observed temperature dependence above the Debye temperature is much weaker than the power law $\beta \sim T^{n-1}$, where n is the order of the multiphonon process, predicted by early multiphonon theories. More recently, however, several new or modified theories have been advanced to satisfactorily explain this weaker temperature dependence.

A study of the frequency and temperature dependence of β allows one to not only study multiphonon processes but also to investigate extrinsic processes as well. In this study we have measured several different samples which exhibit the predicted intrinsic properties as well as some properties which can be clearly attributed to various extrinsic absorption mechanisms such as those due to surfaces and impurities. Specifically, data on β vs. ω and β vs. T will be reported for NaCl, KCl, and ZnSe, over the multiphonon regime and over the temperature range from 100K to several hundred degrees above room temperature. The subsequent analysis will show a correlation with current multiphonon theories when intrinsic processes dominate and provide insight into the nature of the limiting mechanisms when extrinsic processes dominate.

MULTIPHONON THEORIES

Multiphonon theories were at once successful in explaining the exponential variation of β as a function of ω and only more recently at satisfactorily explaining the dependence of β on T. The least sophisticated temperature dependence theory involves merely assuming that the only dependence on temperature comes from Bose-Einstein population factors and that the matrix elements of the interaction as well as other physical parameters are temperature independent. Such considerations lead to the following expression (3a) for β as a function of T when only summation processes are taken into account:

$$\beta_n(T) = \beta_n(0) \left[(\bar{N} + 1)^n - (\bar{N})^n\right] \qquad (1)$$

where $\bar{N} = \left[\exp(\hbar\omega_c/kt)-1\right]^{-1}$ and ω_c is an Einstein oscillator frequency. Eq. 1 has been quite successful in predicting the temperature dependence at low temperatures and energies (n=2 or 3) but has met with only limited success at higher temperatures and energies ($n \geq 4$).

In general the more sophisticated theories take into account the temperature dependence of both matrix elements and phonon dispersion. Several groups (2c-f) have approached the problem from the point of view of adding to their existing multiphonon theories the experimentally obtained temperature dependence of certain critical parameters such as lattice constants, optical mode frequencies, and parameters in the Born–Mayer potential. In this manner the density of states, if included in the theory, is renormalized, the strength of the predicted temperature dependence is weaker than T^{n-1}, and the experimental results are found to be in greater accord with theory. Another approach taken by several investigators (2a, 2g, 2h) begins with a molecular model with an exactly solvable anharmonic (Morse) potential. In this way the temperature dependence is built into the model and is not introduced in an ad hoc manner as is the case with the other approach. These theories also predict a greatly reduced temperature dependence of β; in fact often a little too weak as will be evident from our results. Finally, it is possible to predict a weaker temperature dependence by considering only Bose–Einstein population factors (6). This method, which involves using only a certain percentage of each nth order contribution to β_n (cf. Eq. 1), has been used to satisfactorily explain β vs. T data in NaF for frequencies up into the 3-phonon regime. For higher orders it is not clear how well this procedure will work.

EXPERIMENTAL TECHNIQUES AND METHODS

Room temperature measurements of the absorption coefficient as a function of frequency in the multiphonon region were performed using a Beckman IR-12 spectrometer. The temperature dependence of β was obtained at or near 10.6 μm using a tunable CO_2 laser and calorimetric methods (7). The low temperature data from 300K to 100K were taken using a specially constructed cryostat. The sample is cooled slowly by radiation and a liquid nitrogen temperature shield and cold windows insure that contaminants are condensed on these materials rather than the warmer sample. Just prior to measurement the NaCl and KCl crystals were chemically polished (8) to reduce surface absorption to a minimum.

RESULTS AND DISCUSSION

Deutsch (1) has found that for many materials the log of the absorption coefficient is proportional to frequency over the range of absorption coefficients measurable on an infrared spectrometer, i.e. down to the low $10^{-3} cm^{-1}$ region. It has been customary to

extrapolate these log β vs. ω plots to the CO₂ laser frequency and
call the value of β obtained in this manner the intrinsic absorption coefficient. In Fig. 1 log β vs. ω data are shown for

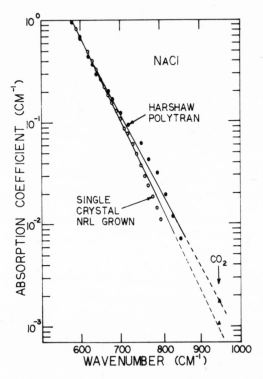

Fig. 1 log β vs. ω for single and polycrystalline NaCl.

two different samples of NaCl at room temperature. The solid lines
drawn through the data are extrapolated (dotted lines) to the 943
cm⁻¹ (10.6 μm) laser frequency where the calorimetric data are
plotted as well. The lower values of β are those of high purity
single crystal NaCl grown by the RAP method (7) at the Naval Research Laboratory (NRL). β_{int} at 10.6 μm is 1 x 10⁻³cm⁻¹ (calorimetry) and approximately 8 x 10⁻⁴cm⁻¹ (extrapolated). This
agrees very well with Deutsch's results. The small difference in
these values is due to surface absorption of approximately 10⁻⁴cm⁻¹
per surface. The higher absorption coefficients belong to a sample
of polycrystalline (Polytran) NaCl obtained from Harshaw Chemical
Co. The calorimetric and extrapolated value at 10.6 μm is 1.7 x

10^{-3}cm^{-1} or slightly above the single crystal value. The discrepancy between the two curves reflects the difference in the
physical properties of the two samples. Polytran has an approximately 10 times greater yield strength than the single crystal.
This difference, which is also seen for two samples of KCl, may be
multiphonon or extrinsic in nature. This point is discussed again
following the KCl results in Fig. 5.

The temperature dependence of β provides yet another method
of distinguishing between intrinsic and extrinsic absorption.
Fig. 2 is a plot of log β vs. log T for NRL grown single crystal
NaCl at 10.6 μm. The low temperature (100-300K) data were taken

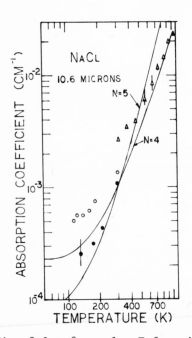

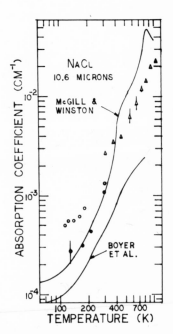

Fig. 2 log β vs. log T for single
crystal NaCl. Solid lines are
Bose-Einstein fits.

Fig. 3 log β vs. log T for single
crystal NaCl. Theoretical curves
are from Ref. 2a and 3a.

on two different samples which clearly reflect a difference in intrinsic levels. The data above room temperature are from Ref. 4.
The improved near intrinsic room temperature values seen for the
low temperature data are due to the better samples obtained from
NRL. The solid curves are Bose-Einstein fits to the data as calcu-
lated from Eq. 1 with $n\omega_c = \omega_L = (943 \text{ cm}^{-1})$, n = 4 or 5 and each
curve is fit to the room temperature intrinsic value of $\beta=1\times10^{-3}$.

It can be seen that neither the n = 4 or n = 5 curve fit the data
too well. Comparing the phonon frequencies corresponding to n = 4
and 5 with the one phonon density of states for NaCl reveals that
n = 5 (ω_C = 189 cm^{-1}) is higher than the TO (164 cm^{-1}) phonon and
in a region of relatively high phonon density but that n = 4 (ω_C =
236 cm^{-1}) is near the LO mode (264 cm^{-1}) in a region of relatively
few phonons. If one neglects the NaCl- open circles in favor of
the more intrinsic material (closed circles) then it would appear
that the population factors with n = 4 are too weak and with n = 5
too strong so that an appropriate n would be somewhere in between.
An analysis based on a fractional n value has been used to fit β
vs. T data for NaF (6). It was decided however, not to pursue
this type of analysis here; instead contact is made with other
available multiphonon theories.

Two multiphonon theories have been chosen for comparison to
our NaCl data. The theory of McGill and Winston (3a) is shown as
a solid line in Fig. 3. Reasonable agreement is obtained, especially at the low temperatures. The theory of Boyer, et al.
(2a), as seen in Fig. 2, falls well below the experimental values.
This is in part due to their use of no adjustable parameters.
Visual inspection, however, reveals that apart from the mismatch
in absolute values of β, the shape of their curve is in reasonable
agreement at low temperatures but predicts too weak a dependence
on temperature above 500K.

Multiphonon data similar to that described above for NaCl
were taken for KCl. As a means of routinely characterizing our
samples, we measured the frequency dependence of the absorption
for several different KCl samples and plotted the results in the
form of log β vs. ω as shown in Fig. 4. The open circles are data
on high purity, RAP processed (7), single crystal KCl grown by NRL.
Upon extrapolation to 10.6 μm, an intrinsic value of about 3 x 10^{-5}
cm^{-1} is obtained. This is in agreement with Deutsch and also the
calorimetric value of 2 x 10^{-4}cm^{-1} when surface absorption of approximately 10^{-4}cm^{-1} per surface is subtracted from this value.
The closed circles are data for RAP processed, single crystal KCl
(9) lightly doped with Sr^{++} (approximately 400 ppm). Again on
extrapolating to the CO$_2$ frequency an intrinsic value of 4 x 10^{-4}
cm^{-1} is obtained. This also agrees well with the calorimetric
value of 7 x 10^{-4}cm^{-1} when surface contributions are removed. From
the data it can be seen that the frequency dependence is the same
for both crystals up to a point after which it deviates to higher
values of β for the doped crystal. As in the case of poly-NaCl,
KCl:Sr is stronger, having a yield strength approximately 10 times
that of the pure crystal. Thus it is expected that other physical parameters such as the compressibility will also differ between
the hosts. When properly taken into account in multiphonon theories this may lead to the correct β vs. ω behavior for the doped
sample. It is possible too, although not as likely, that the ele-

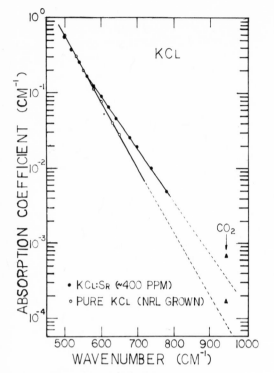

Fig. 4 log β vs. log ω for
single crystal KCl and
KCl:Sr++.

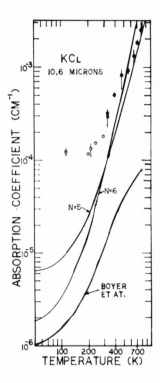

Fig. 5 log β vs. T for KCl.
Theory of Ref. 2a and Bose-
Einstein fits.

vated β vs. ω curve for KCl:Sr is due to the influence of a long,
Lorentzian tail from impurity vibrations set up by the Sr ion.
Such a situation has been considered by Duthler (10) for various
ions in ionic solids.

In measuring the temperature dependence of β in KCl below room
temperature one is faced with an experimentally difficult problem.
Since the intrinsic β at 10.6 μm is $8 \times 10^{-5} cm^{-1}$ and surface absorption is at least $1 \times 10^{-4} cm^{-1}$ per surface there is little hope
of measuring intrinsic absorption as the temperature is decreased
unless surface absorption is somehow taken into account. As a
first measure of β vs. T, data including surface absorption is
shown in Fig. 5 for the single crystal KCl in Fig. 4. The closed
circles are points taken from our earlier work (2a) and the open
circles the data extended to 100K. From the low temperature data
it is clear that, while β is decreasing with decreasing T as expected, the absorption soon levels off at an extrinsic value of
$1 \times 10^{-4} cm^{-1}$. This is mainly surface absorption and work is in
progress to account for this contribution to the total absorption

and thus obtain data which represents just the multiphonon contribution.

Fig. 5 also contains for reference the temperature dependence
of the absorption due to Bose-Einstein factors alone as calculated
from Eq. 1 and the theory of Boyer et al. (2a). The values of
n = 5 and n = 6 predict a temperature dependence too weak and too
strong, respectively, at high temperatures and both fall more rapidly than the data below room temperature. This is due to the
extrinsic surface absorption. The value of the phonon frequency
corresponding to n = 6 is 157 cm^{-1}, which lies above the TO mode
in a region of high phonon density, and for N = 5 is 187 cm^{-1},
which lies nearer the LO mode in a region of relatively few phonons.
Again a fractional n type of analysis similar to that of Pohl (6)
was not attempted primarily because the data at low temperatures
were limited by extrinsic mechanisms. The theoretical calculation of Boyer et al. is, as was the case in NaCl, in poor agreement with the absolute value of the absorption. Visual inspection
indicates that if the theory were shifted upward, the experimental
results would be in reasonable correspondence with the calculation
except at the lowest and highest temperatures. The weaker dependence at the high temperatures is probably due to the neglect
of transitions to the continuum states in the Morse potential
model.

Zinc selenide is extrinsic at 10.6 μm yet its properties are
well studied as it has great potential as a CO_2 laser window
material. Fig. 6 shows the β vs. ω curve for ZnSe at 298K over
the multiphonon edge (left side) and in the extrinsic region
(right side). ZnSe has multiphonon-structure typical of semiconductors evident on the multiphonon tail (12). The extrinsic absorption centered near 6.1 μm may be due to hydride complexes but
its exact nature is unknown.

The temperature dependence of β at 10.6 μm in ZnSe is shown
in Fig. 7. Our room temperature value of 3 x $10^{-3} cm^{-1}$ is certainly extrinsic (the intrinsic value is at least in the $10^{-5} cm^{-1}$
range) and agrees well with the starting value of Skolnik et al.'s
(5) best sample. As expected of extrinsic absorption, there is no
dependence of β on temperature. This could be essentially predicted in advance based on Bose-Einstein factors alone. That is,
knowing the intrinsic absorption at 300K is in the $10^{-5} cm^{-1}$ range
and using $\omega_c \approx \omega_{TO} \approx \omega_{LO} \approx 210$ cm^{-1} and n = 4.5, the multiphonon
absorption cannot increase to even the extrinsic level of 3 x 10^{-3}
cm^{-1} by 500K. Therefore, no increase in β above the extrinsic
value can be expected over the temperature range shown in Fig. 7.
The data do, however, confirm the previous results that the extrinsic absorption is temperature independent.

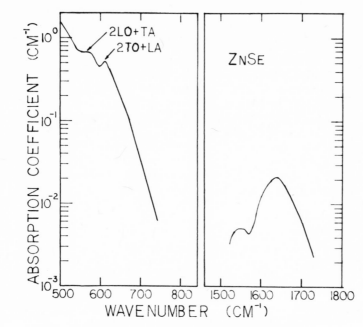

Fig. 6 log β vs. ω for ZnSe (CVD by Raytheon)

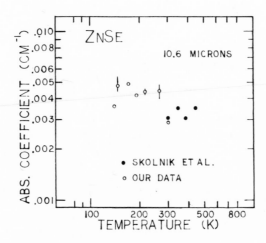

Fig. 7 log β vs. log T for ZnSe

CONCLUSION

The frequency dependence of the absorption coefficient has been found to be in substantial agreement with earlier work for single crystal KCl and NaCl. It has been shown, however, that when more imperfect KCl or NaCl are measured, the β vs. ω curves lie above the corresponding single crystal data. This fact has led us to postulate a fundamental difference in the multiphonon absorption which may be predicted if current theories are modified to account for the difference in physical properties of host crystals. Multiphonon theories have been seen to fit the temperature dependence of the absorption coefficient reasonably well. Extrinsic absorption has been shown to be essentially temperature independent.

*Work supported by U.S. Army Missile Command, Redstone Arsenal, Alabama 35809 under contract DAAH01-74-C-0438.

REFERENCES

1. T. F. Deutsch, J. Phys. Chem. Solids $\underline{34}$, 2091 (1973).
2. (a) L. L. Boyer, J. A. Harrington, M. Hass, and H. B. Rosenstock, Phys. Rev., to be published; (b) J. R. Hardy and B. S. Agrawal, Appl. Phys. Lett. $\underline{22}$, 236 (1973); (c) M. Sparks and L. J. Sham, Phys. Rev. $\underline{B8}$, 3037 (1973); (d) T. C. McGill, R. W. Hellwarth, M. Mangir, and H. V. Winston, J. Phys. Chem. Solids $\underline{34}$, 2105 (1973); (e) B. Bendow, S. C. Ying, and S.P. Yukon, Phys. Rev. $\underline{B8}$, 1679 (1973); (f) K. V. Namjoshi and S. S. Mitra, Phys. Rev. $\underline{B9}$, 815 (1974); (g) D. L. Mills and A. A. Maradudin, Phys. Rev. $\underline{B8}$ 1617 (1973); (h) H. B. Rosenstock, Phys. Rev. $\underline{B9}$, 1963 (1974).
3. (a) T. C. McGill and H. V. Winston, Sol. St. Com. $\underline{13}$, 1459 (1973); (b) M. Sparks and L. J. Sham, Phys. Rev. Lett. $\underline{31}$, 714 (1973); (c) A. A. Maradudin and D. L. Mills, Phys. Rev. Lett. $\underline{31}$, 718 (1973); (d) K. V. Namjoshi, S. S. Mitra, B. Bendow, J. A. Harrington, and D. L. Stierwalt, Appl. Phys. Lett. $\underline{26}$, 41 (1975);
4. J. A. Harrington and M. Hass, Phys. Rev. Lett. $\underline{31}$, 710 (1973).
5. L. H. Skolnik, H. G. Lipson, B. Bendow, and J. T. Schott, Appl. Phys. Lett. $\underline{25}$, 442 (1974).
6. D. W. Pohl and P. F. Meier, Phys. Rev. Lett. $\underline{32}$, 58 (1974); T. F. McNelly and D. W. Pohl, Phys. Rev. Lett. $\underline{32}$, 1305 (1974).
7. M. Hass, J. W. Davisson, P. H. Klein, and L. L. Boyer, J. Appl. Phys. $\underline{45}$, 3959 (1974).
8. J. W. Davisson, J. Mat. Sci. $\underline{9}$, 1701 (1974).
9. Obtained from Dr. Joel Martin, Oklahoma State University.
10. C. J. Duthler, J. Appl. Phys. $\underline{45}$, 2668 (1974).
11. W. G. Spitzer in Semiconductors and Semimetals, Vol. 3, edited by R.K. Willardson and A.C. Beer, Academic Press, NY (1967).

TWO-PHONON ABSORPTION SPECTRA OF III-V COMPOUND SEMICONDUCTORS[*]

E. S. Koteles and W. R. Datars

Department of Physics, McMaster University

Hamilton, Ontario, Canada

High-resolution, two-phonon absorption spectra of
GaAs and InAs were measured with a far-infrared Fourier
transform spectrometer. Prominent Van Hove singularities on the GaAs two-phonon spectrum were identified
with the aid of two-phonon density-of-states spectra
and energy contour plots on symmetry planes. These
were calculated using a shell model, with parameters
derived from neutron diffraction experiments, in a
manner similar to that employed in the investigation
of the two-phonon absorption spectrum of InSb. There
were strikingly similar features in the spectra of GaAs,
InAs and InSb. In the case of GaAs and InSb, energy
contour analysis confirmed that similar features were
produced by phonon combinations of the same type and
from the same Brillouin zone location. This permitted
a tentative assignment of features of the InAs spectrum
in the absence of experimental phonon dispersion curves.
The interpretation of the two-phonon spectra of these
materials demonstrates that not all of the strong peaks
are due to phonon combinations at the high symmetry
points of the Brillouin zone as previously assumed.
Many result from two-phonon processes occurring on the
hexagonal face of the fcc Brillouin zone and at other
locations in the interior of the zone.

[*]Research supported in part by the National Research
Council of Canada

INTRODUCTION

The multiphonon spectra of the III-V compound semiconductors
InSb, InAs, and GaAs have been studied previously by a number of
workers[1-9]. Prominent features in the experimental spectra were
interpreted in terms of combinations of phonons located at high
symmetry points in the Brillouin zone in accordance with group
theoretical selection rules. Recently, we have measured the far-
infrared absorption spectra of these materials with much higher
resolution. The spectral features in InSb were identified by employing experimental phonon dispersion curves, calculated two-
phonon density-of-states curves and calculated two-phonon energy
contours on symmetry planes[10]. It was shown that phonons on the
(111) plane of the Brillouin zone also contribute many prominent
features to the absorption spectrum of InSb.

Now, a similar analysis has been performed on GaAs, which has
experimental phonon dispersion curves available[11]. The dispersion
curves of InAs have not been measured as yet. However, we will
show that many features of the absorption spectra of InAs can be
interpreted by utilizing their resemblance to features in the
spectra of InSb and GaAs.

EXPERIMENTAL

The far-infrared radiation was provided by a Fourier interferometer and detected with a germanium bolometer. Complete details of the experimental apparatus have been given previously[10].
Some parameters of the commercially purchased, nominally undoped
samples are given in Table I. The absorption spectra of un-
oriented single crystals or polycrystalline samples were measured
at liquid helium temperatures.

Table I

	N (cm^{-3})	mobility $(cm^2/V\ sec)$	average thickness (μ)	
			sample 1	sample 2
InSb	10^{14}	6×10^5	2×10^3	4×10^2
InAs	3×10^{16}	2×10^4	6×10^2	2×10^2
GaAs	10^{16}	7×10^3	10^3	4×10^2

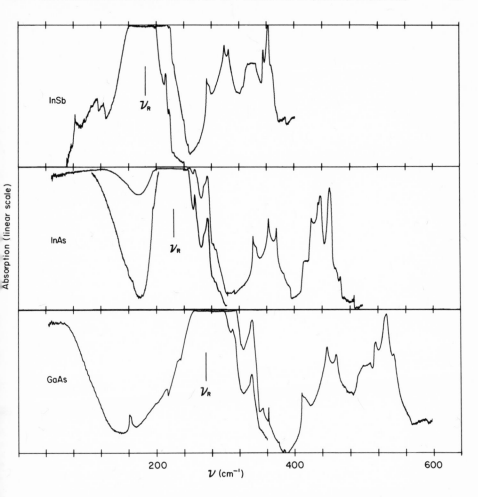

Figure 1: Two-phonon absorption spectra of InSb, InAs and GaAs.
At least two thicknesses of each material were employed in order
to clarify features near the Reststrahlen band (ν_R). The apparent
rise in absorption at low frequencies in InAs and GaAs is due to
the onset of free-electron plasma effects. The sharp peak at
363 cm^{-1} in GaAs is a local phonon mode probably caused by residual Al impurities.

RESULTS AND DISCUSSION

Figure 1 is a computer plot of the far-infrared absorption
spectra of InSb, InAs and GaAs. The high resolution (0.2 cm^{-1})
sharpens many features that are rounded or poorly defined in experiments performed with lower resolution. In most cases the type
and energy of the Van Hove singularities[12] forming the features
are clearly discernable. Since the phonon dispersion curves of

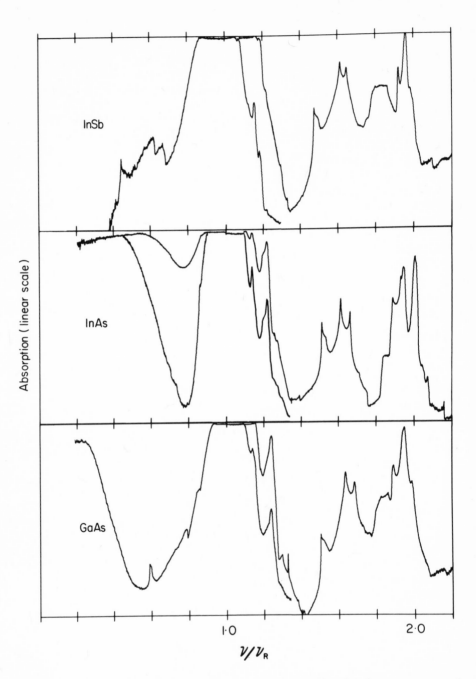

Figure 2: Two-phonon absorption spectra with the frequency scale for each spectrum normalized by the appropriate Reststrahl frequency.

InSb and GaAs have similar shapes it is not surprising that their
absorption spectra should exhibit striking similarities. Further,
many of the features in the spectrum of InAs may be matched with
those in the spectra of InSb and GaAs. This correspondence is emphasized in Figure 2 where the spectra are normalized by dividing
the frequency scale of each by the appropriate Reststrahl frequency.

The resemblance of the dispersion curves of InSb to those of
GaAs lead to the conjecture that similar-looking features in their

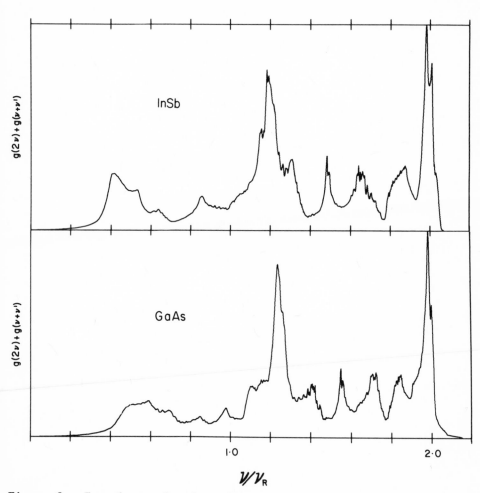

Figure 3: Two-phonon density-of-states for InSb and GaAs calculated using a shell model which had been fit to experimental
phonon dispersion curves. The frequency scales have been normalized as in Figure 2.

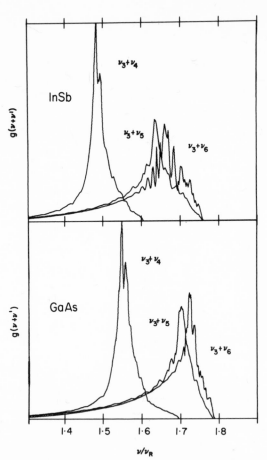

Figure 4: The calculated density-of-states for each of the two-
phonon combinations $\nu_3 + \nu_6, \nu_3 + \nu_5$ and $\nu_3 + \nu_4$ in InSb and GaAs.
The frequency scales have been normalized as in Figure 2.

absorption spectra result from the same two-phonon combination at
the same place in the Brillouin zone. As Figure 3 illustrates,
the theoretical two-phonon density-of-states curve of GaAs has
many features in common with that of InSb. As in the case of
InSb[10], the curve for GaAs was calculated from a 14-parameter
shell model which had been fit to the experimental dispersion
curves. The origin of the peaks arising in the frequency interval
1.4-1.8 in Figures 2 and 3 can be determined from the separate
density-of-states curves for the two-phonon combinations $\nu_3 + \nu_6$,
$\nu_3 + \nu_5$ and $\nu_3 + \nu_4$ in InSb and GaAs shown in Figure 4. (At each
point in the zone, the phonons are numbered sequentially from
lower to higher energy). These are the only two-phonon combinations contributing significantly to the total density-of-states

curve in this interval. The peaks of these three curves form a one-to-one correspondence with the three strong peaks present in the experimental absorption spectra of InSb and GaAs. The location in the Brillouin zone of the particular critical points which give rise to the calculated peaks is determined by studying calculated two-phonon constant-energy contours on the (100),(110) and (111) symmetry planes of the zone. For both InSb and GaAs it is found that the peaks in the $\nu_3 + \nu_6$ and $\nu_3 + \nu_5$ density-of-states curves result from two types of saddle points, with nearly degenerate energies, on the (111) plane. The sharp peak of the $\nu_3 + \nu_4$ curve is also due to a saddle-point-type critical point on the (111) plane. However, for this phonon combination there is a considerable volume of k-space near the Brillouin zone surface in which the energy is relatively constant. This region produces the high energy shoulder which is evident in both the experimental and theoretical peaks. Thus in the frequency interval 1.4-1.8, similar-looking features in the absorption spectra of InSb and GaAs originate from identical two-phonon combinations at identical places in the Brillouin zone.

Having established this connection between spectral features of InSb and GaAs we then relate features in the spectrum of InAs to similar looking features in the spectra of InSb and GaAs. For example, continuing the study of the interval 1.4-1.8, the three prominent peaks present in the spectrum of InAs are identified with those of InSb and GaAs. In order of decreasing energy, these peaks in InAs are ascribed to saddle points on the (111) plane of the two-phonon combinations $\nu_3 + \nu_6$, $\nu_3 + \nu_5$ and $\nu_3 + \nu_4$. These assignments are summarized in Table II. A complete analysis of the two-phonon absorption spectra of InAs and GaAs will be published elsewhere.

Table II

energy of feature (cm^{-1})			critical point		2-phonon combination
			type	Location	
InSb	InAs	GaAs			
273.2	340	410	$P_1 + P_2$	(111)plane & interior	$\nu_3 + \nu_4$
298.8	362	446.5	$P_1 + P_2$	(111)plane	$\nu_3 + \nu_5$
305	373	459.5	$P_1 + P_2$	(111)plane	$\nu_3 + \nu_6$

When the constant-energy contours of all the two-phonon combinations of InSb and GaAs are analysed, some general observations
are possible.

Critical points are present at all the high symmetry points
Γ, X, L and W of this Brillouin zone as predicted by group theory[13].
However, many are found to be either weak or of a type not likely
to produce a prominent feature on the spectrum (e.g. absolute
minima or maxima).

In most cases, the calculated two-phonon combination branches
of GaAs have critical points of the same type and at the same
Brillouin zone location as those of InSb. Where there are significant differences, the branches involved are usually optic overtones or combinations of optic branches with the lower acoustic
modes. Discrepancies in these instances are not too surprising as
shell model theory is known for its inability to fit optic branches
well.

Many prominent features in the density-of-states originate
from critical points on the (111) plane (This does not include the
high symmetry points on this plane which have been discussed previously. This occurs because all of the critical points on this
plane are saddle points, the type of critical point most likely to
produce peak-like features in the spectrum. In addition, there is
generally less dispersion in the phonon branches on this plane
than is usual throughout the zone. Thus, critical points on this
plane give rise to strong features in the spectrum. This latter
point was recognised by some early workers[1,6,7] who attempted to
fit the peaks of the absorption spectra with combinations of only
five "characteristic phonon frequencies". These were defined as
the averages of the frequencies of the phonon branches near the
zone edge.

Thus, in order to interpret the two-phonon spectra of the
III-V compound semiconductors, InSb, InAs, and GaAs, a knowledge
of phonon dispersion curves throughout the Brillouin zone is highly
desirable. Selection rules are useful but, by themselves, provide
insufficient information for a complete analysis since critical
points at non-high-symmetry points contribute significantly to the
spectrum. However, a complete and unambiguous identification of
all the features, large and small, on the experimental lattice
absorption spectra of these compounds awaits the development of a
lattice dynamical model which will fit all the measured phonon
branches with much smaller error.

REFERENCES

1. W. Cochran, S.J. Fray, F.A. Johnson, J.E. Quarrington, and
    N. Williams, J. Appl. Phys. (Suppl.) 32, 2102 (1961)
2. W.G. Spitzer, J. Appl. Phys. 34, 792 (1963)
3. S.S. Mitra, Phys. Lett. 11, 119 (1964)
4. R.H. Stolen, Appl. Phys. Lett. 15, 74 (1969)
5. D.L. Stierwalt and R.F. Potter, Phys. Rev. 137, A1007 (1965)
6. O.G. Lorimor and W.G. Spitzer, J. Appl. Phys. 36, 1841
    (1965)
7. S.J. Fray, F.A. Johnson and R. Jones, Proc. Phys. Soc. Lond.
    76, 939 (1960)
8. F.A. Johnson, Prog. Semicond. 9, 179 (1965)
9. D. Stierwalt, J. Phys. Soc. Jap. Suppl. 21, 58 (1966)
10. E.S. Koteles, W.R. Datars & G. Dolling, Phys. Rev. B9, 572
    (1974)
11. G. Dolling and J.L.T. Waugh, Lattice Dynamics, ed. by
    R.F. Wallis, Pergamon Press (Oxford) p. 19 (1965)
12. L. Van Hove, Phys. Rev. 89, 1189 (1953)
13. J.C. Phillips, Phys. Rev. 104, 1263 (1956)

Section II
Electronic Processes

URBACH'S RULE[*]

John D. Dow

Department of Physics and Materials Research Laboratory
University of Illinois at Urbana-Champaign
Urbana, Illinois   61801

Urbach's Rule states that the fundamental optical absorption
edge exhibits an exponential shape: $K_A(\omega) \propto \exp g(\hbar\omega - \hbar\omega_0)$. The
rule is valid for a variety of non-conducting materials, including
semiconductors and alkali halides (with $g = \sigma/k_B T$ at high temperature, σ being nearly unity and weakly dependent on temperature).
In the case of semiconductors, the origin of the Urbach edge is
well established, largely as a result of a unification of early
theoretical work by Redfield and by Dexter, and as a result of
experiments by Redfield and Afromowitz and by Garrod and Bray.
The semiconductor Urbach rule is attributable to ionization of the
exciton by internal electric microfields. The extent to which the
field-ionization mechanism is responsible for Urbach edges in all
insulators is discussed with emphasis on recent experiments by Williams and Schnatterly. Finally it is suggested that in certain
circumstances the shapes of multiphonon edges and Urbach edges
should be related.

I.   INTRODUCTION

A central theoretical question for the development of ultratransparent lenses and optical fibers is "In an impurity-free
solid, what is the ideal, maximum optical transmission coefficient?"
In the case of the linear optics of solids, this question is equivalent to asking why there is a band gap between multi-phonon

[*]Research supported by the National Science Foundation under grants
NSF-GH-39132 and DMR-7203026.

absorption and electronic absorption, and why the high-energy phonon absorption-edge and the low-energy (Urbach) electronic edge both roll-off into the forbidden band gap as approximately exponential functions of photon energy. Here we shall consider primarily the problem of electronic absorption edges, which has a history of several decades, beginning with Urbach's [1] enunciation of his empirical rule for the shape of the absorption coefficient $K_A(\omega)$ below the edge ($\hbar\omega \ll \hbar\omega_0$):

$$K_A(\omega) = A \exp g(\hbar\omega - \hbar\omega_0) \quad .$$

In Urbach's rule, the parameters A, g, and ω_0 are independent of photon energy $\hbar\omega$ [2].

To understand Urbach's rule, one should first ask why there is a band gap. The band gaps in solids originate from the discrete line spectra of atoms, which, in the condensed state, form broad band spectra. Band gaps, then, are a quantum mechanical diffraction phenomenon, associated with the inability of an electron whose wavelength corresponds to a forbidden energy to meet the Bohr stationary-state constructive-interference condition: that the perimeter of the electron orbit equal an integral number of wavelengths.

The discrete atomic spectral lines in gases are broadened by collisions and by radiative damping, causing weak absorption in the forbidden gap. Normally this broadening is nearly Lorentzian in nature--which, if literally true for energies distant from the line-center (it is not [3]), would imply broad power-law, non-exponential roll-off of the absorption edge into the forbidden gap of the gas. (See Fig. 1.) Without any previous knowledge of the solid-state data, one might naively guess that the fundamental absorption edges in solids should be Lorentzian, not exponential, based on the realization that broadened atomic absorption lines are Lorentzian near their centers. Of course, few solids would be genuinely transparent if this were true.

Theories of Urbach's rule, instead of explaining why fundamental absorption edges do not exhibit broad power-law Lorentzian-type roll-offs, have instead attempted to describe why edges are exponential rather than abrupt. In other words, these theories explain why solids are opaque near (but below) the lowest exciton line, but they do not show why solids are transparent further in the gap. Of course, if we know what does cause a solid to be opaque near its fundamental edge, then we have a good idea of what photoabsorption mechanisms must be forbidden further in the gap, and we need only display the physics that forbids these processes. The natures of

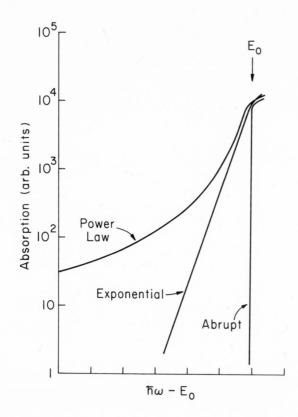

Fig. 1: Illustrating three types of fundamental absorption edges: a power-law roll-off characteristic of a Lorentzian broadened edge, an Urbach-rule exponential edge, and an abrupt edge expected for a solid with perfect translational periodicity.

theories of transparency and opacity are necessarily different, however: to demonstrate opacity requires only one absorption mechanism with sufficient oscillator strength at the photon energy of interest; to demonstrate transparency requires a selection rule or sum rule which forbids all conceivable absorption mechanisms.

The theory of Urbach's rule attempts to explain observed exponential edges in terms of disorder and deviations from the perfect translational periodicity of an ideal insulator; the ideal insulator has an abrupt fundamental absorption edge, which is broadened to the Urbach form by the disorder. Experimentally, the Urbach

slope parameter g exhibits a variety of dependences on temperature
and on impurity concentration, being inversely proportional to temperature (at high temperatures) in alkali halides and II-VI compounds,
but determined by dopant concentrations in III-V semiconductors. The
exciton interaction with phonons (and longitudinal optical phonons, in
particular) has been listed as a primary cause of Urbach's rule in
alkali halides [4], whereas charged impurities seem to shape the edge
of impure GaAs [5]. There are, of course, several different theories
which are capable of producing approximately exponential edges caused
by either impurities or phonons; but if one assumes that the same
essential physics is responsible for all exponential edges, then one
is forced to assume that a single universal mechanism exists.* If a
single mechanism underlies all Urbach edges, a unified theory of the
phenomenon must (i) identify the universal feature of disorder which
causes the exponential tail, and (ii) display this feature in a manner conducive to predicting the outcome of experiments.

II. MICROFIELD MODEL

A. Data

In 1963, Redfield published his electric microfield model of
Urbach's rule, which states that the exponential edge-broadening is
caused by electric fields within the solid [8]. This electric
microfield model was subsequently extended by Dexter [9] and by Dow
and Redfield [2,10]. It has the very attractive feature that it
does not specify the origin of the electric microfields and therefore can offer an explanation of both the impurity-related edges in
III-V compounds and the phonon-related edges in more ionic solids
(the longitudinal optical phonons carry large electric fields).

Two classic experiments support the electric microfield interpretation. The first, by Redfield and Aframowitz [5], demonstrated
that the breadth of the edge is related to the state of charge of
impurities. Redfield and Aframowitz compared the temperature-
dependences of the absorption edges if p-type and compensated GaAs.
The state of charge of the impurities and the observed breadth of
the edge in the compensated sample were both virtually independent
of temperature, whereas the doped sample exhibited increasingly
broad edges as the temperature increased and the impurities ionized
(Fig. 2).

---

*Experiments were once thought to lend strong support to this uni-
versality hypothesis [2,6,7]; recent data [7], however, have removed
that support, but have not contradicted the hypothesis. Thus uni-
versality is an assumption.

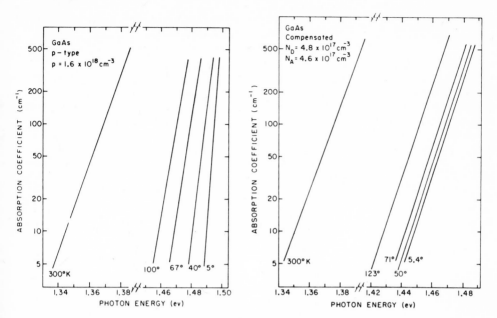

Fig. 2: Absorption edges of p-type and compensated GaAs as functions of temperature (after Ref. 5).

A second experiment by Garrod and Bray [11] consisted in
shooting a pulse of amplified piezoelectric (fast transverse acoustic) phonons down a bar of GaAs and monitoring the domain-induced
change in the absorption edge. The resulting edge was exponential
and has been termed a piezo-Urbach edge [12]. When the incident
acoustoelectric domain struck an interface with a second GaAs crystal rotated by 90°, the transmitted waves were slow transverse
acoustic phonons and did not produce a broad Urbach edge. A similar result held for the longitudinal acoustic waves produced after
reflection. Hence Garrod and Bray have shown that acoustic domains
with large amounts of internal electrical energy cause considerable
Urbach broadening, but that domains with large mechanical energy do
not.

These two experiments lend strong support to the electric
microfield model of Urbach's rule (at least for GaAs); indeed, the
microfield model (with all parameters determined by independent
experiments) has successfully and quantitatively predicted the
breadths of Garrod's and Bray's piezo-Urbach edges [12].

B. Theory

    Theoretically these results can be understood in terms of the
wavefunction of an optically-excited electron-hole pair in the
presence of disorder described by a potential energy function V_d.
When a photon of energy $\hbar\omega$ is absorbed by a non-metal, it excites
an electron from a valence level, leaving a hole behind in its
place. The electron and hole necessarily overlap (or else the
transition is forbidden) and have nearly zero net crystal momentum
$\hbar\mathcal{K} = 0$ (because characteristic lattice constants are negligible in
comparison with optical wavelengths). In terms of the exciton
envelope wavefunction $\Psi(\underset{\sim}{R},\underset{\sim}{r})$, the shape of the absorption spectrum
$K_A(\omega)$ is [13]

$$K_A(\omega) \quad \propto \quad \mathcal{V}^{-1} \lim_{r \to 0} \sum_{\lambda}{}' \left| \int \Psi_\lambda(\underset{\sim}{R},\underset{\sim}{r}) \, d^3R \right|^2 \delta(\hbar\omega - E_{gap} - E_\lambda) \quad ,$$

where $\underset{\sim}{R}$ is the exciton's center-of-mass coordinate, $\underset{\sim}{r}$ is the relative motion coordinate, E_{gap} is the band gap, the integral over
d^3R projects out the $\mathcal{K} = 0$ component of the wavefunction; and the
requirement $r = 0$ insures that the electron and the hole overlap.
If the disorder is static or quasi-static (that is, characteristic
times over which the disorder changes [e.g., phonon periods] are
much longer than electronic relaxation times), then $\Psi(\underset{\sim}{R},\underset{\sim}{r})$ satis-
fies a hydrogenic Schrödinger equation, which we write (for
simplicity) with isotropic effective total and reduced masses M
and μ, respectively:

$$\left(\frac{P^2}{2M} + \frac{p^2}{2\mu} - \frac{e^2}{\epsilon_0 r} + V_d(\underset{\sim}{R},\underset{\sim}{r}) \right) \Psi_\lambda(\underset{\sim}{R},\underset{\sim}{r}) \quad = \quad E_\lambda \Psi_\lambda(\underset{\sim}{R},\underset{\sim}{r}) \quad .$$

In the electric microfield model, the disorder potential in a particular region of the solid is approximated by the potential energy
of a uniform electric field

$$V_d(\underset{\sim}{R},\underset{\sim}{r}) \quad = \quad -eFz \quad .$$

Hence the absorption coefficient is obtained by averaging the theoretical uniform-field absorption coefficient $K_A^\circ(\omega,F)$ (which is an
accurately exponential function of ω for typical microfield strengths
[10]) over a distribution $P(F)$ of internal fields:

$$K_A(\omega) \quad = \quad \int K_A^\circ(\omega,F) P(F) \, dF \quad .$$

Strictly speaking the microfield model is valid only when the characteristic wavelength of disorder is longer than the exciton radius
$a = \hbar^2 \epsilon_0 / \mu e^2$ and the electrooptic length $\ell \equiv (\hbar^2/2\mu|e|F)^{1/3}$; thus
it is best applied to the large radius excitons in III-V compounds
and may require corrections when applied to more ionic materials
such as alkali halides. The physical picture underlying the model,
however, appears to be more general.

To understand the physics and generality of the microfield model, we consider the effect of disorder on energy band edges in real space (Fig. 3) and decompose any perturbation of the perfect solid into an "electrostatic" part (defined as one that causes parallel spatial undulations of valence and conduction band edges) and an "elastic" part (causing antiparallel undulations). This separation is useful if one type of disorder dominates.

For each type of disorder, we ask the question: "In a typical, highly-disordered region of the solid, what is the lowest energy optical transition which contributes significantly to the Urbach edge?"

Recalling that the absorption coefficient is proportional to the square of the electron-hole overlap

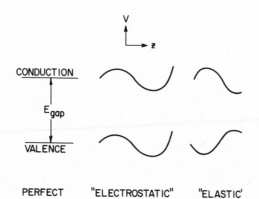

Fig. 3: Energy band edges plotted as a function of position for a perfect solid, "electrostatic" disorder, and "elastic" disorder.

$$K_A(\omega) \quad \propto \quad |\langle \text{electron}|\text{hole}\rangle|^2 \quad ,$$

we see that "elastic" disorder produces a transition from a hole
state at the top of a valence potential energy mountain to an electron state in a potential energy valley in the conduction band
(Fig. 4). Because the electron and the hole overlap so thoroughly,
the primary effect of the disorder is to locally shrink the gap by
an amount Δ. Hence the absorption edge shape may be written crudely
in terms of a shifted theoretical absorption $K_A^o(\omega-\Delta)$, averaged over
$p(\Delta)$, a distribution of shifts

$$K_A(\omega;\text{"elastic"}) \quad \approx \quad \int K_A^o(\omega-\Delta) \; p(\Delta) d\Delta \quad .$$

For $K_A^o(\omega)$ abrupt (a step function) and $p(\Delta)$ Gaussian, we find an

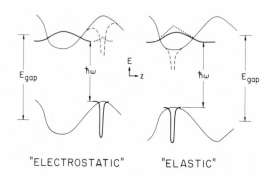

"ELECTROSTATIC"   "ELASTIC"

Fig. 4: Illustrating electron and hole wavefunctions (heavy solid
lines) for low-energy transitions in "electrostatic" and "elastic"
disorder. If the mass of the hole is assumed infinite, the electron
feels the (dashed) Coulomb potential of the hole, which enhances its
wavefunction near the hole (dotted lines). [Ref. 10]

error-function, or roughly Gaussian roll-off. Inclusion of the
electron's interaction with the presumed-massive hole (Fig. 4)
does not alter this conclusion. Hence these crude arguments suggest that "elastic" disorder does not account for the exponential
Urbach edge shape.

"Electrostatic" disorder, however, does seem to produce an
exponential edge. The lowest-energy transition corresponds to a
hole at a valence mountain and an electron in a conduction valley;
the electron-hole overlap is negligible, however, and consequently
this transition between extremely localized hole and electron states
does not contribute significantly to the Urbach edge. A more significant contribution comes from slightly delocalized states where the
electron-hole overlap is significant (Fig. 4). These states overlap
in a region of large disorder potential gradient, not in the region
of large potential. Furthermore, the added Coulomb potential of the
hole permits the electron's wavefunction to overlap the hole significantly. This exciton effect enhances the absorption edge by
orders of magnitude [10].

The important point, then, is that the atypical transitions
which dominate the Urbach edge correspond to the large microfield
regions of the solid; in the regions of electron-hole overlap it
is permissible to replace the "electrostatic" disorder with a quasi-
uniform microfield potential energy $V_d \approx -eFz$ and to compute the
Urbach edge shape using the microfield approximation

$$K_A(\omega) \approx \int K_A^\circ(\omega,F)P(F)\,dF \quad .$$

Physically the disorder electric fields ionize the exciton sweeping
the electron away from the hole and causing it to tunnel into a
nearby potential valley (Fig. 5). Calculations reveal that the
precisely exponential shape of the edge is a general result of
field ionization of the exciton and insensitive to the details of
the microfield distribution for typical microfield strengths [2,10].

The principal features of the microfield model then are (1) it
is capable of providing a unified explanation of Urbach edges; (2)
it relies on internal quasi-static electric microfields; (3) it
involves exciton enhancement of the edge; and (4) it attributes the
Urbach broadening to the internal, relative motion of the exciton.

The microfield mechanism provides a clue to why some solids
are ultratransparent, for it contains a spatial quasi-selection rule
in the electron-hole overlap factor. By attributing the deep Urbach
tail to excited charge-transfer-type states in which the electron is
removed for the hole, the microfield model suggests qualitatively
that only somewhat delocalized intrinsic states contribute to the
Urbach tail.

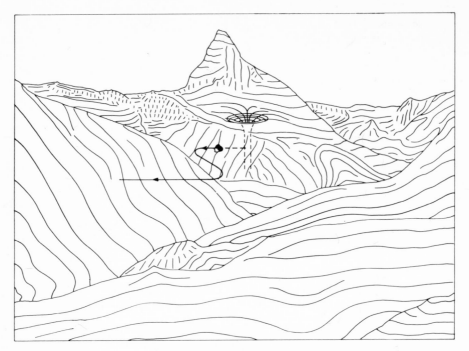

Fig. 5:   Schematic illustration of potential energy of electron in
"electrostatic" disorder, showing the Coulomb potential of the
presumed-massive hole. The electron tunnels out of the Coulomb
well to a nearby potential valley (after Ref. 2).

The electric microfield model has as its principal advantage
the qualitative picture it proposes. Recent work [11,12,14] con-
firms the model's qualitative and quantitative predictions for
covalent materials. Moreover the model appears to be consistent
with absorption edge data for ferroelectrics [15]. However, the
limitations of the model imposed by the requirement of long wavelength disorder demand theoretical contortions in order to describe
the temperature dependences of the Urbach breadths in alkali halides.

III.   OTHER THEORIES

There are two other classes of theories of Urbach's rule, configuration coordinate theories and phonon sideband theories. These
theories rely on phonon <u>dynamics</u> and do not pretend to be universal;
they do not describe impurity-caused or acoustoelectric Urbach edges,

but confine themselves to phonon-related edges in II-VI's, alkali halides, and ferroelectrics.

A crude, but transparent representation of typical configuration coordinate model is given in Fig. 6. The probability that the ground electronic state is in a vibrational state of energy V (above the ground electronic state's vibrational minimum E_0) is given by a Boltzmann factor, with energy-conservation governing the transition.

$$K_A(\omega) \sim e^{-V/k_B T} \delta(E_0 + V - E_f - \hbar\omega) \quad .$$

Integrating over all vibrational levels gives an Urbach shape. A more sophisticated version of this model has been advanced by Sumi and Toyozawa [16], who have applied it to alkali halides. Williams and Schnatterly, in the following paper [7], discuss models of this type.

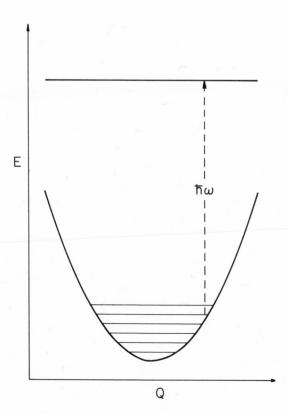

Fig. 6: Schematic configuration coordinate diagram.

Phonon sideband models have been advanced by Segall [17], by Dunn [18], and by Bosacchi and Robinson [19]; the Bosacchi-Robinson model is a polaron model with an exact solution. These models have the attractive feature that they describe the low-temperature quantum structure on the absorption edge. It should be possible to relate the shapes of this multi-phonon-sidebands on the fundamental electronic absorption edge to the shapes of multi-phonon edges in the infrared [20].

IV.   EXPERIMENTS

Experimenters should attempt to determine if the Urbach shape arises from a single mechanism or many different ones. Studies of Urbach edges should concentrate on determining (i) the characteristic wavelength of the disorder; (ii) whether the edge is primarily of dynamic (i.e., depending on ionic mass) or static origin; (iii) the role of perturbations which affect primarily the exciton center-of-mass or relative motion (or the motion of electron or hole); (iv) the dependences of the edges on temperature, impurities, and dielectric constant; (v) the classical or quantum nature of Urbach's rule; and (vi) connections between Urbach edges and multiphonon edges.

ACKNOWLEDGEMENTS

The author is grateful to D. Redfield, who first sparked his interest in Urbach's Rule, for many and continuing conversations. He gratefully acknowledges stimulating conversations with R. Williams and S. Schnatterly, and the influence of many people including D. L. Dexter, J. J. Hopfield, R. S. Knox, and M. H. Reilly. He especially thanks J. J. Hopfield for discussions of the simplified model of Fig. 6.

REFERENCES

1. F. Urbach, Phys. Rev. 92, 1324 (1953).
2. J. D. Dow and D. Redfield, Phys. Rev. B5, 594 (1972).
3. See, e.g., B. H. Winters, S. Silverman, and W. S. Benedict, J. Quant. Spectrosc. and Radiative Transfer 4, 527 (1964). The author thanks Dr. G. Birnbaum for bringing this reference to his attention.
4. Y. Toyozawa, Tech. Rept. ISSP (Univ. Tokyo) A119 (1964).
5. D. Redfield and M. A. Aframowitz, Appl. Phys. Letters 11, 138 (1967).
6. S. E. Schnatterly, Phys. Rev. B1, 921 (1970).
7. R. T. Williams and S. E. Schnatterly (to be published), and these proceedings.
8. D. Redfield, Phys. Rev. 130, 916 (1963).
9. D. L. Dexter, Phys. Rev. Letters 19, 383 (1967).
10. J. D. Dow and D. Redfield, Phys. Rev. B1, 3358 (1970); J. D. Dow, Comments on Sol. St. Phys. IV, 35 (1972).
11. D. K. Garrod and R. Bray, Proc. 11th Internatl. Conf. Phys. Semiconductors, (PWN-Polish Scientific, Warsaw, 1972), p.1167.
12. J. D. Dow, M. Bowen, R. Bray, D. L. Spears, and K. Hess, Phys. Rev. B10, 4305 (1974).
13. R. J. Elliott, Phys. Rev. 108, 1384 (1957).
14. Y. Brada and G. Yacobi, Proceed. 12th Internatl. Conf. Phys. Semicond. (M. H. Pilkuhn, ed., B. G. Teubner, Stuttgart, 1974), p.1212.
15. D. Redfield and W. Burke, J. Appl. Phys. 45, 4566 (1974); S. H. Wemple, Phys. Rev. B2, 2679 (1970); M. DiDomenico and S. H. Wemple, Phys. Rev. 166, 565 (1968); A. Frova, Nuovo Cimento 55B, 1 (1968).
16. H. Sumi and Y. Toyozawa, J. Phys. Soc. Japan 31, 342 (1971).
17. B. Segall, Phys. Rev. 150, 734 (1966).
18. D. Dunn, Phys. Rev. 174, 855 (1968).
19. B. Bosacchi and J. E. Robinson, Sol. St. Commun. 10, 797 (1972).
20. J. D. Dow (to be published).

MAGNETIC CIRCULAR DICHROISM OF THE URBACH EDGE IN KI, CdTe, AND TlCl

R.T. Williams[*] and S.E. Schnatterly

Princeton University

Magnetic circular dichroism in the low-energy
tail of the fundamental absorption edge has
been measured in KI, CdTe, and TlCl, covering
a substantial range of temperatures and span-
ning three to four decades of absorption coefficient. The data will be discussed in relation to simple models of interactions that
have been proposed to account for exponential
broadening of the absorption edges in these
materials.

INTRODUCTION

The shared successes of a number of theories in predicting an exponential tail for the excitonic absorption
in various materials emphasises the need for experimental
data which can be compared to more detailed aspects of
the proposed mechanisms. As indicated in Professor Dow's
review paper on Urbach edges, the relative importance of
electrostatic and elastic interactions serves as a useful
classification of mechanisms for the broadening of fundamental absorption edges. One experimental measurement
which relates to this question is magnetic circular dichroism (MCD) of the absorption edge.

In the materials we shall consider, the effective-
mass envelope function for the lowest exciton state is
essentially s-like, so that the overall degeneracy of that
level is a consequence of the single-particle band symmetries and is encompassed in the cell-periodic part of

the exciton wavefunction. Therefore a perturbation or broadening interaction will lift the 1s exciton degeneracy only to the quantitative extent that it splits the electronic band states from which the exciton is derived. As an example, we note that the importance of electric field effects for the exciton depends almost exclusively on the large radius and closely spaced energies characteristic of solutions to the effective mass equation. It can be readily shown that the Stark effect splitting of the first (1s) exciton level compared to the Stark shift of its center of gravity should go roughly as the square of the ratio of the exciton Rydberg to the band gap, a number typically of order 10^{-3}.

Since the Urbach edge is by nature broad and relatively featureless, the splitting of the lowest exciton level can obviously not be measured directly. However, MCD provides an indirect measure, as illustrated with the aid of Fig. 1. In the left half of the figure, we have represented schematically an excitonic edge which has been broadened without lowering the exciton symmetry (e.g., cubic). In this case the effect of an applied magnetic field may be described as a rigid shift of edge replicas corresponding to the magnetic substates. A simple derivative spectrum of MCD is observed.

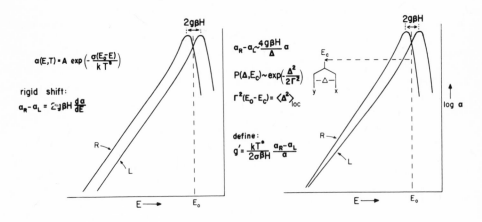

FIGURE 1

On the other hand, if the broadening process splits
the lowest exciton level, taken here to have p-like overall symmetry, then the appropriate description of the
magnetic field's effect is the mixing of nondegenerate
levels separated by energy Δ, instead of the Zeeman splitting of eigenstates of angular momentum. This is represented schematically, for the simple case of zero spin-orbit
coupling, in the right half of Fig. 1. As in Ref. 1, we
can assume Δ to obey a Gaussian probability distribution,
whose width is characteristic of the average energy shift
of the states relative to the exciton peak. Thus the
width Γ defines the rms splitting corresponding to the
average over local interactions, $\langle \quad \rangle_{loc}$, capable of producing a given shift $E_o - E_c$ in the mean 1s exciton energy.

We will also have occasion to characterize the edge
by an effective g-parameter, g', defined in terms of the
two equations at the left in Fig. 1. The parameter g'
is essentially that value which g would have to assume
in order to force consistency between observed MCD and
a rigid shift of the edge, at a given point on the edge.
As a means of comparing data for different temperatures,
g' has the virtue of compensating for the temperature
dependence of the edge slope, and will approach the exciton g-value as the rigid shift conditions are satisfied.

EXPERIMENT

The two quantities to be measured were the optical
absorption coefficient, α; and the difference, $\Delta \alpha$, in
absorption of right and left circularly polarized light
with the sample in a magnetic field. Since both α and
$\Delta \alpha$ in the Urbach edge are sensitive exponential functions of energy and temperature, and since their ratio
is the quantity of primary physical interest for our
investigation, it was imperative to make the two measurements under closely matched conditions. Both were performed in the same experimental apparatus, alternating
on a point-for-point basis between measurements of α and
$\Delta \alpha$ at each wavelength. On the other hand, crystal
thickness cancels in computing the ratio of $\Delta \alpha$ to α
from experimental data, as does the primary dependence
on absolute temperature shared by both measurements.

The probe light was monochromatized by a 3/4-meter
grating spectrometer. MCD was measured by modulating
the circular polarization of the light beam at 50 kHz
with a piezo-optic device, and using lock-in detection.[2]
The sample was placed between the pole pieces of an iron-
core electromagnet operated typically at 12.2 kG, with

the light beam introduced axially through the pole pieces.
Crystals were heated in a vacuum oven designed to fit in
the magnet pole gap, and only fused quartz was allowed
to contact the samples at high temperature. Some of the
alkali halide data were also obtained with a ceramic oven
in air, still with fused quartz crystal support. TlCl
was heated in sealed quartz cells within the oven to
minimize sublimation. Since CdTe is known to undergo
significant changes of stoichiometry when heated above
710 C,[3,4] our experiments in vacuum with CdTe were held
below 350 C, and checks were made for relative optical
changes before and after heating, without finding any.
Care was taken to minimize strains in the mounting of
all samples.

Crystals of KI and TlCl were obtained from Harshaw
Chemical Co., and Samples of CdTe from D.T.F. Marple, at
General Electric Research and Development Center. The
CdTe was from zone-refined ingot No. 7234-213, from
which the edge spectra in Fig. 2 of Ref. 4 were obtained.
All bulk samples were used with cleaved or chemically
polished surfaces. Thin films of KI and TlCl were made
by evaporation onto fused quartz substrates, and in the
case of KI, by solidification between fused quartz plates.

The low-energy tail of optical attenuation may be
regarded as the sum of the intrinsic exponential absorption edge and a background due to impurities and scattering processes. Since we are concerned only with the intrinsic exponential edge in this study, an approximate
subtraction of the background attenuation is effected by
extrapolation from the "transparent" spectral region just
below the exponential edge. Lower temperature data are
also a guide in determining the baseline. It should be
noted that our displays of absorption coefficient refer
to the intrinsic exponential component, and therefore
may extend below the typical level of background absorption. Similar baselines were established from the MCD
background, which can appear as an artifact in transparent
spectral regions, for example, by the combined effects of
strain birefringence and Faraday rotation.

RESULTS

In Fig. 2 are shown a set of experimental results
for KI. The absorption coefficient is given by the open
points, and the MCD ($\Delta_a \alpha$) is given by the solid points,
shown multiplied by 10^3. The approximate straight lines
on this semilogarithmic plot are evidence of the Urbach
rule behavior. The quantity of most interest is the ratio

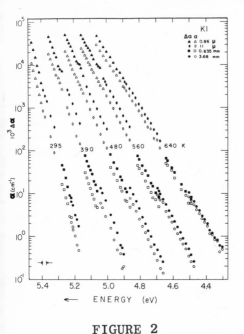

FIGURE 2

$\Delta\ \alpha/\alpha$. In subsequent displays of these data we will
use a single point for the
average of all data obtained
with a given sample at a given
temperature.

We have computed the
parameter $g'=(kT\Delta\ \alpha)/(2\sigma\beta H\alpha)$
from the data in Fig. 2.
The values are plotted in
Fig. 3 against energy measured from the exciton peak
(where $E_o(T)$ is determined
from the optical conductivity
spectra by Tomiki, et al.[5]).
A least-squares fit extrapolates back to a g-value of
0.87 at the peak, which is
the g-value found for the exciton in KI from magneto-
reflectance[6] and Faraday rotation.[7] Further down the
tail, g' drops consistently
below the exciton value, indicating that the edge does
not shift rigidly in a magnetic field. It will be noted
that the points measured at
room temperature (open circles)

are consistently lower than others at corresponding spectral energies. We attribute this to residual strain
effects, which are partly relieved, and partly covered
over by the intrinsic edge, as temperature is raised.

Also, we have used the experimental values of
$\Delta\ \alpha/\alpha$ to compute the average splitting of the exciton
states as discussed briefly in connection with Fig. 1.
In KI, however, the valence band spin-orbit splitting
is quite sizeable, (0.89 eV) and the earlier discussion
does not necessarily apply quantitatively. The relevant
treatment of a $P_{3/2}$ quadruplet subject to splitting in
a tetragonal deformation field, with subsequent mixing
of the nondegenerate orbital substates in a magnetic field
has been given.[1] It leads to essentially the same result
as that of Fig. 1, although the contribution to MCD
from Zeeman splitting of the Kramers doublets when there
is nonzero spin-orbit coupling was neglected. The latter

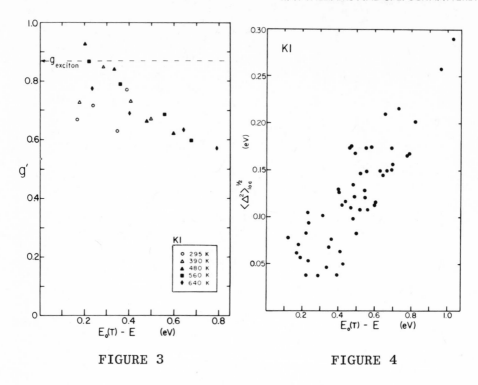

FIGURE 3   FIGURE 4

contribution is essentially a rigid-shift term which may
diminish the apparent energy-dependence of $\langle \Delta^2 \rangle_{loc}$ as
computed from the equations in the right half of Fig. 1.

    In Fig. 4 each point represents a complete run with
a given sample at one temperature. The collection of
points includes all data taken for KI, with sample thicknes
from 8600 A to 3.68 mm, and temperatures from 295 K to
930 K. The average splitting of the exciton extrapolates
approximately to zero at the exciton peak, and has a slope
of 0.27 in the plot against energy relative to the peak.
This can reasonably be taken as evidence that the edge-
broadening interaction is about 27% noncubic in KI.
That is, if the mean energy of a pair of p-like exciton
states is reduced by 1 eV relative to the unperturbed
exciton, those states will on the average have been split
0.27 eV apart in the process.

We exhibit in Fig. 5 data for CdTe corresponding
in format to Fig. 2, where again the open points are absorption coefficient, and the solid points are MCD times
10^3.   Recall that the ratio of $\Delta\alpha$ to α in KI decreased
with energy down the edge, presumably because of the increasing strength of the noncubic part of the broadening
interaction.   The trend in CdTe is just the opposite; i.e.
$\Delta\alpha/\alpha$ decreases toward the exciton peak.   In fact, the
sign of the MCD is observed to reverse for the highest

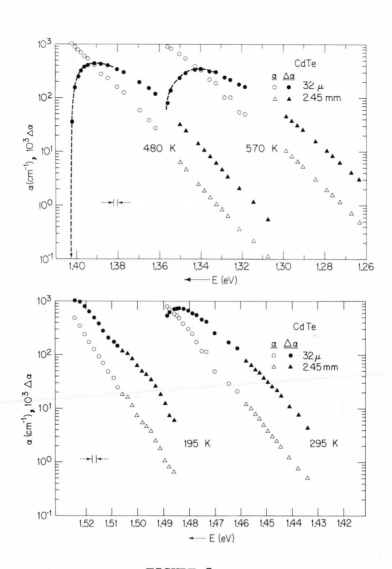

FIGURE 5

energy point at 480 K.   The low energy part of the edge
displays something like rigid-shift behavior.   It therefore seems reasonable to suppose that this anomalous
MCD spectrum is the sum of a rigid shift component and
some unspecified contribution of opposite sign that
dominates at higher energies.   To isolate the higher
energy component, we re-construct the rigid-shift MCD
spectrum as an extrapolation of the low-energy data,
proportional to the derivative of the absorption spectrum, and subtract from that the observed MCD.   The
result is shown in Fig. 6.

The solid line is the reconstructed rigid-shift
spectrum, and the points with broken line represent the
difference spectrum; that is, the presumed negative MCD
component of the edge.   There is thus a rather strong
suggestion that the Urbach edge in CdTe is the sum of
two exponential edges, differing in slope and being

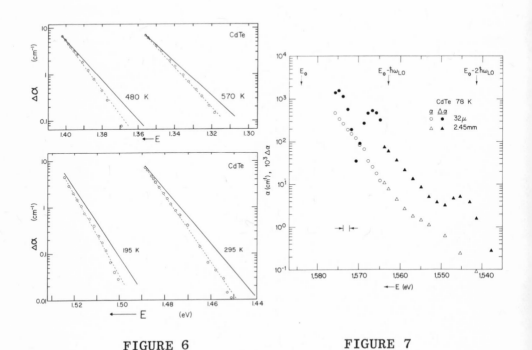

FIGURE 6   FIGURE 7

characterized by opposite net circular polarization in
a magnetic field.

Figure 6 also argues against the possibility that
an impurity causes the "anomalous" MCD. Transport
measurements, as reported for the zone-refined CdTe
from which our samples were taken,[4] indicated a donor
concentration of 1.05×10^{15} cm^{-3} and an acceptor concentration of 6.7×10^{14} cm^{-3}. Since neither the g-
value nor the oscillator strength for an impurity transition in CdTe is likely to be radically greater than unity,
the only way for trace impurities to seriously affect
the intrinsic MCD spectrum at the 10^3 cm^{-1} level of absorption is for the impurity band to be extremely narrow.
Although such a narrow band is in principle a reasonable
explanation, the data of Fig. 6 rule out that possibility.
The "anomalous" MCD component is exponential with a slope
comparable to that of the intrinsic edge, and therefore
the associated absorption must have approximately an exponential distribution. But application of a generalized
Smakula equation[8] to such a postulated impurity edge,
along with values of unity for g and the oscillator strength,
shows that an impurity concentration greater than 10^{17}
cm^{-3} would be required to account for the observed spectra

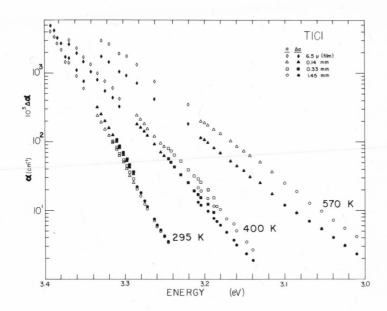

FIGURE 8

on an impurity basis.

At 78 K, the appearance of the MCD in CdTe changes rather dramatically, developing the bumps shown in Fig. 7. For comparison, we have marked the energies corresponding to the exciton peak, and to one- and two-phonon assitance thresholds. These bumps probably are properties of the intrinsic edge, since measurements at 10 K did not reveal any residual structure at corresponding energies.

Data for TlCl are shown in Figs. 8 and 9. Figure 9, especially, should be compared to Fig. 3 for KI. The g-value of 0.8 indicated in Fig. 9 was deduced from Faraday rotation in the neighborhood of 4300 A, assuming a rigid shift of the dispersion. If this characterizes the

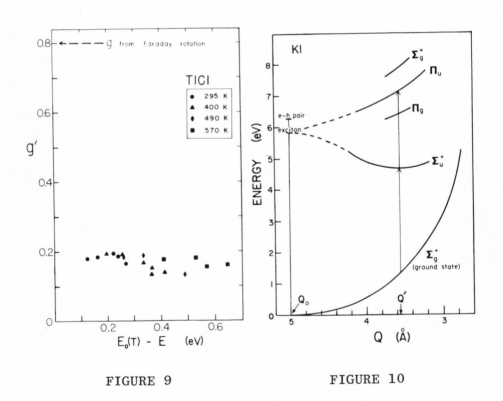

FIGURE 9   FIGURE 10

principal absorption components at the band edge, then
our MCD data suggest that the low-energy exponential
tail in TlCl is not simply a smear of all transitions that
constitute the fundamental edge. The broadening process
appears to favor a particular state or combination of
states characterized by an unusually small g-value.

DISCUSSION

We have remarked that an electric field is not likely
to produce a large (on the scale of edge broadening)
splitting of the lowest exciton level. In considering our
results for KI, it will be helpful to make comparisons as
well to a model which does predict large splittings. The
model we will describe falls in the general category of
the deformation potential models, and has to do with an
unusually strong interaction between the exciton and a
particular localized vibrational mode. As two neighboring
halide ions in an alkali halide crystal are compressed
together, some of the ion electronic states form net bonding orbitals, so that certain of the excited-state potential curves of the system are unstable against lattice
distortion. Thus excitons in alkali halides spontaneously
relax, or "self-trap", in a configuration characterized
as a hole localized on a pair of nearest-neighbor halide
ions separated by a distance Q, and surrounded by an
electron in a fairly diffuse orbital.[9] The normal lattice
configuration Q_o can thus correspond to a local maximum
of the first excited state energy as well as to the minimum of the ground state energy, so that an expansion of
the exciton-lattice interaction in the coordinate Q would
have no linear term. This led Toyozawa[10] and Keil[11] to
suggest that the quadratic exciton-lattice interaction
which is required to obtain an exponential edge within a
deformation-potential theory may correspond to an odd-
parity (110) vibrational mode which produces compression
of the neighboring halide ions.

If one imagines that the lattice vibrational energy
causes a momentary local distortion to the configuration
Q', it is obvious from Fig. 10 that a transition from
the ground state of the crystal to the first excited state
(exciton) takes less energy than the lowest exciton transition in the undistorted crystal. At the same time, the
exciton states are split by the uniaxial deformation,
leading to odd-parity states formed from the halide p-
orbitals oriented along the deformation axis (Σ_u^+), and
perpendicular to it (Π_u). The orbital notation we are
using is only approximate, referring to the point group
$D_{\infty h}$ of a diatomic molecule, but it describes the im-

portant features adequately for the present discussion.
The important energies shown in Fig. 10 are fairly accurate
in a quantitative sense. (See, e.g., Ref. 9). Thus for
the configuration Q', we can measure the splitting Δ between the states Σ_u^+ and Π_u, while the cubic shift (E_o-E_c)
can be measured from the ground state Σ_g^+ to the midpoint
between Σ_u^+ and Π_u. The ratio $\Delta/(E_o-E_c)$ found in this
way for KI is 2. This is not expected to hold accurately
for a general point on the diagram (e.g., $Q \approx 3.9$ A should
correspond to a value of E_o-E_c which is observable at
930 K in a thick crystal of KI). However it does suggest
a reasonable limit for the fractional noncubic character
of this kind of edge-broadening mechanism in alkali halides

In presenting the CdTe data, we have already suggested
that the Urbach edge in CdTe is a composite of two exponential edges characterized by opposite magnetic circular
polarizations. This is not altogether without precedent,
if we recall that a number of II-VI,[12, 13] III-V,[13] and
group IV[14] semiconductors, including CdTe,[12, 15] display
a sign reversal of Faraday rotation as the band edge is
approached. The authors of Ref. 15 suggested that the
reversal in CdTe can be explained by taking the exciton
to have a negative g-value, with the interband transitions
characterized by positive g. In that case, we should
ascribe the Urbach edge to comparable broadening of both
the exciton peak and the interband edge; furthermore,
the behavior of MCD which we have observed would suggest
that the interband edge (with positive g) is most effec-
tively broadened.

An alternative view could be based on a theoretical
treatment by Boswarva and Lidiard[16] of Faraday rotation
in GaAs, GaSb, Ge, InAs, and InSb. Those authors suggested that the reversals in sign of Faraday rotation in the
materials treated are the results of competition between transitions originating in the heavy- and light-
mass valence bands, which contribute with opposite signs
to the Faraday rotation below the band edge. If we are
to account for Fig. 7 by analogy to the theory of
Boswarva and Lidiard, then the exponential edges associated
with the light- and heavy-mass states must have different
slope parameters. In fact, the hole mass is an important
parameter in several of the theories of Urbach's rule.

CONCLUSION

Characterization of the process(es) giving rise to
Urbach edges as either fully "electrostatic" or fully
"elastic" is not supported as a generally applicable

rule by the foregoing results.

Consider first the case of KI. The theory of Dow and Redfield[17] models the interaction of excitons with long-wavelength phonons in terms of the electric micro-fields set up by the phonons. Such internal electric microfields certainly exist in alkali halides, and on that basis we must conclude that excitons in KI are broadened exponentially as predicted by the Dow-Redfield theory. Whether electric-field broadening accounts sub-stantially for the observed edge, however, is entirely a matter of its magnitude relative to contributions from other interactions.

For example, short-wavelength phonons do not couple efficiently to the exciton in terms of the field-ioniza-tion model, and so contribute very little to that aspect of the broadening.[17] The same is true for nonpolar phonon modes of any wavelength. However, when the interaction of excitons with such phonons is modeled on the exciton self-trapping phenomenon known to occur generally in halides,[9] an essentially quadratic deformation potential is obtained, and this will lead to the exponential broadening as proposed by Toyozawa[10] and Keil.[11] Corres-pondingly, the long-wavelength optical phonons do not couple efficiently in this model.

Thus, while asking whether long- or short-wavelength phonons are responsible for Urbach's rule, we should al-so consider the possibility that both can have comparable importance in the same observed edge, even though neither model presently being considered accounts adequately for both limits. Likewise, both polar and nonpolar phonons may participate significantly, though the interactions will be described by "electrostatic" and "elastic" models, respectively. This is basically the interpretation we choose to apply to our MCD results in KI, where the ob-served magnitude of splitting falls approximately mid-way between model predictions based on electric micro-fields and quadratic deformation potentials.

This hybrid view does not exclude an eventual unified theory. We believe, however, that such a theory should include features encompassing both the facts that internal electric microfields ionize the exciton, and that in many halide crystals there exist low-energy "self-trapped exciton" states which can be reached optically from the ground state with the aid of short-wavelength phonons. For non-halide crystals, short-wavelength phonons should

be included as well, but the essential physics probably
will not have any resemblance to self-trapped excitons,
which may well be peculiar to halide and rare-gas crystals.

Thus in non-halide crystals, the inappropriateness
of a self-trapping model requires a dynamic model of the
interaction between excitons and short-wavelength phonons.
Whether this contribution is to be considered as an addition to the edge broadened by long-wavelength electric
microfields, or whether a dynamic model can include both
limits of phonon wavelength adequately is perhaps still
an open question. (The solution could be either along
the lines of Professor Dow's suggested "A.C. microfield
model," or a version of the treatment by Sumi and
Toyozawa,[18] or perhaps an expanded phonon-sideband model.)
For the present, we will consider the observed edge in a
material such as CdTe to be potentially the sum of a
phonon-assistance edge and a microfield-broadened edge.

The first significance of our MCD results in CdTe
is as a demonstration that the simple ideas (Fig. 1) for
splitting of an essentially localized exciton, which
seemed to work for MCD in KI, are not broadly applicable
for comparisons between arbitrary materials. We suggested a possible explanation of the apparent sum of distinct
exponential edges in CdTe in terms of competition between light-and heavy-hole states, coupled with the hole-
mass dependence of each of the proposed models. It is
also possible to invoke two different models for the two
edge components. That is, Segall's phonon-assistance
model[19] favors the heavy-hole states, and so may contribute an edge with some characteristic slope, σ, and a
dominant right magnetic circular polarization. The electric-microfield mechanism favors the lighter reduced exciton masses, and therefore the light-hole band, giving
an edge with a different slope, σ', and possibly a net
left MCD. The exaggeration of structure in the MCD of
CdTe at 78 K as compared to structure in absorption might
also be a result of looking at the sum of two components
in absorption, one bumpy and one smooth, but looking at
their difference in MCD. There remains the possibility
that the features of both model contributions may be encompassed in an eventual unified theory.

The large exciton radius and substantial crystal
ionicity in TlCl should greatly enhance the importance
of electric-microfield broadening.[17] Conversely, excitons do not appear to self-trap in TlCl. Our MCD results in TlCl are rendered somewhat uncertain by   am-

biguities in the appropriate exciton g-value. Even so, the absence of substantial indications of exciton splitting in proportion to the edge broadening (in spite of the large non-cubic deformation potentials characteristic of $TlCl^{20}$), can be viewed as implicitly consistent with dominance of electrostatic interactions.

REFERENCES

*Present address: Naval Research Laboratory, Washington, D.C. NRL support of R.T.W. during this work is gratefully acknowledged. Research supported in part by NSF.

1. S.E. Schnatterly, Phys. Rev. B1, 921 (1970).

2. S.N. Jasperson and S.E. Schnatterly, Rev. Sci. Instrum. 40, 761 (1969).

3. D.de Nobel, Philips Res. Repts. 14, 361 (1959).

4. D.T.F. Marple, Phys. Rev. 150, 728 (1966).

5. T. Tomiki, T. Miyata, and H. Tsukamoto, Z. Naturforsch. 29a, 145 (1974).

6. R.K. Abrenkiel and K.J. Teegarden, Phys. Stat. Sol. (b) 51, 603 (1972).

7. G. Baldini and G. Canossi, Sol. St. Commun. 10, 373 (1972).

8. D.Y. Smith and D.L. Dexter, Progress in Optics, Vol. X (North Holland, Amsterdam, 1972), p. 167.

9. R.T. Williams and M.N. Kabler, Phys. Rev. B9, 1897 (1974); M.J. Marrone, F.W. Patten, and M.N. Kabler, Phys. Rev. Lett. 31, 467 (1973); M.N. Kabler and D.A. Patterson, Phys. Rev. Lett. 19, 652 (1967).

10. Y. Toyozawa, Tech. Rept. ISSP (Univ. Tokyo) A119, (1964).

11. T.H. Keil, Phys. Rev. 144, 582 (1966).

12. P.S. Kireev, L.V. Volkova, and V.V. Volkov, Soviet Phys.-Semicond. 5, 1816 (1972); M. Zvara, F. Zaloudek, and V. Prosser, Phys. Stat. Sol. 16, K21 (1966).

13. V.V. Karmazin and V.K, Miloslavskii, Soviet Phys.-
    Semicond. 5, 866 (1971).

14. A.K. Walton and T.S. Moss, Proc. Phys. Soc. 78, 1393
    (1961).

15. V.V. Karmazin and V.K. Miloslvaskii, Soviet Phys.-
    Semicond. 5, 928 (1971); V.V. Karmazin, V.K.
    Miloslavskii, and V.V. Mussil, Soviet Phys.-Semicond.
    7, 639 (1973).

16. I.M. Boswarva and A.B. Lidiard, Proc. Phys. Soc.
    London, A278, 588 (1964).

17. J.D. Dow and D. Redfield, Phys. Rev. B5, 594 (1972).

18. H. Sumi and Y. Toyozawa, J. Phys. Soc. Japan 31, 342
    (1971).

19. B. Segall, Phys. Rev. 150, 734 (1966).

20. E. Mohler, G. Schlögl, and J. Treusch, Phys. Rev.
    Lett. 27, 424 (1971).

THEORY OF NEW TRANSIENTS AND OPTICAL PHENOMENA IN SPATIALLY DISPERSIVE MEDIA[*]

Michael J. Frankel and Joseph L. Birman

Physics Department, The City College of the CUNY

New York, New York 10031

Abstract

An integral representation for the linear transients of coupled propagating modes in a spatially dispersive medium is investigated. New "exciton precursors" are predicted as well as altered signal velocities. Numerical estimates are made for CdS and ZnSe.

It is well known that allowing the polarization to depend upon the electric field non-locally through the linear relation

$$\vec{P}(\vec{r},\omega) = \int \chi(\vec{r},\vec{r}',\omega) \; \vec{E}\,(\vec{r}',\omega) \; d^3\vec{r}' \tag{1}$$

with $\chi(\vec{r},\vec{r}',\omega)$ a non-local susceptibility introduces a $\vec{k}$-vector dependence in the susceptibility $\chi(\omega,\vec{k})$.

We consider a semi-infinite medium occupying the region $z>0$ with vacuum in $z<0$. Employing a model non-translationally invariant susceptibility[1]

$$\chi(\vec{r},\vec{r}',\omega) = [\chi_0 \; \delta(\vec{r}-\vec{r}') + \chi_1 \; G_+(\vec{r}-\vec{r}') - \chi_1 G_+(\xi)]\theta(z)\theta(z')$$

with

$$G_+(\vec{r}-\vec{r}') = \frac{e^{ik_+(\omega)|\vec{r}-\vec{r}'|}}{|\vec{r}-\vec{r}'|} \quad , \quad G_+(\xi) = \frac{e^{ik_+(\omega)\xi}}{\xi}$$

[*] This work supported in part by NSF, AROD, and CUNY-FRAP.

and with

$$\xi = [(x-x')^2 + (y-y')^2 + (z+z')^2]^{\frac{1}{2}}$$

$$k_+(\omega) = [\frac{(\omega_o^2-\omega^2)^2 + \omega^2\Gamma^2}{b^2}]^{\frac{1}{4}} \; e^{\frac{i(\theta + \pi)}{2}} \; ,$$

$$\tan\theta = -\frac{\omega\Gamma}{\omega_o^2 - \omega^2} \; , \quad -\pi < \theta < 0 \; . \tag{2}$$

it can be shown that the propagating waves in the medium must
satisfy the dispersion

$$k^2 = \frac{\omega^2}{c^2} [\; 1 + 4\pi \; \chi^{PROP}(\omega,k) \;] \tag{3}$$

where

$$\chi^{PROP} = \chi_o + \frac{4\pi\alpha_o\omega_o^2}{\omega_o^2 + bk^2 - \omega^2 - i\omega\Gamma} \tag{4}$$

The solutions to equation (3) give rise, for $\omega > \omega_\ell$, to two
propagating solutions $k_1(\omega)$ and $k_2(\omega)$ which are respectively
photon like and exciton like while for $\omega < \omega_o$ there is the usual
photon like branch. The introduction of such an "extra" propagating mode for $\omega > \omega_\ell$ gives rise to the problem of the correct
additional boundary conditions (abc) to be employed, a subject
of much discussion in the literature.

The problem we wish to consider here is that of a truncated
laser signal of carrier frequency ω_1 normally incident on the
boundary (z=0) of our medium at t=0. The applied field is then
taken to be of the form $f(t) = \sin\omega_1 t \; \theta(t)$ with $\theta(t)$ the Heavy-
side function. We now consider the nature of the disturbance seen
at some (z,t) in the medium. In our work we generalize a complex
integral representation due to Sommerfeld[4] who investigated the
response of a medium considering only frequency dispersion.

We then take for $f(z,t)$

$$f(z,t) = \frac{Re}{2\pi} \int_c \sum_{j=1}^2 \frac{A_j(\omega)e^{ik_j(\omega)z-i\omega t}}{\omega - \omega_1} \; d\omega \tag{5}$$

The $A_j(\omega)$ are determined by solving the boundary value problem and including the (abc) aré given by

$$A_1(\omega) = \frac{(n_2^2 - \varepsilon_o)}{(n_2-n_1)(n_1+n_2+n_1n_2+\varepsilon_o)} \quad ,$$

$$A_2(\omega) = \frac{(n_1^2 - \varepsilon_o)}{(n_1-n_2)(n_1+n_2+n_1n_2+\varepsilon_o)} \quad , \tag{6}$$

where

$$n_j^2 = \frac{c^2 k_j^2}{\omega^2} \quad , \quad \varepsilon_o = 1+4\pi\chi_o \quad . \tag{7}$$

The contour c is along the real axis and closed in the upper or lower half plane for the appropriate retarded time $\hat{t}$ less than or > 0 respectively. This choice is made to successfully reproduce the coupled polariton field $f(0^+,t)$ inside the crystal boundary 0^+.

It can be shown [2,3] that choice of the specific model non-local susceptibility fixes the form of the additional boundary conditions and the $A_j(\omega)$.

The representations of eq.(5) can now be used to consider a number of problems concerning wave propagation in spatially dispersive media. We consider the problem of the transient response in time between z/c and the signal arrival time (approximately $z/v_g(\omega_1)$) where $v_g(\omega_1)$ is the group velocity at ω_1. This problem has been considered by Sommerfeld[4] and Brillouin[5] for non-spatially dispersive media and it had been predicted that there would arise two precursor regimes to the main signal, one characterized by very high frequencies and the second by very low frequencies.

For spatially dispersive media the integrals in eq.(5) can be evaluated asymptotically for small times t. For times less that than the signal arrival time it is found that the signal amplitude can be written as a sum of three contributions

$$f(z,t) = \sum_{\alpha=1}^{3} f_\alpha(z,t) \ \theta \ (t-t_\alpha) \tag{8}$$

eq.(8) is approximate. It indicates that contributions $f(z,t)$ are the Sommerfeld high frequency precursors (S.P), a Brillouin low-frequency precursor (B.P) and a third new high-frequency exciton-like precursor (E.P) peculiar to spatially dispersive media. The t_α are given explicitly by:

$$t_{S.P} = \sqrt{\epsilon_0}\, z/c \quad,$$

$$t_{B.P} = \sqrt{\epsilon_0}\, \frac{\sqrt{1+4\pi\alpha_0}}{\epsilon_0}\, z/c \quad,$$

$$t_{E.P} = z/\sqrt{b} \quad. \tag{9}$$

The forms for $f_\alpha(z,t)$ are obtained from a steepest descent evaluation and are given for the high frequency precursors: Sommerfeld Precursor and new Exciton Precursor, by:

$$f_\alpha = B_\alpha(\omega_\alpha z)(t-t_\alpha)^{\frac{1}{4}} \exp(-\Gamma_\alpha(t-t_\alpha))\cos(\omega_\alpha(t-t_\alpha) + \frac{\pi}{4}$$

with

$$B_{S.P} = \frac{4\omega_1}{\sqrt{\pi}}\left(\frac{c\sqrt{\epsilon_0}}{8\pi\alpha_0 z}\right)^{3/4} \frac{A_1}{\omega_0^{3/2}} \quad;\quad B_{E.P} = \frac{4\omega_1}{\sqrt{\pi}}\left(\frac{\sqrt{b}}{2z}\right)^{3/4} \frac{A_2}{\omega_0^{3/2}} \tag{10}$$

For the Brillouin precursor:

$$f_{B.P} = B_{B.P}(\omega_B z)(t-t_B)^{-\frac{1}{4}} \exp(-\Gamma_{B.P}(t-t_{B.P}))\cos(\omega_{B.P}(t-t_{B.P})+\frac{\pi}{4})$$

with

$$B_{B.P} = \frac{1}{\omega_1\sqrt{\pi}}\left(\frac{3\sqrt{\epsilon_0 b}}{4zc}\right)^{\frac{1}{4}}\omega_0^{\frac{1}{2}} \tag{11}$$

The functional dependence of the ω_α on the time is given by

$$\omega_\alpha = \gamma_\alpha\omega_0 / \sqrt{t-t_\alpha} \qquad \alpha = S.P, E.P$$

and

$$\omega_{B.P} = \gamma\omega_0\sqrt{t-t_{B.P}} \tag{12}$$

The Γ_α , γ_α and γ are constants dependent upon the material parameters.

A more exact analysis[3,5] gives the Brillouin precursor in the form of an Airy function. An observer at z thus receives three partially overlapping precursor packets. Each precursor represents a superposition of harmonic frequency components with relatively large group velocities i.e. from the three regions of larger slope in the dispersion curve illustrated in Fig. 1.

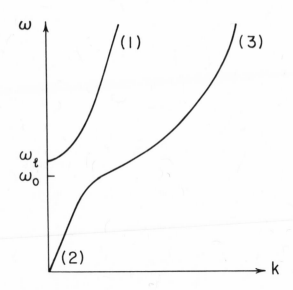

Fig. 1: Dispersion curve of eq.(3) (with $\Gamma = 0$ for convenience). ω_1 incident laser frequency, degenerate propagating modes in the crystal for (ω,k_1) and (ω,k_2). (1)(2)(3) represent three regions of large v_g.

An illustration of the sequence of precursor events as seen by an observer at z is given in Fig. 2 below.

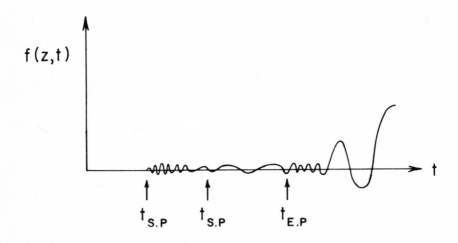

Figure 2: Representation of the signal precursors including spatial dispersion effects. There are three precursors, Sommerfeld Precursor, Brillouin Precursor, and the new Exciton Precursor to the main signal; instead of the two as in the classic calculation[5,6].

In Table I we characterize the precursors by an appropriate delay time t_α , a band of frequencies in which the precursor

TABLE I

Precursor	S.P	B.P	E.P
t_α	9.48×10^{-12} sec	9.49×10^{-12} sec	1.34×10^{-9} sec
Major frequencies in precursor	$\omega_\ell < \omega < 1.2\,\omega_\ell$	$.8\omega_1 < \omega < \omega_1$	$\omega_\ell < \omega < 1.2\,\omega_\ell$
Amplitude	3×10^{-3}	1.5×10^{-4}	2×10^{-4}
Duration	3×10^{-15} sec	3×10^{-8} sec	6×10^{-10} sec

Illustrative precursor values. The material parameters chosen are appropriate to ZnSe[7]; $\hbar\omega_o$ = w.8 e.v , $4\pi\alpha_o = 5.5 \times 10^{-3}$, $\varepsilon_o = 8.1$, $m^* = .88\,m_e$. ω_1 is taken as 2×10^{15} sec^{-1} and $z = .1$ cm. The amplitude is normalized to the incident signal and the major frequencies refer to a region over which the amplitude varies by a factor <10; and $\omega_\ell = \omega_o (1 + 4\pi\alpha_o/\varepsilon_o)^{\frac{1}{2}}$.

attains some appreciable amplitude, some average amplitude and a duration.

The representation (5) has also been used to consider the problem of signal velocity calculation when the damping is appreciable. In such a case the group velocity is no longer a good measure of signal velocity and another criterion is needed. The criterion used here is a generalization of a formalism due to Baerwald[6] wherein the signal is defined to arrive when the amplitude of the steady state response becomes equal to the transient. This involves setting an asymptotic, steepest descent calculation of the integrals in (5) giving the transient amplitude equal to the steady state response as calculated by residues. The results of this procedure for a choice of material parameters appropriate to CdS is illustrated in Table II. In general, the velocities in spatially dispersive media are enhanced over those to be expected in a calculation ignoring such non-local effects and also display much less dispersion in the anomalous absorption region $\omega_o < \omega < \omega_\ell$.

TABLE II

$$\frac{C}{S}$$

velocity frequency	$b \neq 0$	$b = 0$
ω_o	1.9×10^4	7.3×10^4
$\dfrac{\omega_o + \omega_\ell}{2}$	1.7×10^4	2.7×10^4
ω_ℓ	1.0×10^4	7.1×10^4

Comparison of the inverse signal velocity at some points in the interval $\omega_o \leq \omega \leq \omega_\ell$ with $\omega_\ell = \sqrt{1 + 4\pi\alpha_o}$ over ε_o. The case $b = 0$ represents the velocities as calculated by ignoring spatial dispersion effects. Here S is the signal velocity and C the vacuum velocity of light parameters appropriate to CdS[8] were used. $\hbar\omega_o = 2.5528$ e.v , $4\pi\alpha_o = .0125$, $\varepsilon_o = 8$, $m^* = .9\ m_e$, $\hbar\Gamma = 2 \times 10^{-4}$ e.v.

REFERENCES

(1) R. Zeyher, J.L. Birman and W. Brenig, Phys. Rev. B6, 4613 (1972).

(2) J.L. Birman and R. Zeyher, Proc. of Taormina Conf. on Polaritons ed. E. Burstein (Pergammon, 1974)

(3) M.J. Frankel, Ph.D. Thesis, New York University, 1975.

(4) A. Sommerfeld, Ann. der Phys. 44, 177 (1914).

(5) L. Brillouin, Ann. der Phys. 44, 203 (1914) and also in "Wave Propagation and Group Velocity" (Academic Press, N.Y. 1960).

(6) H.G. Baerwald, Ann. der Phys. 6, 295 (1930) and Ann. der Phys. 7, 731 (1931).

(7) A.A. Maradudin and D.L. Mills, Phys. Rev. B7, 2787 (1973).

(8) J.J. Hopfield and D.G. Thomas, Phys. Rev. 122, 41 (1961).

DISPERSION OF THE ELASTO-OPTIC CONSTANTS OF POTASSIUM HALIDES[*]

K. VEDAM, E. D. D. SCHMIDT and W. C. SCHNEIDER

THE PENNSYLVANIA STATE UNIVERSITY

UNIVERSITY PARK, PENNSYLVANIA 16802

ABSTRACT

The variation of the refractive indices of potassium halides with pressure to 14 kbars have been determined by an interfero-metric method. In every case the refractive index was found to increase with pressure with a pronounced non-linearity. However, the same data exhibit perfect linear relationship with strain when the latter is evaluated using the non-linear theory of elasticity. Combining these results with the data using uniaxial pressure measurements, the individual elasto-optic constants have been determined. The observed dispersion of the elasto-optic co-efficients in the spectral region 366 - 589 nm, is attributed mainly to the electronic contribution to the elasto-optic effect.

INTRODUCTION

Even though the photoelastic behavior of crystals was formu-lated by Pockels [1] at the close of the last century, this subject has remained essentially dormant till very recently. With the advent of high power laser systems it is being recognized that the photoelastic behavior of the various optical components in the light path contribute significantly to the distortion of the laser beam and thus compromise the performance of the laser systems and their associated applications [2]. As a part of a large program to understand the origin and thus possibly overcome such distortion effects, Bendow et al. [3] have recently derived theoretical

[*] Research supported by the National Science Foundation.

expressions for the photoelastic effect for a number of materials and have also given numerical estimates of the various elasto-optic coefficients and their dispersion. However, it is found that some of these theoretical values particularly on the dispersion of the elasto-optic constants of alkali halides, are not in complete agreement with the only experimental data [3] reported in the literature. Since the present authors have recently developed an experimental method of measuring some of the photoelastic coefficients to a high degree of precision and accuracy [4], it was felt desirable to carry out a systematic measurement on the photoelastic behavior of alkali halides. Some of the results obtained on the dispersion of the elasto-optic constants of potassium halides are presented below.

EXPERIMENTAL

The experimental procedure employed and the computations involved in piezo-optic measurements at high pressures have already [5,6] been described in detail and hence only a brief outline of the technique is given here. As the specimen (in the form of a thin plate) is subjected to hydrostatic pressure its thickness and refractive index change. The variation of the refractive index of the crystal with pressure can then be determined from the shift of the localized interference fringes across the specimen kept in a high pressure optical vessel. A schematic diagram of the experimental arrangement is shown in Fig. 1. The change in the refractive index Δn was evaluated from the interference formula

$$\Delta n = \frac{k\lambda - 2n \Delta t}{2t_o} \tag{1}$$

where k is the number of fringes shifted, t the initial thickness of the specimen, and Δt the change in thickness of the specimen due to the applied pressure. Since the pressure employed was high, Δt was evaluated with the help of the nonlinear theory of elasticity developed by Murnaghan [7], Birch [8] and others [9-11].

The volume change of a solid due to an applied hydrostatic pressure P can be expressed as

$$\frac{\Delta V}{V_o} = - aP + bP^2, \tag{2}$$

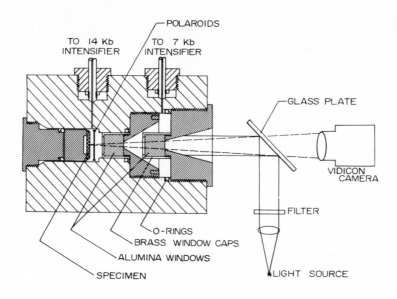

Fig. 1.   Schematic Diagram of the Experimental Arrangement.

where a and b are temperature sensitive constants. These coefficients are related to the bulk modulus B and its pressure coefficient according to

$$a = \frac{1}{B} \, ,$$

$$b = \frac{1}{2B^2} \left[\left(\frac{\partial B}{\partial P} \right) + 1. \right] \tag{3}$$

According to Barsch [12], superscript S or T should be added
to all these quantities to indicate that the compression is made
adiabatically or isothermally, respectively. As pointed out by
Barsch in any precision high pressure measurement it is necessary
to distinguish between the three different sets of pressure derivatives. These are (i) the isothermal pressure derivatives of the
isothermal elastic constants, (ii) the isothermal pressure derivatives of the adiabatic elastic constants, and (iii) the adiabatic

pressure derivatives of the adiabatic elastic constants. The third
case is relevant in, for example, shock wave propagation and hence
will not be discussed here. The isothermal-isothermal derivatives
are pertinent because these are the thermodynamic boundary conditions in the present experimental arrangement and also for
Bridgman's a and b values [13]. The ultrasonically measured
pressure dependence of the elastic constants is constrained by the
isothermal-adiabatic boundary conditions. In order to compare the
a and b values that were obtained by entirely different methods, a
thermodynamic conversion of the ultrasonic data must be made.
Overton [14] and Barsch and Chang [15] have made such corrections
for many materials of cubic symmetry. Table I lists such values
of a and b for potassium halides as evaluated by Barsch and Chang,
and used in the present study to calculate Δt the change in the
thickness of the specimens.

RESULTS AND DISCUSSION

Figure 2 shows the pressure dependence of the Δn evaluated
with the help of Eqn. (1) for KBr and KI to 14 kbar and for KCl to
7 kbar at a wavelength of 5893Å. These data are presented first to
establish the general trend of Δn versus pressure as well as volume-
strain relationships, irrespective of the wavelength of light
employed in the present study. It is seen that all three materials
exhibit linear relationship between Δn vs P at low pressures, but
at high pressures (i.e., above 4.0 kbars, 3.0 kbars and 2.5 kbars
for KCl, KBr, and KI respectively) Δn increases nonlinearly with
pressure. The same data are plotted in Fig. 3 as a function of
volume strain, and it is seen that all the three materials
exhibit a linear relationship even for volume strains as high as
9% when the strain is computed using the nonlinear theory of
elasticity. This clearly indicates that the nonlinearity observed

Table I

Compressibility Data of Potassium Halides at T = 23°C.

	a $(10^{-6}/bar)$	b $(10^{-12}/bar^2)$
KCl	5.76_0	106.9
KBr	7.06_0	165.3
KI	9.08_0	306.0

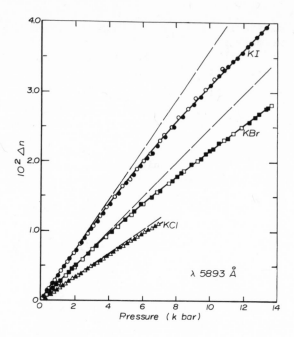

Fig. 2.   Variation of the Refractive Indices of Potassium Halides
with Pressure.

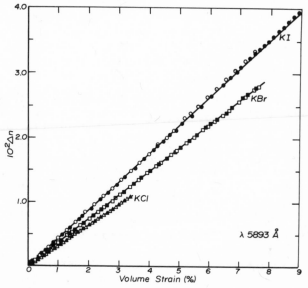

Fig. 3.   Variation of the Refractive Indices of Potassium Halides
with Volume Strain.

in Fig. 2 is a manifestation of the nonlinear stress–strain relation in these materials. Thus it appears that the strain-optical constants relating Δn and the strain are more fundamental than the corresponding stress–optical constants.

A somewhat similar linear relationship between the isothermal volume strain and the melting temperature of various materials has been found by Kraut and Kennedy [16] even for strain in some cases as large as 60%. However, this does not mean that such a linear relation between the change in refractive index Δn and the Lagrangian strain will be valid universally. On the contrary, pronounced nonlinearity between Δn and the strain was observed even at fairly low strain levels in α–ZnS and CdS for $\lambda 5893$Å indicating the influence of the nearness of the absorption edge in these cases [17].

From the slopes of the curves in Fig. 3, the values of

$$\rho(dn/d\rho) = - \Delta n \bigg/ \left(\frac{\Delta V}{V_o}\right) \tag{4}$$

can be obtained which gives a linear combination of the elasto-optic constants p_{11} and p_{12} for these materials. Combining these results with the data from the uniaxial pressure measurements reported in the literature [18,19] allows the determination of the values of the individual elasto-optic components p_{11} and p_{12}.

The above measurements were repeated to 7 kbar on these same materials for six wavelengths in the range 5893Å and 3663Å. The results are listed in Table II giving the wavelength and for each materials, the corresponding index of refraction, the slope of the Δn versus pressure relationship, and the slope of $\rho(dn/d\rho)$ computed by the least squares method.

Table III lists the values of the p_{11} and p_{12} at these wavelengths for these three materials. The values for these coefficients reported by Iyengar [4] are also listed in this table in parentheses for comparison. The values are in fair agreement for KBr and KI but a larger discrepancy (about 5%) is noticed for KCl. This can be considered as quite satisfactory since many reported values of the p_{ij}'s in the literature disagree even as to the sign of the coefficient.

It is seen from Table III that the values of both p_{11} and p_{12} increase as the wavelength is increased from 3663Å to 5893Å for all the three materials. Very recently Bendow et al. [3] have performed a theoretical computation of the dispersion of the elasto-optic constants p_{ij} in the transparent regime for a number of materials including potassium halides. These computations

Table II

Summary of Results on dn/dP and $\rho(dn/d\rho)$ as a Function of Wavelength for KCl, KBr, and KI at T = 23°C.

λ(Å)	KCl			KBr			KI		
	n	dn/dP ($\times 10^2$/kbar)	$\rho(dn/d\rho)$	n	dn/dP ($\times 10^2$/kbar)	$\rho(dn/d\rho)$	n	dn/dP ($\times 10^2$/kbar)	$\rho(dn/d\rho)$
5893	1.4904	0.179_6	0.324_1	1.5598	0.246_3	0.365_1	1.6647	0.385_4	0.436_1
5461	1.4931	0.178_7	0.323_1	1.5639	0.245_3	0.363_6	1.6715	0.384_2	0.434_2
5085	1.4962	0.178_7	0.321_6	1.5685	0.238_7	0.361_2	1.6792	0.382_8	0.432_3
4358	1.5047	0.178_1	0.319_8	1.5815	0.242_0	0.359_1	1.7017	0.376_7	0.425_4
4047	1.5100	0.175_9	0.316_9	1.5898	0.240_1	0.355_5	1.7164	0.372_1	0.419_7
3663	1.5189	0.174_5	0.313_1	1.6038	0.233_6	0.347_7	1.7423	0.364_8	0.415_9

Table III

Elasto-Optic Constants for KI, KBr and KCl at T = 23°C.

λ(Å)	KCl		KBr		KI	
	p_{11}	p_{12}	p_{11}	p_{12}	p_{11}	p_{12}
5893	0.228	0.179	0.228	0.174	0.205	0.181
	(0.215)	(0.159)	(0.212)	(0.165)	(0.203)	(0.164)
5461	0.226	0.178	0.225	0.172	0.202	0.178
	(0.211)	(0.156)	(0.208)	(0.162)	–	–
5085	0.223	0.176	0.222	0.170	0.198	0.175
4358	0.216	0.172	0.210	0.167	0.186	0.166
	(0.182)	(0.134)	(0.194)	(0.151)	–	–
4047	0.213	0.169	0.205	0.163	0.179	0.160
3663	0.207	0.165	0.192	0.157	0.166	0.153

indicate that in the case of potassium halides as the wavelength is increased from uv to the visible region of the spectrum, p_{11} should increase whereas p_{12} should decrease contrary to the present experimental results. In these calculations these workers have assumed the electronic contribution to p_{ij} to be constant - an assumption certainly valid in the infrared region However, in the visible as well as the near uv region of the spectrum covering the range of the present measurements, such an assumption is not quite justified. This will become evident from Table III by comparing the relative dispersion of both p_{11} and p_{12} at the two extreme ends of wavelength range studied. It is seen that the magnitude of the dispersion decreases as one proceeds from uv to the yellow region whereas the opposite behavior would have been noticed if the lattice contribution had played the dominant role in the spectral region studied. However, if these measurements are extended into the IR region we can expect the dispersive behavior predicted by Bendow et al.

A detailed discussion of the observed values of $\rho(dn/d\rho)$ as well as of the p_{ij} of the potassium halides in terms of the various model as well as phenomenological theories [18,20-23] will be reported elsewhere along with similar results obtained on other alakli halides.

REFERENCES

1. F. Pockels, Ann. Phys. Chem. 37(4), 144, 269, 372 (1889).
2. Second and Third Conferences on High Power Infrared Laser Window Materials, ed. by C. A. Pitha, et al. (1973): AFCRL-TR-73-0372 and AFCRL-TR-74-0085, respectively; and Proceedings of the 1972 Electronic Materials Conference, February and May issues of J. Elect. Matls. (1973).
3. B. Bendow, P. D. Gianino, Y. Tsay and S. S. Mitra, Appl. Opt. 13, 2382 (1974); and Tech. Rept. No. AFCRL-TR-74-0533 dated October 24 (1974).
4. K. S. Iyengar, Nature (London) 176, 1119 (1955).
5. K. Vedam and E. D. D. Schmidt, Phys. Rev. 146, 548 (1966).
6. K. Vedam, E. D. D. Schmidt, J. L. Kirk and W. C. Schneider, Mat. Res. Bull. 4, 573 (1969).
7. F. D. Murnaghan, Finite Deformation of an Elastic Solid. John Wiley & Sons, Inc., New York (1951).
8. F. Birch, Phys. Rev. 71, 809 (1947).
9. S. Bhagavantam and E. V. Chelam, Proc. Ind. Acad. Sci. 52A, 1 (1960).
10. A. A. Nran'yan, Sov. Phys. Solid State 6, 936 (1964).
11. K. Brugger, Phys. Rev. 133A, 1611 (1964).
12. G. R. Barsch, Phys. Stat. Sol. 19, 129 (1967).
13. P. W. Bridgman, "The Physics of High Pressure". G. Bell and Sons, Ltd., London (1958).
14. W. C. Overton, Jr., J. Chem. Phys. 37, 117 (1962).
15. G. R. Barsch and Z. P. Chang, Phys. Stat. Sol. 19, 139 (1967).
16. E. A. Kraut and G. C. Kennedy, Phys. Rev. 151, 668 (1966).
17. K. Vedam and T. A. Davis, Phys. Rev. 181, 1196 (1969).
18. A. Gavini and M. Cardona, Phys. Rev. 177, 1351 (1969).
19. R. Srinivasan, Z. f. Phys. 155, 281 (1959).
20. H. Mueller, Phys. Rev. 47, 947 (1935).
21. J. Yamashita and T. Kurosawa, J. Phys. Soc. Japan, 10, 610 (1955).
22. J. Van Vechten, Phys. Rev. 182, 891 (1969).
23. S. H. Wemple and M. Didominico, Phys. Rev. 1B, 193 (1970).

OPTICAL TRANSMISSION IN IODINE TRANSPORTED α-HgS

M. M. KREITMAN* and S. P. FAILE

University of Dayton, Dayton, OH 45469 and

C. W. LITTON and D. C. REYNOLDS

Aerospace Research Laboratories, WPAFB, OH 45433

Optical transmission and X-ray measurements confirm the growth of α-HgS (cinnabar) crystals by an
iodine transport method.

INTRODUCTION

Mercury sulfide is a wide bandgap semiconductor which is of
considerable interest because its unique properties offer excellent potential for non-linear optics and laser technology devices.
It is the most optically active of all known mineral compounds,
possessing the greatest birefringence, a remarkable rotary power
and strong piezoelectric properties.

Sapriel[1] has reported that α-HgS has the highest acousto-
optical figures of merit of any known crystal, making it an efficient material for acousto-optical applications, such as transducers, light modulators, deflectors, delay lines, etc. Ultrasonic attenuation and the elastic and photoelastic constants in
α-HgS have been measured.[2] Sapriel and Lancon[3] have pointed out
that, since the electromechanical coupling coefficients of cinnabar are approximately two times larger than those of Quartz,
piezoelectric α-HgS transducers bonded on a parallelpiped cinnabar crystal would realize perfect accoustical matching and therefore yield wide bands for acousto-optical deflection. Thus α-HgS
should serve well as a good deflector of laser light. For example, if one assumes a frequency bandwidth of 250 MHz for a
transducer made of cinnabar and an acoustic beam 1/2 mm square,
it may be shown that only 210 mW of acoustic power is necessary

to deflect all of the incident light of a He-Ne laser beam into 62,500 different positions with an access time of 1 μs.

Measurements[4] on the red colored α phase of synthetic and good natural crystals indicate a transparency of more than 50% between 0.6 and 14 μ at 300 K. Risidual absorption in the transparent region is known to be less than 1 cm^{-1} for good quality natural crystals.[5,6] Large indices of refraction have been measured[7] over the range 0.62-11 μ. The proximity of the upper limit of the transparent region to the 6328Å He-Ne laser line, and the large temperature dependence of the HgS absorption edge ($\sim 9.10^{-4}$ eV/C), has already been exploited[8] for thermal modulation of the laser line (by the thermoabsorption of cinnabar).

Photoelectronic properties of α-HgS have been studied, chiefly in the more available, good quality, natural crystals and these studies reveal a close simularity to those of other II-VI compounds. Some results[9] in synthetic crystals yield a peak photoconductivity at ~6000 Å (300 K) for HgS ($\sim 10^{12}$ Ωcm); the position, within the energy gap, of localized levels of electrons (0.62 eV) and holes (1.08 eV) from the conduction band; and thresholds (0.67 eV and 1.08 eV) for optical quenching. In other studies[10] the conductivity of illuminated natural crystals has been observed to increase by 6 orders of magnitude. Furthermore the mobility in a direction perpendicular to the c-axis has been determined[11] to originate from electrons at 10 cm^2 $V^{-1}s^{-1}$ while others[10] have noted a photo-Hall mobility of 38 cm^2 $V^{-1}s^{-1}$ in a crystal which may not have been oriented.

The strong non-linear properties of cinnabar are also important for integrated optical circuits, parametric oscillator upconversion, difference frequency generation, tunable lasers, and other applications, For example, Martin and Thomas[12] has suggested that the stimulated Raman effect be combined with the non-linear electrooptic effect[13] in HgS in order to conveniently generate pulses of radiation between 9 and 15 μ. They utilize the non-linear electrooptic effect to generate a beat frequency between a Nd^{3+}/glass laser and the laser activated, stokes Raman emission. The difference frequency is equal to the molecular vibrational frequency of the Raman process.

The successful future utilization of all of these devices, and the exploration and study of these and additional properties of α-HgS rests heavily on the growth of good quality single crystals, which until recently, has been very difficult. It is the purpose of this paper to present details on the growth of α-HgS by an iodine transport method and to report optical and x-ray measurements which confirm the yield of good cinnabar crystals.

GROWTH OF CINNABAR BY VAPOR TRANSPORT

Mercury sulfide exists in two modifications, cinnabar (α-HgS) and metacinnabar (β-HgS). The α phase crystalizes in an unusual, dihedrally coordinated, red colored, low-symmetry (D_3^4) structure, the bandgap of which has been measured by Zallen[5] as 2.1 eV. This trigonal form is stable at 20 C and is strongly piezoelectric with[14] a=4.149 Å and c=9.495 Å at 26 C. Metacinnabar, however, is black, metastable at 20 C, and cubic (T_d^2)(a=5.817 Å)[14] at 26 C, with Eg= -0.15 eV.

An early technique, reported by Hamilton[15], on the synthesis of single crystals of the sulphides of Zn, Cd and Hg was based on a vapor phase growth method utilized by Reynolds and Czyzack[16] for the growth of ZnS and CdS crystals. Hamilton employed temperatures below 550°C for the crystallization of HgS in a pyrex tube containing a pressure of 35 cm of H_2S prepared by triple vacuum distillation. A temperature of 490 C was applied to the powder source while the region of normal growth was maintained at 440°C with a value of 8°C/cm for dT/dx and a minimum run time of 50 hours to produce a growth of crystals with a volume of 20 mm^3. The deposited crystals were found to grow only in the form of hexagonal columns a few mm long and 1-2 mm thick, with crystal faces so poorly developed that it was difficult to determine structure. There was no evidence of dendrites in this growth. A rough theoretical estimate of the surface free energy according to the method of Harkins[17] (assuming nearest neighbor interactions only and covalent bonding) was found to be in excellent agreement with an experimental value of 300 ergs/ cm^2 obtained for HgS assuming two-dimentional nucleation.

Curtis[18], while reproducing and extending Hamilton's experiment, observed a phase transformation at 335 C above which mercury sulfide deposited as a black, β-HgS, cubic phase and in cooling below, transferred to a red, α-HgS phase. He concluded that it was unlikely that the vapour phase growth techniques would lead to true single crystal cinnabar because of the phase transformation which served to prevent the growth.

However, Carlson[19] attempted the growth of cinnabar below the phase transformation. He obtained crystalline HgS above and below a 344 C transition temperature, but below, only polycrystalline material or small single crystals were produced. He used 99.0% pure HCl gas (at concentrations of 0.02 moles/liter) as the transporter gas, and 0.4 g of HgS pelletized charges synthesized from elements each with a purity of 99.999%. Pyrex tubes having 11 and 8 mm bores, were used to obtain equant crystals with maximum dimentions of about 1.8 mm and crystals with plate like habit with dimentions up to 2.5 x 2.5 x 1.0 mm^3. He also observed a prismatic habit with dimentions as large as 3.0 x 0.5 x 0.5 mm^3 at high HCl concentrations.

Curtis also had observed a large hysteresis in the phase transformation. Although the temperature associated with the change from α to β phase for a pure sample has been reported[20] as 344 C, Ohmiya has found an upward shift for samples of higher purity covering a temperature range of from 100 to 362 C on heating and from 200 to 295 C on cooling. He has also recently measured the thermal expansion coefficients of α-HgS as $\alpha_a=1.81\ 10^{-5}$ and $\alpha_c=1.88\ 10^{-5}$ for 20-200 C, and of β-HgS as $\alpha=4.3\ 10^{-6}$ for 211-348 C.

As noted above, a chief difficulty in growing cinnabar is that temperatures lower than approximately 330 C are needed to prevent the material from transforming to the black cubic phase. However, the low vaporization temperature of the metal iodide species allows one to overcome the transformation temperature limitation. In the past the iodine transport method[22] has been used at high temperature because many metal iodides are not very volatile. For ZnI_2 the vapor pressure is one atmosphere at 623 C, and for CdI_2 it is at 796 C in contrast to 354 C for HgI_2. We have observed that the transport of HgS by iodine transport occurs as low as 200 C. Some initial results on the growth of cinnabar crystals by iodine transport were reported recently[23,24].

Generally the experimental runs involved the use of 50-150 grams of technical grade HgS with a fairly high impurity content as a source material. The Na content was about 80 ppm, the K content 2.5 ppm and the Cd content 6.7 ppm. About 5 mg/cc of iodine was added to a pyrex tube. See Fig. 1. To prevent the loss of iodine during

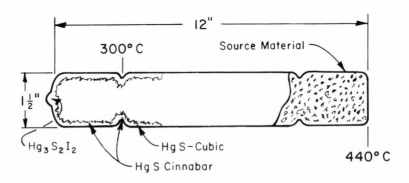

Fig. 1   Iodine Transport of HgS. The source material in the pyrex tube is composed of 150 gm HgS and 3 gm I_2.

evacuation of the air from the tube during vacuum pumping, one end of the tube was immersed in liquid nitrogen. After the tube was sealed off, trace amounts of fine HgS powder at the deposition end of the tube were removed by using a reverse temperature gradient for a few days.

The run is started by having the source end of the tube at a higher temperature than the deposition end. The transport species, HgI_2 and S_8, react to form HgS several weeks after a few grams of $Hg_3S_2I_2$ have been deposited. For a tube 1.5 inches in diameter about 60 grams of material is transported in approximately 5 weeks to form cinnabar at temperatures less than approximately 330 C and metacinnabar above 330 C.

One of the more successful runs involved the use of an 8 inch long, 1.5 inch diameter pyrex tube which was run for 4 weeks over temperatures of from 315 C to 260 C and then run for 5 weeks from 400 to 290 C. Single crystals as large as $5 \times 4 \times 1/2$ mm^3 and $7 \times 2 \times 1.5$ mm^3 were obtained.

The duration of the runs for I_2 transport of HgS can be reduced, either by increasing the diameter of the container tube, or by including a compound (containing hydrogen) in the starting material which reacts to form the transport gas H_2S. When the diameter of the 8 inch long tube is increased from 1 to 1.5 to 2.75 inches the transport of material increases respectively from about 3 to 1.5 months to about 3 weeks. This result is obtained for approximately the same depth of HgS powder (1 inch) initially at the warmer end of the tube. For the 1.5 inch diameter tube this corresponds to about 75 grams of HgS. The hot end of the tube was held at 400 C while the cool end was maintained at 300 C. For runs involving the use of containers with a diameter greater than 1.5 inches and an end temperature difference greater than 100 C, there was a tendency for small crystals of cinnabar to form at the cool end, followed later on by larger crystals of metacinnabar. For a 2.75 inch diameter tube this occured even for deposition temperatures below 200 C.

For more rapid transport rates a higher proportion of impurities may be deposited resulting in a lowering of the transformation temperature. (In a proceeding paragraph we have already noted Ohmiya's[17] observations of the lowering of the transmission temperature by an increased impurity content.) Towards the end of a typical run this effect should become more pronounced because, during the early stages, a higher ratio of HgS to impurities is transported. As the HgS is depleted relative to the impurities at the source end, the concentration of the impurities in the transport gas should increase. According to atomic absorption analyses, the source material contains an excess of 70 ppm Na while initial amounts of deposited

HgS crystals were found to contain approximately 7 ppm Na.

By adding less than 4 mg/cc of a compound containing hydrogen, such as an organic, to the starting materials, the rate of transport can be speeded up, for a 1 inch diameter tube, from about 3 months to less than 3.5 weeks and, for a 1.5 inch diameter tube from 1.5 months to 7.5 days. If the starting HgS powder is very fine the compound will have reacted at 300 C in 24 hours to form H_2S and other materials that will transport faster than the transport species would have if there had not been a reaction with HgS. The increased transport rate, in part, is probably due to the conversion of the S_8 species at a pressure about 100 mm to H_2S at a few atmospheres, resulting in a higher diffusion rate and a higher concentration of sulfer in the vapor phase. In fact even when the organic, 1-dodecanol, $CH_3(CH_2)_{11}OH$, is used without any iodine there is an indication of some crystal growth, such as needles of centimeter lengths.

Encouraging results have been obtained using HgS powder with a purity of 99.999% for a run in a 1 inch diameter tube where larger than normal cinnabar crystals were produced. It should be noted, however, that the use of the technical grade HgS in this rapid transport method nearly always produces metacinnabar crystal growth.

The iodine transported HgS has also been found to be useful as a substrate for hydrothermally grown cinnabar. A dendritic formation develops against the pyrex surface as a layer of polycrystalline cinnabar is deposited during iodine transport. This formation of small thickness and generally less than 7 mm width is converted during hydrothermal growth to a wider and thicker mass yielding crystals approximately 1 cm in length.

CINNABAR GROWTH BY OTHER TECHNIQUES

Many laboratories have attempted to grow cinnabar by various methods but not all of these attempts have been successful in producing good crystals of significant size. For example, we have tried cinnabar growth from Na_2S-sulfur fluxes and achieved almost the same dendritic results (1-5 mm) as was reported by Garner and White.[25] Others[26] have achieved cinnabar crystal growth (the largest crystal being 3 mm on edge) by the decomposition of a chemical complex involving excess sulfide ions in solution. We have also tried, like others[27], without success, to grow HgS from Hg-rich solutions.

One of the most successful techniques which has been reported to date is the hydrothermal growth of cinnabar, which has sometimes yielded[28-30] single crystals up to 20 g, when utilizing basic solutions, and crystal platelets of $4 \times 4 \times 0.4$ mm^3 dimentions, when using strong acid solutions.[31] The use of basic solutions appears to be the most successful method in obtaining crystals approaching centimeter dimensions.

X-RAY DATA

The results of an x-ray diffraction analysis verified that the material was α-HgS with no evidence of the β phase. A summary of these results are given in Table I. Although the measurements established clearly that the material was cinnabar, little could be said about the purity of the samples. Since it was known that the ASTM standard[14] contained a rather high level of impurities, it could not be used as a standard for purity. Furthermore in most materials 5000 ppm is the limit of impurities necessary to produce effects on the lattice parameters, and this is far above tolerable levels in semiconductors.

OPTICAL RESULTS

Some of the larger cinnabar crystals grown by the iodine vapor transport methods described above were used as samples for optical investigations in the visable and infrared spectral regions. A typical measured value of crystal thickness was 330 μ. In one study a 2 meter spectrograph was employed to take photographic plates (4 Å/mm) of the edge absorption of a number of crystals on which a tungsten source was focused. In this arrangement the crystals were mounted on a rod within a fixed temperature dewar and measurements were carried out at room temperature and at 4.2 K. A densitometer analysis of the photographic plates enabled values of the wavelength at zero absorption to be tabulated with the exposure times and this led to an extrapolation of the wavelength corresponding to zero exposure time and a value of 6115 Å for the approximately $\parallel$ absorption edge at 300 K with a decrease of approximately 2.0 Å/K on cooling to 4.2 K. Additional transmission data taken on a Cary spectrophotometer was used to verify this value for the edge. A value of the reflection at 0.63 was calculated as 26%. The observed temperature dependence is in rough agreement with a literature value[32] of

Table I   X-Ray Measurements in HgS

α-HgS		β-HgS	
$a_\alpha(\text{Å})$	$c_0(\text{Å})$	$a_\alpha(\text{Å})$	Reference
4.172 +0.003	9.546 +0.006		Present
4.1488+0.0002	9.5039+0.0009	5.8514+0.0003	Ohmiya[21]
4.149 +0.002	9.495 +0.002		Toudic[29]
4.149	9.495	5.8517	Swanson[14]

1.8 Å/K, the difference being easily attributed to the pronounced
dichroism of the α-HgS edge. In fact, because of the small size and
delicate nature of the crystals, use was made of this anisotropy to
check the orientation of the crystal after mounting.

Although we did not thoroughly search for it at very low absorption levels[5], we did not observe any discrete structure near the
edge at 4.2 K. Typical values of resistivity for our samples were
approximately 2×10^9 Ωcm which may be compared to 10^{12} Ωcm obtained
by Roberts, Lind and Davis[9] for their undoped synthetic cinnabar.
Their high resistivity crystals of mm dimentions were grown at temperatures just below the polymorphic transition temperature by a
reaction of the HgS elements in sealed quartz ampoules containing
200-400 mm Hg pressure at room temperature.

Infrared measurements were performed with the crystals at temperatures of 32, 92, and 300 K in a metal cryostat and mounted on
a cold finger instrumented with Ge, Pt, and carbon thermometry. The
tail and barrel of the dewar was enclosed within a vacuum spectrophotometer having a focal length of 60 inches. The spectrometer has
Czerny-Turner optics and utilizes interchangeable 7" x 9", kinemati-
cally mounted reflection gratings, blazed at 1, 3, 9, 18, 30, 45,
60 and 100 μ with associated spectral reststrahlen filter plates,
filters, choppers and I.R. sources together with a Golay cell detector. Saphire, ZnSe and Polyethelene dewar windows were employed for
the low temperature measurements and at 30 μ a typical value of the
inverse linear dispersion was experimentally computed to vary between
0.028-0.030 μ/mm for a range of grating angles.

Measurements of the infrared transmission were carried out in
vacuum over a wavelength range which extends from the fundamental
absorption to beyond the restrahlen edge beginning at approximately
27 μ. We believe our crystal temperature to be approximately 32 K
for some typical experimental data representing some of the more
prominent absorption peaks which we have enclosed in Table II for
comparison with the work of others.

Table II Low Temperature I-R Data on α-HgS Phonon Frequencies

E		A	E		T(K)	Source
T	L	T	T	L		
285	290	333.5	340	348	32	Present ARL
283	290	340	344	350	300	Barcelo, et. al.[33]
277	287	333	338	346	90	Zallen, et. al.[34]
284		345			77	Riccius, et. al.[35]
280		335	343		300	Poulet, et. al.[36]

CONCLUSION

From the data comparisons summarized above it may be concluded that the optical and x-ray measurements verify the growth of α-HgS crystals and that the iodine transport method and its associated techniques merit further study and development. Furthermore it appears that when the cinnabar growth methods have been developed to the extent where large single crystals are available in sufficient supply, that α-HgS, because of its unique properties, will be in great demand as a device material.

ACKNOWLEDGEMENT

We would like to acknowledge Mr. R. S. Harmer and Prof. J. Schneider of the University of Dayton for providing x-ray data, and useful discussions; and Capt. J. Gorrell, Lt. Wayne Anderson and Dr. L. Greene of the ARL Laboratories for making equipment available and for helpful technical discussions.

FOOTNOTES

* Work performed at Aerospace Research Laboratories, Air Force Systems Command, under Contract F 33615-71-C-1877.

1. J. Sapriel, Appl. Phys. Letters, 19, 533 (1971).
2. J. Sapriel, L. Rivoallan, and J. L. Ribet, J. de Physique, 33, C6-150 (1972).
3. J. Sapriel and R. Lancon, Proc. IEEE, 61, 678 (1973).
4. Y. Toudic and R. Aumont, J. Crystal Growth, 10, 170 (1971).
5. R. Zallen, in II-VI Semiconducting Compounds, edited by D. G. Thomas (Benjamin, New York, 1967), p. 877.
6. G. G. Roberts and R. Zallen, J. Phys. C: Solid St. Phys., 4, 1890 (1971).
7. W. L. Bond, G. D. Boyd, and H. L. Carter, Jr., J. Appl. Phys., 38, 4090 (1967).
8. M. Abkowitz, G. Pfister, and R. Zallen, J. Appl. Phys., 43, 2442 (1972).
9. G. G. Roberts, E. L. Lind, and E. A. Davis, J. Phys. Chem. Solids, 30 833 (1969).
10. C. Verolini, and H. Diamond, J. Appl. Phys. 36, 1791 (1965).
11. M. Tabak and G. G. Roberts, J. Appl. Phys., 39, 4873 (1968).
12. M. D. Martin and E. L. Thomas, IEEE J. Quant. Electron, QE-2, 196 (1966).
13. E. H. Turner, IEEE J. Quant. Electron, QE-3, 695 (1967).
14. H. E. Swanson, R. K. Fuyat, and G. M. Ugrinic, NBS Circular 539, IV, 17 (1953).
15. D. R. Hamilton, Brit. J. Appl. Phys., 9, 103 (1958).
16. D. C. Reynolds and S. J. Czyzack, Phys. Rev., 79, 543 (1950); ibid, and R. C. Allen, D. C. Reynolds, J. Opt. Soc. Amer., 44,

864, (1954); ibid., J. Opt. Soc. Amer., 45, 136 (1955); S. J. Czyzack, D. J. Craig, C. E. McCain and D. C. Reynolds, J. Appl. Phys., 23, 932 (1952).

17. W. D. Harkins, J. Chem. Phys., 10, 268 (1942).
18. O. L. Curtis, J. Appl. Phys., 33, 2461, (1962).
19. E. H. Carlson, J. Crystal Growth, 1, 271 (1967).
20. F. W. Dickson and G. Tunell, Amer. Min., 44, 471 (1959).
21. T. Ohmiya, J. Appl. Cryst., 7, 396 (1974).
22. K. E. Spear, J. Chem. Ed., 49, 81 (1972).
23. S. P. Faile and P. W. Yu, Bull. Am. Phys. Soc., 19, 821 (1974).
24. M. M. Kreitman, S. P. Faile, R. S. Harmer, and D. G. Wilson, Bull. Am. Phys. Soc., 20, in press, (1975).
25. R. W. Garner and W. B. White, J. Crystal Growth, 7, 343 (1970).
26. A. F. Armington and J. J. O'Connor, J. Crystal Growth, 6, 278 (1970).
27. E. Cruceanu and N. Nistor, J. Crystal Growth, 5, 206 (1969).
28. Y. Toudic, A. Regreny, M. Passaret, R. Aumont and J. F. Bayon, J. Crystal Growth, 13/14, 519 (1972).
29. Y. Toudic and R. Aumont, Compt. Rend. (Paris) 269, 74 (1969).
30. A. Pajaczkowska, J. Crystal Growth, 7, 93 (1970).
31. H. Rau and A. Rabenau, Solid State Commun. 5, 331 (1967).
32. Y. O. Dovgii and B. F. Bilenkii, Soviet Physics-Solid State 8, 1280 (1966).
33. J. Barcelo, M. Galtier, and A. Montaner, Comp. Rend. (Paris), 274-B, 1410 (1972).
34. R. Zallen, G. Lucovsky, W. Taylor, A. Pinczuk, and E. Burstein, Phys. Rev. B-1, 4058 (1970).
35. H. D. Riccus and K. J. Siemsen, J. Chem. Phys., 52, 4090 (1970).
36. H. Poulet and J-P. Mathieu, Comp. Rend. (Paris), 270-B, 708 (1970).

Section III
Impurity Effects

IMPURITY INDUCED ABSORPTION IN TRANSPARENT CRYSTALS*

A. A. Maradudin

University of California

Irvine, California 92664

It is well known that the introduction of impurities or defects into a crystal perturbs the atomic vibrations and, consequently, also any physical properties of the crystal in which the vibrations play a central role.

The effects of point defects on the atomic vibrations of crystals can be of several different kinds. First of all they can introduce exceptional modes into the vibration spectrum of the crystal. Thus, for example, if an impurity atom is a good deal lighter than the atom it replaces in a crystal, or if it is coupled much more strongly to the surrounding host crystal than the atom it replaces, it can give rise to certain exceptional vibrational modes called localized modes. These are characterized by the fact that their frequencies lie above the maximum frequency of the unperturbed host crystal and that displacement amplitudes of the atoms, when the crystal is vibrating in these modes, decay faster than exponentially with increasing distance from the impurity site.

Similarly, if the crystal has a gap in its frequency spectrum between the acoustic and optical branches, as is the case, for example, in NaI, NaBr, KI, KBr, LiCl, CsF, and LiH then it is possible to have localized vibration modes whose frequencies lie in the gap. They can arise either because a heavy, or weakly bound, impurity has forced a mode into the gap from the bottom of the optical branches of the frequency spectrum, or because a lighter or more stiffly bound impurity has pushed a mode into the gap from the top of the acoustic band of the phonon spectrum. These modes

* Work supported in part by Air Force Office of Scientific Research under Grant No. AFOSR 71-2018.

are also localized in the sense that the amplitudes of the atoms
when the crystal vibrates in such a mode decay faster than exponentially with distance from the impurity site. However, to
distinguish between the localized modes whose frequencies are in
the gap and localized modes whose frequencies lie above the maximum frequency of the unperturbed crystal, I refer to the former as
gap modes. However, I emphasize that there is no fundamental distinction between a localized mode and a gap mode.

A third kind of exceptional modes introduced into the crystal
by the presence of impurities are the so-called resonance modes.
A very crude picture of a resonance mode might be obtained in the
following way. We know that in the very low frequency vibration
modes of a crystal, i.e., in the low frequency acoustic branches,
all the atoms are moving in phase, in parallel with each other,
independently of their masses. This is a consequence of infini-
tesimal translational invariance. This means that if we have a
very heavy impurity atom, or an impurity atom which is coupled much
more weakly to the surrounding host lattice than the atom it replaces, it will be vibrating in phase with its, let us say, lighter
neighbors. However, as the frequency of the vibrations increases,
the heavy impurity cannot keep up with its lighter neighbors and
begins to lag behind them until a point is reached where the heavy
impurity finds itself vibrating 180° out of phase with its light
neighbors. The frequency at which this switchover in the character
of the vibrations takes place (from the case in which the heavy
impurity vibrates in phase with its neighbors to the one in which
it is moving out of phase with its neighbors) is called the frequency of the resonance mode. It is, then, a localized optical
type of vibration in the crystal and is characterized by the fact
that the mean square amplitude of the impurity as a function of
frequency undergoes a large increase at the frequency of the
resonance mode. However, unlike the localized modes, which are
associated with tightly bound or light impurities in the crystal,
resonance modes are not true normal modes of the perturbed crystal.
They are a kind of collective mode which is the superposition of
all the unperturbed modes of the crystal. Because its frequency
falls in the range of frequencies allowed the normal modes of the
unperturbed crystal, the resonance mode can decay into the continuum of band modes and consequently acquires a width, or finite
lifetime, even in the harmonic approximation.

In addition to introducing exceptional vibrational modes into
the spectrum of normal modes of a perturbed crystal impurities can
also alter the frequencies and vibration patterns of the remaining
normal modes of a crystal, which do not have an exceptional
character but rather are band-like in nature.

Finally, by breaking down the translational periodicity of a

perfect crystal, impurities can break down selection rules governing
physical processes such as the absorption and scattering of light
which have their origins in the translational periodicity of a
perfect crystal.

The effects of impurities on the vibrational properties of
crystals have been studied by a wide variety of experimental techniques, such as specific heat measurements, tunneling between super-
conductors, the Mössbauer effect, neutron scattering, and the
temperature dependence of the relaxation time for spin flips when
there are paramagnetic impurities in a crystal. However, in keeping with the spirit of this conference, I would like to discuss the
effects on the optical properties of crystals caused by the presence
of impurities which have been studied extensively by the techniques
of infrared lattice vibration absorption, and Raman scattering. I
should also like to consider only effects associated with localized
modes on the optical spectra of transparent crystals since it is
only these modes whose frequencies extend into the highly transparent region of the optical spectra of these crystals.

In order that absorption of infrared radiation by these exceptional vibration modes can be observed in one phonon processes,
that is processes in which the incident light excites or de-excites
one vibrational quantum of frequency equal to that of the localized
modes, several selection rules have to be satisfied. The first is
the conservation of energy, which has the consequence that the frequency of the incident light must equal the frequency of the excitation being studied. In addition, the exceptional mode must
transform according to the polar vector irreducible representation
of the point group of the impurity site, and must possess a dipole
moment which, by virtue of its oscillation, can couple to an external electromagnetic field to extract energy from it.

The Raman scattering of light by localized vibration modes has
its origin in the fact that just as a vibrating dipole radiates
energy, so do the electrons surrounding an atom radiate energy when
they are set into vibration by an external electromagnetic field.
Because the nucleus which these electrons surround is itself vibrating, the re-radiated energy is doppler shifted with respect to the
frequency of the exciting light, and the shift in the frequency of
the re-radiated light is equal to the frequency of the vibrating
nucleus. In a localized vibration mode it is only the impurity
atoms which are vibrating to a good approximation, and the intensity
distribution of the light scattered by a crystal possessing localized
modes shows a sharp maximum at a frequency shift equal to the frequency of the localized mode. To be Raman active the localized
mode must transform according to the irreducible representation of
the point group of the impurity site according to which a symmetric
second rank tensor transforms.

There is no conservation law in either case analogous to the
law of conservation of linear momentum, which obtains in the absorption or scattering of light by perfect crystals, because that
conservation law has its origin in the periodicity of a crystal,
and that is destroyed by the presence of defects.

I turn now to a discussion of infrared absorption by localized
vibration modes. In Fig. 1 is depicted the displacement field
associated with a localized mode due to a light substitutional
impurity ion in a one-dimensional diatomic crystal.[1] (We are,
in fact, concerned with the longitudinal vibrations of the chain.
However, for ease of depiction the displacements have been drawn
transverse to the chain.) It is seen that the displacements of the
ions decrease rapidly with increasing distance from the impurity
site. If one takes into consideration the fact that the charges
on the ions alternate along the chain, one finds that this mode
has a dipole moment, so that it can couple to incident electromagnetic radiation and extract energy from it.

The first experimental observation of infrared absorption by
localized modes is due to Schaefer,[2] who measured the infrared
absorption spectra of alkali-halide crystals doped with hydride
impurities present substitutionally in the anion sublattice.
Such impurity centers are known as U-centers. A plot of the
transmissivity of KCl containing 3×10^{17} hydride U-centers per
cubic centimeter is shown in Fig. 2, and shows a pronounced dip in
the vicinity of 20 microns (500 cm^{-1}), a frequency well above the
maximum frequency of pure KCl (210 cm^{-1}). The origin of this dip
was attributed by Schaefer to absorption by the localized vibration
modes to which the hydride U-centers give rise. Group theoretic
arguments show that localized modes due to impurities at sites of
cubic symmetry, O_h in the case of alkali-halide crystals and T_d in
the case of crystals of the zincblende structure, are three-fold
degenerate and are infrared active. The vibrational origin of
this transmission minimum was confirmed by subsequent investigators[3] who replaced the hydride U-centers by deuteride U-centers.
It was found that the frequency at which the transmission minimum
occurred shifted downward when this was done by very nearly a
factor of $2^{-\frac{1}{2}}$, as would be expected for the frequency of an
harmonic oscillator when its mass is doubled. An example of this
effect is shown in Fig. 3, which shows the absorption coefficient
of KBr containing both hydride and deuteride U-centers.[4] The peak
at about 445 cm^{-1} is due to the hydride localized mode, while the
peak at about 320 cm^{-1} is due to absorption by the deuteride localized mode. The ratio of these two frequencies is 1.39, which suggests that the impurity ion in this case can be regarded to a very
good approximation as an independent, three-dimensional harmonic
oscillator.

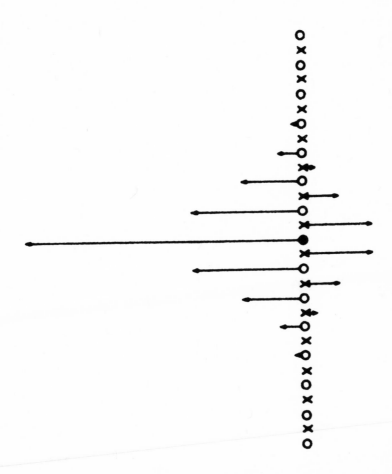

Fig. 1.   Ionic displacements in an infrared active localized mode
          in a diatomic linear chain.   (Ref. 1).

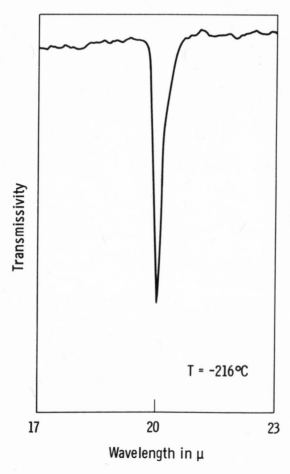

Fig. 2.   An infrared transmission spectrum for KCl containing
           3×10^{17} hydride U-centers per cubic centimeter. The
           strong absorption at a wavelength of 20µ is the absorption
           resulting from the localized vibration modes. (Ref. 2).

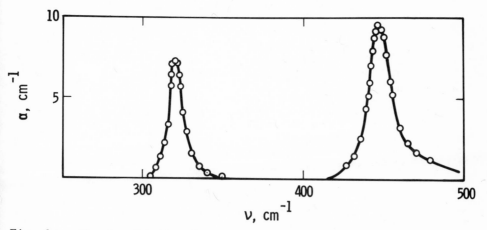

Fig. 3. The localized mode peaks in the infrared absorption spectrum of KBr containing H^- and D^- ions substitutionally in the Br^- sublattice. (Ref. 4).

After the experimental discovery of localized modes a period
began during which a wide variety of different properties of these
localized modes were studied both theoretically and experimentally.
Chief among these are properties which are anharmonic in nature.
These are effects associated with the cubic, quartic,...terms in
the expansion of the potential energy of a crystal in powers of
the displacements of the atoms from their equilibrium positions.
The existence of such terms can either modify effects which already
exist in the harmonic approximation, or they can give rise to entirely new effects which are absent when anharmonicity is neglected.

Because of its very light mass, and the consequent high frequency localized mode to which it gives rise, which implies a
high degree of localization of the atomic displacements about the
impurity site, it is a good approximation to assume that in a
localized vibration mode only the hydride or deuteride impurity
itself is vibrating, and all the other ions in the crystal are at
rest at their equilibrium positions. With this approximation it
is easy to show that the vibrational Hamiltonian for the U-center
can be written in the form[5]

$$H = \frac{p^2}{2M} + \frac{M\Omega^2}{2} (x^2 + y^2 + z^2) + $$

$$+ Lxyz + M_1(x^4 + y^4 + z^4) + M_2(x^2y^2 + y^2z^2 + z^2x^2) . \tag{1}$$

Here M is the mass of the U-center, p is its momentum, x,y,z, are
the components of the displacement of the ion from its equilibrium
position, and Ω, L, M_1, M_2 are constants. This form of the
Hamiltonian obtains if the impurity site is one possessing T_d
symmetry; in the case that the impurity site possesses O_h symmetry,
L = 0. Equation (1) gives the most general Hamiltonian through
quartic anharmonic terms compatible with the point symmetry of the
impurity site. By treating the terms on the second line of Eq. (1)
as a perturbation on the harmonic Hamiltonian represented by the
first line of this equation, it is easy to calculate the vibrational
energy levels of such a three-dimensional anharmonic oscillator by
second-order perturbation theory, to yield results of second order
in L and of first order in M_1 and M_2.[5] These are illustrated in
Fig. 4 for an impurity possessing T_d symmetry and for an impurity
possessing O_h symmetry. Group theoretic arguments show that electric
dipole transitions are allowed between the ground state of Γ_1 symmetry
to excited states of Γ_5 symmetry in the case of T_d site symmetry,
and from the ground state of Γ_1^+ symmetry to excited states of Γ_4^-
symmetry in the case of O_h site symmetry.[6] In the harmonic approximation only transitions from the ground state to the first excited
states would be allowed by the usual harmonic oscillator matrix

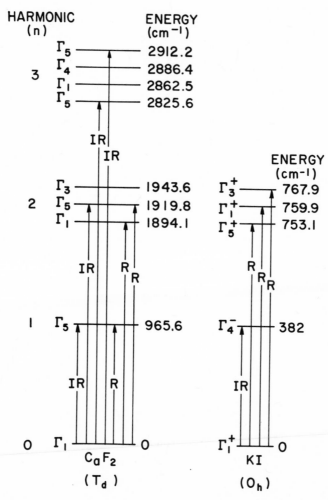

Fig. 4.   The lower vibrational energy levels and their symmetries
of a U-center at sites of T_d and O_h symmetry. The levels
which are infrared and Raman active are indicated.(Refs. 5
and 7).

element selection rules, but in the presence of anharmonicity
transitions from the ground state to the second, third, and higher
harmonics, become allowed. Consequently, one should expect to see
in addition to absorption at the fundamental frequency additional
absorption at one second-harmonic frequency and two third-harmonic
frequencies, in the case of an impurity at a site of T_d symmetry,
but no absorption at second harmonic frequencies in the case of O_h
impurity site symmetry. Similarly, group theoretic arguments permit
Raman active transitions from the ground state of Γ_1 symmetry to
excited states of Γ_1, Γ_3, and Γ_5 symmetry for impurities possessing T_d site symmetry, and from the ground state of Γ_1^+ symmetry to
excited states possessing Γ_1^+, Γ_3^+, and Γ_5^+ symmetry for impurities
possessing O_h site symmetry.[6] Thus, in the former case one would
expect to see in the Raman spectrum of a hydride U-center two
second harmonic peaks, in addition to the peak corresponding to
scattering by the fundamental, while in the case of O_h site
symmetry one would expect to see three second harmonic peaks in
the Raman spectrum. Such experiments have been carried out,[5],[7]
and the predicted overtones observed in both infrared absorption
and Raman scattering measurements. From the experimental values
for the peak positions and the theoretical expressions for the
energy levels of the Hamiltonian, Eq. (1), one can work backwards
to determine the values of the parameters Ω, L, M_1, and M_2, appearing in Eq. (1).

A second class of effects to which anharmonicity can give rise
is the broadening of the localized mode peaks in absorption and
scattering spectra, and shifts in peak positions with increasing
temperature. In the harmonic approximation, the peak in the absorption or scattering spectrum would be a δ-function, whose
position would not shift with temperature. The temperature effects
arise because of the coupling of the localized vibration mode to the
continuum of band modes through the anharmonic terms in the crystal
potential energy. From considerations of energy conservation it
follows that because the frequency of the hydride localized mode
is more than twice the maximum frequency of the unperturbed host
crystal, quartic anharmonic terms are the lowest that can contribute
to the broadening and shift in position of the peak in the absorption
spectrum. In the case of deuteride U-centers whose localized mode
frequencies are lower than twice the maximum frequency of the unperturbed host crystal, cubic anharmonic terms as well as quartic
contribute. The mechanism contributing to the broadening and frequency shift of the localized mode through cubic anharmonic terms
is the decay of the localized mode into two band modes. In the
case of the quartic anharmonic terms, two kinds of processes are
possible: a decay of the localized mode into three band modes, or
a process in which one localized mode is scattered by a second
localized mode. It is found that the latter is the dominant

mechanism for both the deuteride and hydride localized modes, which
has as one consequence the result that the greater width of the
higher frequency hydride localized mode, seen in Fig. 3, is due to
the larger mean square displacement of the lighter hydride ion.[8]
In Fig. 5 are plotted the temperature dependences of the linewidths
of the fundamental, second harmonic, and one of the two third
harmonics of the localized mode due to a hydride impurity at a
fluorine site in CaF_2[5]. At elevated temperatures these curves
are closely proportional to the square of the absolute temperature,
as is predicted by theory.

In his first experiments on infrared absorption by hydride and
deuteride U-centers in alkali-halide crystals, Fritz[3] noticed
satellites or sidebands to the fundamental absorption peak. His
results for KCl doped with hydride U-centers is shown in Fig. 6.
Fritz suggested that these sidebands represent summation and difference bands in which one of the phonons participating is a localized mode phonon. If this is the case, then according to theory
these side bands should provide a weighted density of states, or
frequency spectrum of the unperturbed host crystal. This suggestion
was confirmed by subsequent work, such as that of Timusk and Klein[9]
shown in Fig. 7, which gives a high resolution spectrum of the high
frequency side band to the hydride localized mode peak in KBr. The
remarkable feature of this spectrum is that it reproduces among
other features of the frequency spectrum of the unperturbed host
crystal the gap in the spectrum between the optical and acoustic
branches, which is indicated in this figure.

There are three dominant mechanisms which give rise to these
sidebands. They are illustrated graphically in Fig. 8. The first
is a process in which the incident light interacts with the localized mode through its first-order dipole moment, whereupon the
localized mode breaks up into a second localized mode and a band
mode through cubic anharmonic terms in the crystal potential energy.
Energy conservation requires that the frequency of the incident
light equal the frequency of the localized model plus or minus
the frequency of the band mode. This is the mechanism that was
suggested by Fritz[3] to explain his experimental observations.
The second mechanism is one in which the incident light excites
two normal modes directly through the crystal second-order dipole
moment.[10] One of these modes is a localized mode and the other
is a band mode. Again, energy conservation requires that the frequency of the incident light equal the frequency of the localized
mode plus or minus the frequency of the band mode. The third
mechanism for the production of sidebands is the interaction between the first two processes. In this case the incident light
interacts directly with a localized mode through its first order
dipole moment, the localized mode then breaks up into a localized
mode and a band mode through the cubic anharmonic terms in the

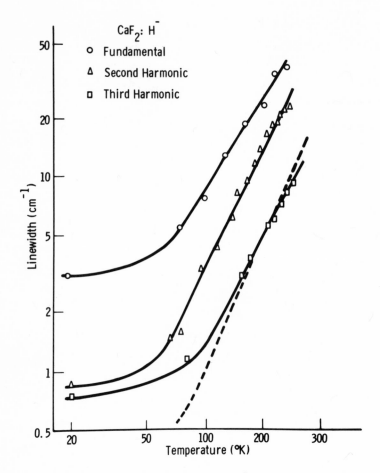

Fig. 5.   The temperature dependence of the widths of the funda-
           mental, second harmonic, and one third harmonic of the
           localized mode due to substitutional H⁻ ions in CaF₂.
           (Ref. 5).

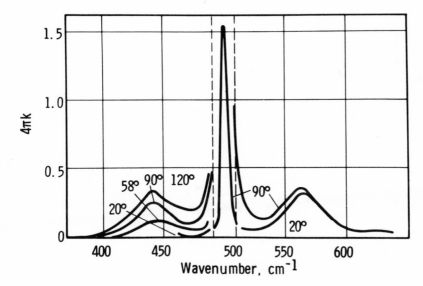

Fig. 6.   Sidebands to the localized mode peak in the infrared
           absorption spectrum of KCl containing substitutional H
           impurities.   The scale is reduced by a factor of 10 in
           the center of the figure.   (Ref. 3).

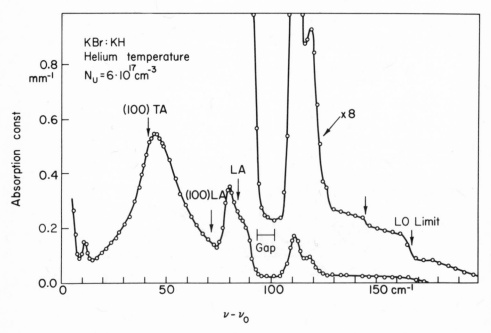

Fig. 7.   The high frequency sideband to the localized mode peak
in the absorption spectrum of KBr containing substitutional
H^- U-centers. (Ref. 9).

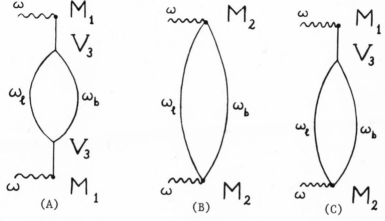

Fig. 8.   The three dominant mechanisms which give rise to side-
          bands to localized mode peaks in infrared absorption
          spectra.

crystal potential energy, and then this process interferes con-
structively with the process in which the light directly excites
a localized mode and a band mode. For each mechanism the shape
of the sideband which is obtained is a weighted frequency spectrum
of the perturbed host crystal to which only modes of Γ_1^+, Γ_3^+,
and Γ_5^+ symmetry contribute, when the symmetry at the impurity
site is O_h. It is just because these sidebands provide a means
of studying at least a part of this frequency spectrum that they
are so interesting to study. The consensus at the present time
is that the dominant mechanism responsible for the sidebands of the
localized mode peak is the one which proceeds through cubic anharmonic processes and a first order dipole moment.

Other properties displayed by localized modes due to hydride
and deuteride U-centers in ionic crystals are their morphic effects.
Morphic effects are effects induced in a crystal through the lowering of its symmetry by the application of an external force, such
as a stress or electric field, for example. In Fig. 9 are shown
the effects on the fundamental and second harmonic of the localized
mode due to hydride impurities in CaF_2 when a stress is applied
along the [100], [110], and [111] directions. When the applied
stress is along the [110] direction, the three-fold degeneracy of
the localized mode, and of its second-harmonic, is completely
split, while a stress applied along either the [100] or the [111]
directions splits the three-fold degenerate mode into a doublet
and a singlet. These results are in complete agreement with the
predictions of group theoretic arguments, based on the lowering
of the symmetry at the impurity site by the applied stress. From
such measurements one can also obtain information about the magnitudes of the anharmonic force constants entering the interaction
between the U-center and the host crystal.

The application of a stress, however, is not the only way in
which the site symmetry of a U-center in an alkali-halide crystal
can be lowered. This can also be achieved by adding cation impurities to the host crystal, for example by adding a small amount
of RbCl to NaCl. A certain fraction of the Na^+ sites are now
occupied by Rb^+ ions. This means that there is a certain probability that one of the nearest neighbors to a U-center in the Cl^-
sublattice is occupied not by an Na^+ ion but by a Rb^+ ion, and the
site symmetry of the impurity site is reduced from O_h to C_{4v}. In
this case the three-fold degeneracy of the U-center localized mode
is split into a singlet, which corresponds to motion of the U-center
along the line joining it to the additive Rb^+ impurity, and a
doublet arising from the vibrations of the U-center perpendicular
to this direction. Such measurements were in fact carried out by
Schaefer in his earliest studies of hydride U-center localized
modes,[2] and one of his results is shown in Fig. 10. This shows

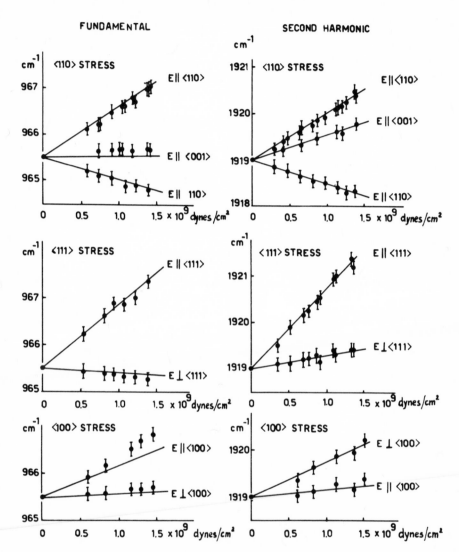

Fig. 9.   The effects of a uniaxial stress on the frequencies of
the threefold degenerate localized mode, and its second
harmonic, due to substitutional H⁻ U-centers in CaF₂.
(Ref. 11).

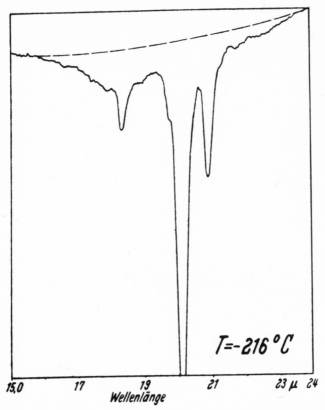

Fig. 10.   The localized mode peaks in the infrared absorption
           spectrum of NaCl containing 2% RbCl in addition to
           substitutional H⁻ U-centers.   (Ref. 2).

the transmission spectrum of NaCl containing 2% RbCl, in addition
to the hydride impurities present in the anion sublattice. The
presence of the two peaks, one on each side of the fundamental
absorption peak, is evidence of the splitting of the three-fold
degeneracy of the localized mode just described. The higher
frequency peak represents the doublet, and the lower frequency
peak corresponds to the singlet mode.

A very interesting property of localized mode absorption is
the strong temperature dependence (decrease) of the integrated
absorption. In Fig. 11 the areas under the localized mode peaks
for absorption by both hydride and deuteride U-centers in KCl
are plotted against the square of the absolute temperature on a
semi-logarithmic plot.[12] The results indicate that the integrated absorption decreases exponentially with increasing temperature with the exponent varying as the square of the absolute
temperature. This is just the kind of temperature dependence one
would expect in the high temperature limit if the absorption coefficient contained a temperature factor analogous to the Debye-
Waller factor familiar in the theory of x-ray or neutron diffraction by the atomic vibrations of crystals. A great deal of
theoretical effort has shown that just this kind of temperature
dependence is to be expected for the absorption by U-center
localized modes.[13]

All of the localized modes, and the effects assoiated with
them, described so far have been induced by either hydride or
deuteride U-centers, on account of their very light masses. Lest
it be thought that hydrogen atoms can give rise to localized modes
only in their ionic form, I should like to point out that atomic
hydrogen present interstitially in CaF_2 also gives rise to localized vibration modes. The existence of such modes had been predicted several years ago on the basis of the temperature dependence of the spin lattice relaxation time for paramagnetic[14]
hydrogen and deuterium atoms present interstitially in CaF_2.
However, such modes have now been observed directly by infrared
absorption by Shamu et al.,[15] whose results are shown in Fig. 12.
Notwithstanding the fact that the hydrogen atom is a neutral entity,
it acquires a dynamic effective charge in its interstitial site due
to the difference between its electronic polarizability and those
of its neighbors in the host crystal. The symmetry of the localized
mode to which such an interstitial hydrogen atom gives rise is one
which possesses a nonvanishing dipole moment, and the localized
mode is therefore infrared active. Its frequency can be obtained
with much greater position by optical means than indirectly
through the temperature dependence of the spin lattice relaxation
time.

If one is dealing with a host crystal which is made up of

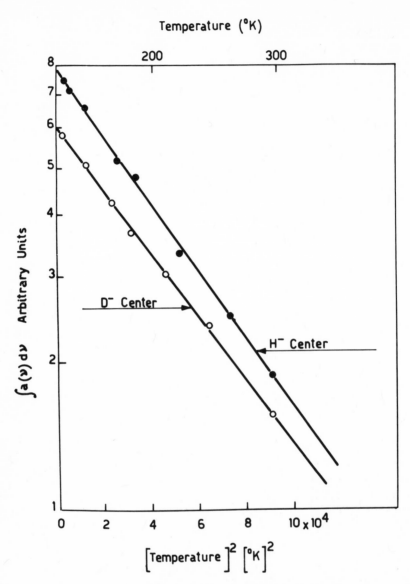

Fig. 11. Temperature dependences of the areas under the localized
          mode peaks in the absorption spectrum of KCl containing
          substitutional H⁻ and D⁻ U-centers. (Ref. 12).

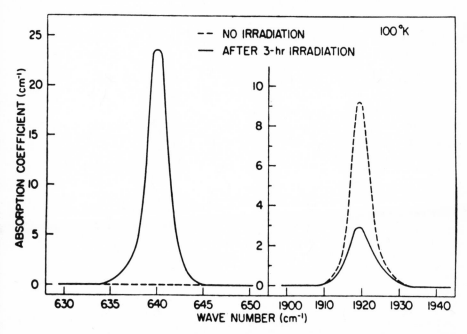

Fig. 12.   Localized mode peak in the infrared absorption spectrum
           of CaF_2 containing interstitial H atoms (left), and the
           second harmonic of the localized mode due to H^- ions
           substitutionally in CaF_2 (right).   (Ref. 15).

electrically neutral atoms, such as the homopolar semiconductors
diamond, silicon, and germanium, one can induce localized modes
which are infrared active by the simple expedient of introducing
impurities which are charged. This observation is due originally
to Dawber and Elliott.[16] If these charged impurities give rise
to localized modes which possess a dipole moment they can be
studied by infrared absorption techniques. In Fig. 13 is depicted
the absorption spectrum of silicon containing boron atoms as
substitutional impurities and compensated with lithium atoms which
occupy interstitial sites.[17] Since the maximum normal mode frequency of silicon is 518 cm^{-1}, all of the structure depicted in
this figure is due to absorption by the localized modes due to the
boron and lithium impurities. The peaks at 532 cm^{-1} and 522 cm^{-1}
are due to localized modes associated with Li^6 and Li^7 impurities,
respectively. The relative areas under these two peaks are a reflection of the relative abundances of Li^6 and Li^7 in naturally
occurring lithium. The peaks at 644 cm^{-1} and 620 cm^{-1} are due to
the localized vibration modes induced by the substitutional B^{10}
and B^{11} impurities, also present in their natural abundance.
Because the substitutional boron impurities occupy sites of T_d
symmetry, the localized modes to which they give rise are triply
degenerate. However, when an interstitial lithium impurity is
adjacent to a substitutional boron impurity, the site symmetry of
the boron impurity is lowered from T_d to C_{3v}. Just as with the
additive cation impurities described above, the three-fold degeneracy of the boron localized mode is split into a singlet and
a doublet when its site symmetry is reduced to C_{3v}, with the singlet
corresponding to motion of the boron impurity along the three-fold
symmetry axis, and the doublet associated with motions of the boron
impurity perpendicular to this axis. In Fig. 13 the peaks at
584 cm^{-1} and 683 cm^{-1} are the singlet and doublet respectively,
associated with the B^{10} localized mode when a lithium impurity is
adjacent to it in an interstitial position, while the peaks at
564 cm^{-1} and 657 cm^{-1} are the singlet and doublet, respectively,
associated with the B^{11} localized mode when a lithium impurity is
adjacent to it in an interstitial site.

It is not necessary, however, that the impurity in a homopolar
semiconductor be charged in order to give rise to a localized mode
which is infrared active. Carbon present substitutionally in
silicon gives rise to localized modes which can be observed by
infrared absorption. Although carbon is a neutral impurity in
silicon, since they are both from the same column of the periodic
table, the carbon impurity atoms nevertheless acquire dynamic
effective charges, again because of the difference between the
electronic polarizabilities of the carbon and silicon atoms. In
Fig. 14 are presented the experimental results of Newman and
Willis[18] for the absorption spectrum of silicon containing C^{12}
and C^{13} impurities. The heavier C^{13} atoms give rise to a localized

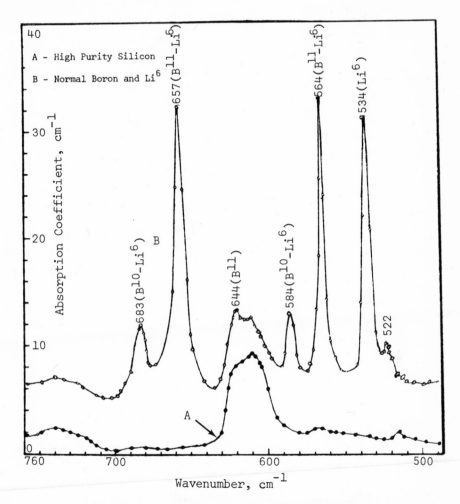

Fig. 13. Localized mode peaks in the infrared absorption spectrum
          of Si containing substitutional B^{10} and B^{11} impurities,
          compensated by interstitial Li^6 and Li^7 impurities.
          (Ref. 17).

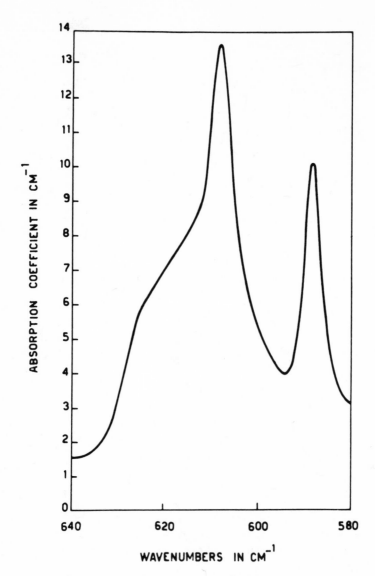

Fig. 14.   Localized mode peaks in the infrared absorption spectrum
           of Si containing substitutional C^{12} and C^{13} impurities.
           (Ref. 18).

mode of a frequency somewhat lower than that due to C^{12} impurities.

In Fig. 15 are shown the experimental results of Parker, et al.,[19] for the Raman spectrum of germanium containing increasing concentrations of silicon atoms as substitutional impurities. The site symmetries of atoms in homopolar crystals of the diamond structure, or in polar semiconductors of the zincblende structure, is T_d. One of the consequences of this fact is that the localized modes associated with subsitutional impurities in such crystals give rise to a first order Raman scattering of light. The maximum frequency of germanium is 350 cm^{-1}. The peak in the Raman spectrum at 400 cm^{-1} is therefore identified as the localized mode peak due to an isolated silicon impurity. A second peak at 462 cm^{-1} is tentatively identified as due to the localized mode arising from a nearest neighbor pair of silicon impurity atoms. This work was the first observation of localized vibration modes by Raman spectroscopy, and there have been several subsequent studies of localized modes by Raman scattering in succeeding years.

All of the localized modes studied so far have been induced by impurities in cubic crystals, occupying positions of cubic symmetry. However, there have been a few studies of localized modes in crystals of lower symmetry. In Fig. 16 are presented experimental results by Manabe et al.,[20] for the absorption spectrum of CdSe containing Be atoms substitutionally in the Cd sublattice. The symmetry of the impurity site in this case is C_{6v}. The localized mode in this case is no longer triply degenerate but consists of a singlet, corresponding to motion of the Be impurity along the six-fold axis, and a doublet, corresponding to the motion of the Be impurity parallel to the basal plane. These can be observed by using infrared radiation polarized parallel and perpendicular to the axis respectively, and this is illustrated in Fig. 16. One can construct a local oscillator Hamiltonian for an impurity at a site of C_{6v} symmetry, similar to that in Eq. (1) for an impurity at site of T_d symmetry. One can again work out the energy levels and group theoretic selection rules for infrared absorption and Raman scattering for a particle moving in such an anharmonic potential. One of the results of such an analysis is that there are four infrared active second harmonics, two associated with each of the two fundamental vibration modes. These two sets of two second harmonics can also be observed in infrared absorption measurements by using incident radiation polarized parallel and perpendicular to the hexagonal axis of the crystal. Experimental results demonstrating this are also shown in Fig. 16.

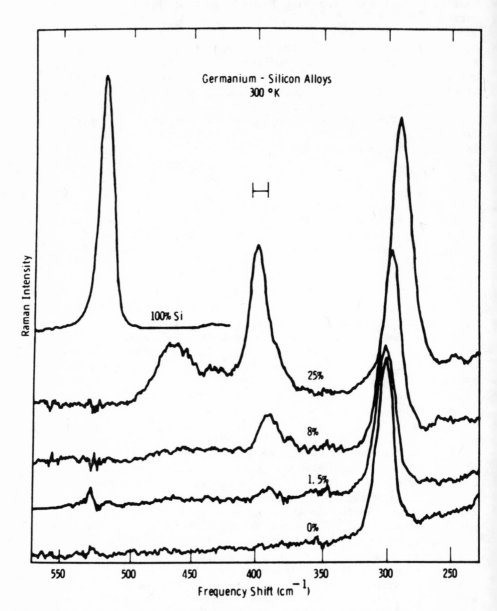

Fig. 15. First order Raman spectrum of Ge containing subsitutional
          Si impurities. (Ref. 19).

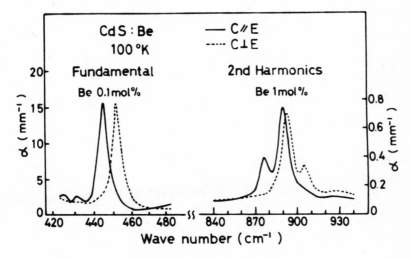

Fig. 16. Localized mode peaks and their second harmonics in the infrared absorption spectrum of CdSe containing substitutional Be impurities. (Ref. 20).

In this brief survey I have tried to indicate something
of the richness of the optical properties of localized modes
induced by impurities in transparent crystals. An understanding of these effects will lead in turn to an understanding
of the mechanisms responsible for absorption in the highly
transparent portion of the optical spectra of such crystals.

References

1.   K. F. Renk, Z. Physik 201, 445 (1967).

2.   G. Schaefer, J. Phys. Chem. Solids 12, 233 (1960).

3.   B. Fritz, in Lattice Dynamics, edited by R. F. Wallis
     (Pergamon Press, New York, 1965), p. 485.

4.   D. N. Mirlin and I. I. Reshina, Fiz. Tver. Tela 6, 3078 (1964).
     [English translation: Soviet Physics–Solid State 6, 2454 (1964)].

5.   R. J. Elliott, W. Hayes, G. D. Jones, H. F. Macdonald and
     C. T. Sennett, Proc. Roy. Soc. A289, 1 (1965).

6.   A. A. Maradudin, E. W. Montroll, G. H. Weiss, and I. P.
     Ipatova, Theory of Lattice Dynamics in the Harmonic Approxi-
     mation (Academic Press, New York, 1971), p. 416; the notation
     used for labeling the irreducible representations is that of
     G. F. Koster, J. O. Dimmock, R. G. Wheeler, and H. Statz,
     Properties of the Thirty-Two Point Groups (M. I. T. Press,
     Cambridge, Mass., 1963).

7.   J. A. Harrington, R. T. Harley, and C. T. Walker, Solid State
     Comm. 8, 407 (1970); G. P. Montgomery, W. R. Fenner, M. V.
     Klein, and T. Timusk, Phys. Rev. B5, 3343 (1972).

8.   W. Hayes, G. D. Jones, R. J. Elliott and C. T. Sennett, in
     Lattice Dynamics, edited by R. F. Wallis (Pergamon Press,
     New York, 1965), p. 475; M. A. Ivanov, M. A. Krivoglaz, D. N.
     Mirlin, and I. I. Reshina, Fiz. Tver. Tela 8, 192 (1966)
     [English translation: Soviet Physics–Solid State 8, 150
     (1966)] ; I. P. Ipatova and A. A. Klochikhin, Zhur. Eksper.
     i Teor. Fiz. 50, 1603 (1966) [English translation: Soviet
     Physics–JETP 23, 1068 (1966)].

9.   T. Timusk and M. V. Klein, Phys. Rev. 141, 664 (1964).

10. Nguyen Xuan Xinh, Solid State Comm. 4, 9 (1966); Phys. Rev.
     163, 896 (1967).

11. W. Hayes, H. F. Macdonald, and R. J. Elliott, Phys. Rev.
     Letters 15, 961 (1965).

12. A. J. Sievers and S. Takeno, Phys. Rev. Letters 15, 1020 (1965).

13. I. P. Ipatova, A. V. Subashiev, and A. A. Maradudin, Proc. Int. Conf. on Localized Excitations in Solids, edited by R. F. Wallis (Plenum Press, New York, 1968), p. 93. A. E. Hughes, Phys. Rev. 173, 860 (1968); I. P. Ipatova, A. V. Subashiev, and A. A. Maradudin, Annals of Physics (N.Y.) 53, 376 (1969).

14. P. G. Klemens, Phys. Rev. 125, 1795 (1962); D. W. Feldman, J. G. Castle, Jr., and J. Murphy, Phys. Rev. 138, A1208 (1965); P. G. Klemens, Phys. Rev. 138, A1217 (1965).

15. R. E. Shamu, W. M. Hartmann, and E. L. Yasaitis, Phys. Rev. 170, 822 (1968).

16. P. G. Dawber and R. J. Elliott, Proc. Phys. Soc. 81, 453 (1963).

17. W. G. Spitzer and M. Waldner, J. Appl. Phys. 36, 2450 (1965).

18. R. C. Newman and J. B. Willis, J. Phys. Chem. Solids 26, 373 (1965).

19. D. W. Feldman, M. Ashkin, and J. H. Parker, Jr., Phys. Rev. Letters 17, 1209 (1966).

20. A. Manabe, A. Mitsuishi, H. Komiya, and S. Ibuki, Solid State Comm. 12, 337 (1973).

RAMAN, PHOTOCONDUCTIVE, AND ACOUSTOELECTRIC PROBES OF RESIDUAL DEEP IMPURITIES AND ABSORPTION IN GaAs*

D. A. Abramsohn, G. K. Celler[#] and Ralph Bray

Physics Department, Purdue University

West Lafayette, Indiana 47907

We present several techniques to test and characterize GaAs for residual deep impurities and weak optical absorption at 1.06 μ. 90° Raman scattering intensities were measured as a function of position along a sample to probe the attenuation of a collimated YAG laser beam and to obtain the absorption coefficient. The absorption at 1.06 μ may be due to native defects. Their concentration was obtained from saturation photoconductive studies with intense Q-switched YAG laser radiation. A 25 fold increase in conductivity yielded a defect concentration of $\approx 10^{16}$ cm^{-3}. In addition we have used propagating acoustoelectric domains, ≈ 1 mm wide, to scan the sample for inhomogeneities in both dark and photoexcited carrier concentration.

I. INTRODUCTION

We shall describe here three quite different techniques for testing and characterizing various aspects of the residual, weak absorption of 1.06 μ YAG laser light in n-type GaAs. We measured 90° Raman scattering intensities as a function of position along a sample to probe the attenuation of a collimated YAG laser beam, and thus to determine the absorption coefficient. This technique eliminates the need to take into account surface reflection and absorption effects.

* Work supported by NSF Grant GH 43409 and by NSF-MRL grant GH 33574-A3.

Formerly: J. K. Wajda

As a probe technique, it can detect spatial variations of the
absorption coefficient arising from such factors as inhomogeneity
of the material or non-linear effects at high light intensity. The
residual absorption at 1.06 μ appears to be due to native defects.
The second technique is a photoconductive study of intense laser
excitation from deep lying levels associated with these defects.
From the near saturation of the photoconductive effect at the
highest light intensities, it is possible to obtain a measurement
of the concentration of these defects. The third technique involves
the use of propagating acoustoelectrically-amplified phonon domains
to scan the length of the sample for spatial variations in both the
dark and photoexcited carrier concentrations. The information appears in the form of variations in the time trace of the acoustoelectric current through the sample.

     A study of the literature reveals a rather remarkable disagree-
ment in the magnitude of the parameters describing basic non-linear
optical effects. For example, in GaAs the values quoted [1,2] for
the two-photon absorption coefficient differ by factors greater than
200. These discrepancies may be related to problems with defining
the spatial distribution of the laser beam intensity, and failure
to take into account the material characteristics of the specific
samples. Thus it has been suggested [2] that residual absorption
can induce transient thermal self-focusing effects, which would
exaggerate the strength of the two-photon absorption. In either
case, the Raman probe technique offers the possibility of investigating such effects by scanning the distribution of light intensity
in the sample, both laterally and longitudinally.

II. RAMAN PROBE TECHNIQUE

     The experimental arrangement for a point by point probe of Raman
scattered light from the sample is shown in Fig. 1. Collimated cw
YAG laser radiation is passed through the 2 cm length of the sample,
in our case along a <110> direction. The material used has an elec-
tron concentration $n_o = 6 \times 10^{14}$ cm^{-3} and conductivity $\sigma_o = 0.67$ (ohm
cm)$^{-1}$ at room temperature. The 1.17 eV photon energy is well below
the intrinsic absorption edge of GaAs. Light which is scattered in
the bulk of the sample is collected at 90° by a light-tight cone
with a 1 mm x 3 mm aperture placed against the side of the sample.
The scattered light is focussed into a double monochromator, and
detected then recorded by a cooled S-1 photomultiplier and photon
counting equipment. The monochromator is set at a Raman wavelength
with a resolution of $\approx$ 4 cm^{-1}, the room temperature half-width of
the individual Raman lines. The sample is mounted on a translation
table so that any position along the length of the sample can be set
in front of the cone aperture. The high index of refraction of GaAs
assures us that light is only collected from scattering which occurs
directly in line with the aperture.

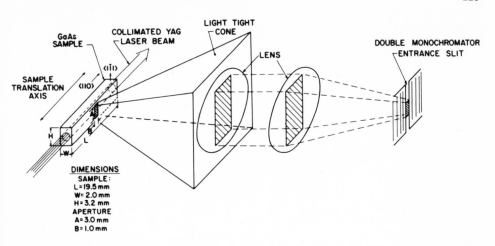

Fig. 1.   Experimental arrangement for probe of 90° Raman scattered
          light.

     The Raman scattered light serves very nicely as a probe to give
an accurate measure of the intensity of the laser beam at each point
along the sample. Fig. 2(a) shows data for unpolarized light incident on a sample cut obliquely at the far end to prevent multiple
internal reflections. The data were taken for the transverse and
longitudinal Stokes and anti-Stokes Raman lines. The relative intensities are shown as measured, corrected only by subtraction of a
background signal of 2-3 dark counts/sec. The spatial resolution of
the data is determined by the 1 mm cone aperture. Greater spatial
resolution is easily attainable at the expense of counting times
longer than the 300 sec at each point for the $\approx$ 1 watt of laser power
used in the present runs. The exponential variation of the scattered
light intensity for each Raman line gives the absorption coefficient
α, at 1.06 μ. The average value of α is 1.39 cm^{-1}. The absence of
appreciable scatter of the data points and the self-consistency of
the slopes suggest that the method can be used to measure coefficients well below 0.1 cm^{-1}. It is shown in Fig. 3 that a scan of
the sample using parasitic or Rayleigh scattered light does not provide an adequate probe, presumably because of an irregular distribution of parasitic scattering centers.

     Our method has some advantages over more standard measurements
of absorption coefficient. First, it can determine the spatial variation of the absorption coefficient along the sample, of whatever origin, rather than its average value over the entire sample. Secondly,
the reflection of incident laser power from the first surface of the
sample does not enter into the absorption measurement. A correction

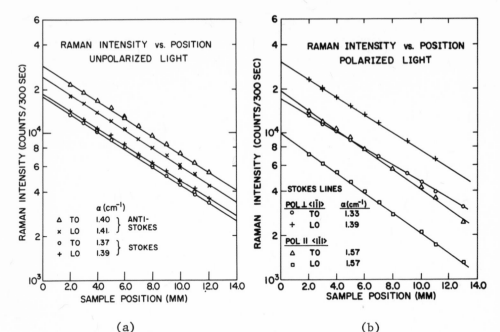

(a)   (b)

Fig. 2.   Raman line intensities for (a) unpolarized and (b) polarized
incident light as a function of position along the length of the GaAs
sample.   The slope gives the absorption coefficient α.

can easily be made for multiple reflections within a crystal with
parallel end surfaces; however, the sample can be cut to prevent
reflection of the light back along the sample and to eliminate the
need for such corrections. This allows the direct measurement of
the absorption coefficient, independent of the reflectivity and other
surface effects. This method also has the advantage that the measurement is done at a wavelength where stray light and parasitic
radiation from the laser are not detected.

Since the Raman probe technique worked so well with unpolarized
light, it came as a surprise to discover appreciable deviations when
measurements on the same sample were repeated with polarized light.
As illustrated in Fig. 2(b), there is a 15% variation in the slopes
depending on the polarization direction of the incident light. This
seems well outside the scatter of the data and not to be due to experimental error. A more detailed study gave a fairly systematic
dependence of absorption coefficient on the direction of polarization. These initial observations of optical anisotropy and some
smaller day-to-day variations in absorption coefficient (possibly
with laser power or sample temperature) will be investigated more
thoroughly.

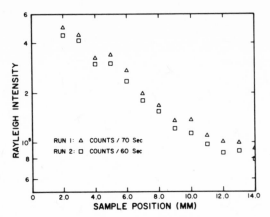

Fig. 3. Rayleigh or parasitic scattered light intensity as a function of position along the sample.

Optical anisotropy in cubic crystals can be due to internal strains and spatial dispersion. Spatial dispersion-induced bire-fringence for light propagation along a ⟨110⟩ direction has been reported extensively in GaAs [3] and other cubic materials [3,4]. We have found no equivalent information for absorption. For the latter, the anisotropy would be dependent on the specific source of absorption rather than on the host crystal as in the case of birefringence. Birefringence, whether due to spatial dispersion or strain, can be an important source of error in the present absorption measurements when polarized light is used. From the GaAs data of Yu and Cardona [3], it can be deduced that effective depolarization of the incident beam occurs in a 2 cm long sample. This depolarization, if coupled with a strong dependence of Raman scattering efficiency and instrumental response on polarization, can influence the measurement of Raman scattering intensities along the sample and thus distort the determination of the absorp-tion coefficient. However, such an effect can not explain the difference in slopes seen for the TO lines at the two polarizations shown in Fig. 2(b), since the scattering strength in this case is essentially independent of the polarization.

It is interesting to compare values of the absorption coef-ficient in the literature for GaAs at 1.06 μ, and to inquire into the source of the absorption. Extensive measurements in a large variety of samples have been reported by Bois and Pinard [5]. After corrections for free carrier absorption they find α to vary between 0.7 and 2.0 cm^{-1} at 1.17 eV, in samples with electron concentrations ranging from 10^{15} to 10^{18} cm^{-3}. This result is consistent with other investigators [2,6], and appears to corre-spond to an absorption band centered near 1.2 eV at temperatures between 4 K and 300 K. This band was observed in all bulk GaAs samples tested by Bois and Pinard [7]. Transient capacitance

measurements [8,9] reveal the same center to be present in high
purity epitaxial samples. The center responsible for this absorption band appears to be a universally present impurity or a native
defect, possibly Ga vacancies [7]. Estimates of the concentration
of this defect are mid 10^{14} cm^{-1} in all liquid epitaxial grown
crystals [9]. This is also a lower limit estimated in [7] and [8]
for bulk and epitaxial samples. The importance of the residual
absorption in GaAs for the measurement of non-linear optical properties was emphasized in Sect. I.

III. SATURATION PHOTOCONDUCTIVITY ANALYSIS OF DEFECT CONCENTRATION

Extrinsic photoconductivity at very high light intensity is a
very useful tool for studying the deep levels associated with the
residual absorption process, and in particular for determining the
density of these defects.

Although the YAG laser light is only weakly absorbed in GaAs,
it produces very large increase in conductivity, $\Delta\sigma \gg \sigma_0$ at high
Q-switched laser power. We made all our measurements at room temperature with laser pulses at intensities up to ≈ 3.5 MW/cm^2. To
achieve such intensities, the illumination had to be concentrated
in a spot of about 2 mm in diameter. In order to measure directly
the photo-induced change in conductivity, it was necessary to illuminate the sample uniformly and for this we had to make small
samples whose illuminated face was equal to or smaller than the
cross section of the concentrated laser beam. The thickness of the
sample was about 0.5 mm, so that the attenuation of the beam in
passing through the sample was negligible. With the sample thus
quite uniformly illuminated, the change in conductivity was determined from the pulse change of current for a nearly constant voltage
of 5 V applied across the samples. The incident light intensity,
as monitored on a Si pin diode, and the current modulation pulse
were separated by a 500 nsec delay line and measured on a storage
oscilloscope.

In Fig. 4 we show log-log plots of $\Delta\sigma/\sigma_0$ vs. light intensity
for a GaAs sample from the same slice as the one used for the absorption measurements. This is undoped material from Monsanto
with $n_0 = 6 \times 10^{14}$ cm^{-3} at 300 K and it shows little electron freeze-
out at 77 K. We see that $\Delta\sigma/\sigma_0$ increases sublinearly over all of
the measured range, and approaches a saturation level [10] at the
highest light intensities. The data for pulses of 60 and 200 nsec
half-width are in substantial agreement except at the lowest light
intensities. Where they correspond, the carrier lifetime must be
smaller than 60 nsec and the data must represent steady state photoconductivity. At the lowest intensities, the electron lifetime
may be quite long and the photoconductivity not in steady state.

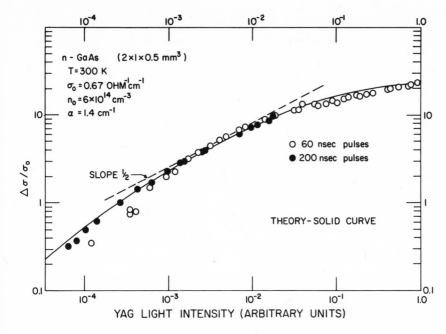

Fig. 4. Fractional change in conductivity versus light intensity, showing saturation of photoconductivity at maximum light intensity of 3.5 MW/cm^2. Theoretical curve is obtained with $N_T = 1.5 \times 10^{16}$ cm^{-3} and $r = 8.5 \times 10^{-9}$ cm^3/sec.

The photoconductive response may be attributed completely to laser excitation of electrons to the conduction band from deep levels, presumably the ones responsible for the absorption band near 1.2 eV found in all GaAs [5-7]. The saturation of $\Delta\sigma/\sigma_0$ at a value of about 25 implies that the levels are completely exhausted at that stage and that their concentration N_T is $\approx 25 \times n_0$ or 1.5×10^{16} cm^{-3}. This very direct determination gives a defect concentration substantially higher, by a factor of $\sim 10^2$, than the estimates quoted earlier for epitaxial and bulk samples. We have assumed here that there is no change in electron mobility due to changes in the charge status of the defects. This is a fair approximation at 300 K, where charged impurity scattering contributions are not very important.

The strong tendency of $\Delta\sigma/\sigma_0$ to saturate guarantees that there is no appreciable contribution from non-linear processes, even though the maximum intensity in Fig. 4 is estimated to be a few MW/cm^2. The same evidence implies that there is no restoration of the supply of electrons on the deep levels by direct optical activation by the YAG laser or by thermal activation from the valence band. The absence of such effects may be due to the relatively high cross section for

hole capture. According to Henry and Lang [9] the hole capture cross
section at 300 K, on so-called type A impurities at 1.2 eV below the
conduction band is $\approx 10^3$ times greater than the electron capture
cross section.

The solid line in Fig. 4, calculated from a very simple model,
can account for the whole shape of the curve of $\Delta\sigma/\sigma_0$ vs. light intensity. We assume N_T defects, all at a single energy level, serving
both as the sole source of absorption of light and of retrapping of
the excited electrons. The data indicate that steady state conditions prevail over nearly the whole range of light intensity. In
the dark, the deep levels, lying well below the Fermi level (determined by the shallow donors), can be assumed completely occupied
by electrons.

In steady state,

$$\alpha\,(1 - \frac{\Delta n}{N_T})\frac{L}{\hbar\omega} = r\,(n_0 + \Delta n)\Delta n$$

where n_0 is the free electron concentration in the dark, Δn is the
concentration of electrons optically excited to the conduction band
from the deep levels, $L/\hbar\omega$ is the number of photons/(cm^2sec) entering
the sample, α is the absorption coefficient, and r is the recombination coefficient. N_T, as determined from the saturation of the photoconductivity, is 1.5×10^{16} cm^{-3}. The best fit of the theoretical
curve to the experimental data is obtained assuming that
$r = 8.5 \times 10^{-9}$ cm^3/sec, which corresponds to a free electron lifetime at low excitation levels $\tau_0 = 1/(r\,n_0) \approx 200$ nsec.

According to this model the change of conductivity is linear
at low light intensities where $\Delta n \ll n_0$. The sublinear response
at higher light intensities is caused by a reduction of the photoexcited carrier lifetime $\tau = 1/r(n_0+\Delta n)$. Thus, when $\Delta n \gg n_0$
but still $\Delta n \ll N_T$, the slope of the $\Delta\sigma/\sigma_0$ versus light intensity
curve is 1/2. At still higher light intensities, where $\Delta n \approx N_T$,
the absorption goes to zero and the photoconductive curve saturates.

IV.   ACOUSTOELECTRIC PROBE

Piezoelectrically-active shear phonons in n-type GaAs can be
amplified by 10 orders of magnitude above the thermal background
by applying a high electric field along the <110> crystallographic
axis [11]. These highly amplified phonons are localized in narrow domains [12], creating a very high resistance region moving
along the sample with the velocity of sound, $v_s = 3.35 \times 10^5$ cm/sec.

The fact that most of the voltage drop across the sample is
concentrated in the acoustoelectric domain enables the latter to

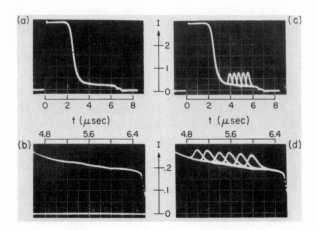

Fig. 5. (a) Current pulse showing characteristic acoustoelectric
drop to a saturation value; (b) expanded portion of the saturation
current revealing spatial variations of n_o; (c) a multiple exposure
photograph of current traces with light pulses shifted in space and
time to probe different parts of the sample; (d) expanded portion
of (c) with light pulses of lower intensity.

scan the conductivity of a sample as a function of position, with
the spatial resolution determined by the domain's size (typically
about 1mm). The utilization of the domain to scan fluctuations
of the free electron concentration in GaAs crystals, which is representative of the variation in shallow donor concentration, was
reported earlier by Spears and Bray [13]. We now show how
similar measurements can be made with the domain area illuminated
by extrinsic light, to provide information about the variation of
the concentration of the deep impurities as a function of position.

In Fig. 5(a) a current pulse trace characteristic of the
domain formation is shown. After a domain incubation time of
≈ 2 μsec, the current drops from an ohmic value to a low saturation
level, given by $fn_o ev_s A$, where A is the sample cross sectional area
and f is a small number ≈ 2, varying with carrier concentration and
temperature. In Fig. 5(b) a portion of the saturation current is
expanded to reveal deviations of the current from a straight line.
Such deviations, representing spatial variations in n_o, are quite
small here, indicating that this sample is uniform. Examples for
less uniform samples are shown in [13].

To probe the variation in defect concentration, we focussed
200 nsec laser pulses on successive portions of the sample and
scanned the sample with the propagating domain. When the domain
passes an illuminated region, a current spike is generated (Fig.

5(c)) representing the local concentration of photoexcited carrie
[14]. An expanded picture of a series of such spikes is shown in
Fig. 5(d). The small increase in spike amplitude in this figure
as the domain propogates (1 μsec corresponds to 3.35 mm) indicate
a gradual increase of the density of defects along the sample. I
is also possible in principle to illuminate the whole sample uniformly with cw light and let the domain scan the change in carrie
concentration. However, it is difficult to obtain such uniform
excitation with sufficient intensity to give measurable response.

We wish to express our appreciation to Dr. E. Palik for the
loan of a pulsed YAG laser and S. Mishra for assistance with the
use of the laser for the photoconductive measurements.

REFERENCES

1. C. C. Lee and H. Y. Fan, Phys. Rev. B 9, 3502 (1974).
2. D. A. Kleinman, R. C. Miller and W. A. Nordland, Appl. Phys.
    Lett. 23, 243 (1973).
3. P. Y. Yu and M. Cardona, Solid State Commun. 9, 1421 (1971);
    M. Bettini and M. Cardona, Proc. 11th Intern. Conf. Phys.
    Semicond. (Polish Scientific Publishers, Warsaw 1972) p. 1072
4. J. Pastrnak and K. Vedam, Phys. Rev. B 3, 2567 (1971).
5. D. Bois and P. Pinard, Phys. Stat. Sol. (a) 7, 85 (1971).
6. M. D. Sturge, Phys. Rev. 127, 768 (1962).
7. D. Bois and P. Pinard, Phys. Rev. B 9, 4171 (1974).
8. D. Bois and M. Boulou, Phys. Stat. Sol. (a) 22, 671 (1974).
9. C. H. Henry and D. V. Lang, Proc. 12th Intern. Conf. Phys.
    Semicond., (B. G. Teubner, Stuttgart 1974) p. 411.
10. It is interesting to note that an analogous saturation in
    optical absorption has been seen by A. E. Michel and M. I. Na
    Appl. Phys. Lett. 6, 101 (1965), under different circumstances
    Their observations were made with GaAs injection laser radiati
    (10^5 W/cm^2) on compensated p-GaAs with $\approx 5 \times 10^{17}$ atoms/cm^3 o
    manganese as the absorbing impurity and absorption coefficient
    of about 140 cm^{-1} at low light intensities.
11. D. L. Spears, Phys. Rev. B 2, 1931 (1970).
12. For reviews on domain formation and propagation see: N. I.
    Meyer and M. H. Jorgensen, Feskörperprobleme 10, 21 (1970); M.
    B. N. Butler, Rep. Prog. Phys. 37, 421 (1974).
13. D. L. Spears and R. Bray, J. Appl. Phys. 39, 5093 (1968).
14. J. K. Wajda and R. Bray, Proc. 12th Intern. Conf. Phys. Semi-
    cond. (B. G. Teubner, Stuttgart 1974) p. 877.

IDENTIFICATION OF Fe^{4+}, Fe^{5+} and Fe^{4+}-V_O PHOTOCHROMIC ABSORPTION BANDS IN $SrTiO_3$*

K.W. Blazey, O.F. Schirmer, W. Berlinger and K.A. Müeller

IBM Laboratory, Rüschlikon, Switzerland

ABSTRACT

$SrTiO_3$Al exhibits photochromism due to the presence of iron as a minority impurity. The wavelength dependence of the optical absorption due to Fe^{5+} agrees with the wavelength dependence of the bleaching of Fe^{5+} EPR.[1] After the decay of the Fe^{5+} and an initial decay of Fe^{3+} EPR, the intensity of the latter is enhanced with the same wavelength dependence as the optical absorption due to Fe^{4+}. During these processes, a hole center is also created. This correlation of the optical absorption with the EPR bleaching and enhancement curves, shows that the photochromic absorption bands must be due to electron transfer from the O^{2-} valence states to the Fe^{5+} and Fe^{4+} ions. The bands occur at 2.09 and 2.82 eV for Fe^{4+}, and 1.99 and 2.53 eV for Fe^{5+}; they show a shift to higher energies as the charge state of the impurity decreases.[2] A weak EPR, spectrum axially symmetric around <100>, with $g_{\|} \sim 8$, $g_{\perp} \sim 0$ was observed in Al and Mg-doped crystals. This spectrum is due to Fe^{4+}-V_O. The angular dependence of the resonance fields is given by

$$H \cdot 4g_{\|} \cdot \cos \delta = [(h\nu)^2 - a^2]^{1/2}$$

similar to Mn^{3+}-V_O.[3] From X- and K-band measurements, $g_{\|} = 2.007 \pm 0.001$, $a = 0.1504 \pm 0.0008$ cm^{-1}. Optical charge transfer of valence O^{2-} electrons to Fe^{4+}-V_O, converts the center to well-known Fe^{3+}-V_O; the respective EPR quenching and creation rates are identical in the investigated energy range 1.85 - 3.07 eV, this allowed a determination of the charge transfer bands.[4] $Fe^{4+}-V_O$ is a shallow acceptor and observed only in low Fermi-level $SrTiO_3$ at temperatures below $77°K$.

*Text not available. Details will be published in refs. 2 and 4.

REFERENCES

1. K. A. Müller, Th. von Waldkirch, W. Berlinger and B. W. Faughnan, Solid State Commun. $\underline{9}$, 1097 (1971), and references quoted therein.

2. K. W. Blazey, O. F. Schirmer, W. Berlinger and K. A. Müller, to be published in Solid State Comm.

3. R. A. Serway, W. Berlinger and K. A. Müller, Proc. 18th Congress Ampere, Nottingham, September, 1974.

4. O. F. Schirmer, W. Berlinger and K. A. Müller, to be published in Solid State Communications.

PICOSECOND SPECTROSCOPY OF TRANSIENT ABSORPTION IN PURE KCl

R.T. WILLIAMS, J.N. BRADFORD and W.L. FAUST

NAVAL RESEARCH LABORATORY

WASHINGTON, D. C. 20375

We present a brief review of various metastable and stable absorbing species generated during electron-hole recombination in alkali halides, emphasizing transient components which are the dominant features on the time scale of short laser pulses. The F-band in KCl falls in this category, since it is produced with high efficiency (~ 15% of e-h recombinations) but decays to 10% of its initial value ($\tau \lesssim 20$ μsec) even at liquid helium temperature. We report experiments in which a single 20 psec pulse of 2660 Å light produces dense coloration in a pure KCl crystal at flux levels too low to cause damage from other processes such as avalanche breakdown. We have used delayed pulses of 5320 Å light to monitor the rise of the absorption on a picosecond time scale, as well as its decay (at room temperature) on a nanosecond time scale.

Because of two-photon absorption processes, an intense laser pulse with frequency equal to at least half the band gap in a "transparent" material shares many of the damaging effects normally associated with such forms of ionizing radiation as X-rays, hard ultraviolet, and energetic charged particles. A well-known effect of ionizing radiation in insulators is of course the generation of point defects, most notably F centers. In regard to the transparency of optical materials, defect generation is frequently regarded as a chronic, cumulative degradation proceeding with moderate-to-low efficiency, confined to a few specific optical bands, and susceptible to correction by occasional annealing.

In the time regime of short pulses, however, the situation is quite different, as illustrated in Fig. 1. These are absorption

spectra measured in pure alkali halides following excitation by a
5 nsec pulse of energetic electrons.[1] The spectra have been
decomposed by decay time, as noted in the figures. First, we
notice that the absorption is certainly not confined to just a few
bands, but extends more or less continuously from near-ultraviolet
to near-infrared. The absorption due to transient effects effec-
tively overwhelms the stable components of familiar color center
bands. One of the strongest stable defect absorption components
apparent in Fig. 1 is the F band in KCl (2.3 eV). Even though
measured within one second of irradiation, the KCl F-band component
labeled "STABLE" in Fig. 1 is only 1/10 as large as its short-time
counterpart. Thus for practical purposes in this time regime, the
F band itself is a transient defect, decaying by recombination of
the close vacancy-intersitial pairs, which are apparently produced
in much greater numbers than well-separated pairs.

Aside from the transient F and H bands, most of the other
absorption components in Fig. 1 have no stable counterparts and
are associated with relaxed, or "self-trapped," excitons. Without
going into detail, we briefly mention that the prominent low-
energy bands comprise the hydrogenic series of the relaxed exciton,
and the broad near-UV absorption arises from excitation between
levels involving different self-localized hole states.[1] Very
strong transient absorption bands observed just below the fundamental edge in alkali iodides are attributed to excitons perturbed
by self-trapped excitons.[2] These latter features are expected to
occur near the edge in other alkali halides as well, though we
have not attempted to work that far into the UV.

To provide an approximate scale of production efficiency for
these spectra, we point out that about one out of every six
electron-hole pairs created in pure KCl (and possibly as many as
one out of three in KBr) results in a transient F center,[1,3]
while most of the remaining electron-hole pairs reach the lowest
self-trapped exciton states, contributing to the characteristic STE
absorption.

Having reviewed the scope of transient absorption generated in
alkali halides by ionizing radiation, we again consider the propagation of a very intense laser pulse (with $h\nu \geqq E_{gap}/2$) through
a pure alkali halide crystal. If the two-photon band-gap excitation leads to the same transient defect states that we have observed after electron pulse excitation, and if the process occurs
in a short enough time, then the propagating laser pulse will
introduce and may _itself_ be attenuated by transient defect states
in the crystal. This is in addition to the normal losses by two-
photon absorption, and to absorption by the ordinary stable defects.

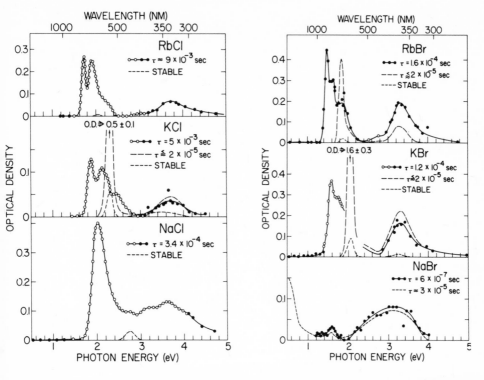

FIGURE 1:   Optical absorption in several alkali halides at a
temperature of 10 K, following excitation by a 5 nsec pulse of
energetic electrons.   The spectra have been decomposed by decay
times, as noted in the figure.

        It is therefore of interest to know how quickly the absorbing
species appear after band-gap excitation.   Furthermore, since the
annealing or destruction of these transient defects is hastened at
higher temperatures, the measurement of decay times at room temperature will be useful.   There is already strong evidence(4,5) that
one of the transient defects (self-trapped exciton) is a precursor

to the formation of another (the F center). The process is
comparable to a molecular photochemical rearrangement reaction.
The mechanism remains unresolved between several models,[3,4,6-8]
so that kinetic data on the time scale of vibrational relaxations
will be quite useful.

As noted above, the techniques of optical spectroscopy with
electron pulse excitation are limited to times of the order of
nanoseconds, which is too slow to see the defect formation processes. We have therefore turned to picosecond laser techniques.
The first experiments, described here, deal with F center production in KCl.

Our first experiment was a simple visual one, to ascertain
whether in fact a ~ 20 psec pulse of fourth harmonic light (2660 Å)
from our Nd:YAG laser would produce measurable concentrations of
point defects in KCl at flux levels too low to cause damage from
other processes (e.g., avalanche ionization). Starting with photon
energies less than half the band gap, we focused first the fundamental, then the second, then the third harmonic beam in the bulk
of a KCl crystal. In each case, a line of fractures was generated
at sufficiently high powers, but no coloration was observed. On
the other hand, the fourth harmonic (2660 Å) pulse generated dense
purple coloration (characteristic of the F band, which absorbs
green light) throughout the focal region at power levels considerably below those which would produce fracture damage. Watching
through a microscope focused in the interior of the crystal, we
could observe the color appear following a single pulse. As we
remarked earlier, of course, the transient coloration is expected
to be much stronger. Similar qualitative observations were made
with the third harmonic pulse in KBr and KI.

We also noticed that for relatively low power levels, such
that multiple pulses were required to produce significant coloration, the color extended rather uniformly through the crystal. As
the power was increased, the distribution of color peaked up gradually toward the entrance face of the crystal. This behavior is
interpreted as evidence of the role of two-photon absorption in the
defect production. The tendency of the F centers to fade rapidly
defeated our initial efforts to obtain a depth profile with a
densitometer, which would have allowed a quantitative check of the
expected quadratic dependence on intensity.

In order to make time-resolved measurements, we introduced a
known delay between the fourth harmonic damaging pulse and the
second harmonic interrogation pulse, the latter falling conveniently on the F band peak at low temperature. Actually, three green
pulses were used in each measurement, all derived from the same
pulse by appropriately placed partial reflectors. One green pulse

preceded the UV pulse to establish a reference probe intensity; the second green pulse arrived at a known delay after the UV; and the third green pulse arrived ~ 10 nsec later to sample the "final" absorption (i.e. before significant recombination set in at low temperature). The green probe pulses were attenuated by about 10^{-3} in order to avoid serious bleaching.

All pulses were combined colinearly before being focused on the sample through a single fused silica lens, making use of the lens dispersion to bring the probe pulses to a well-defined focus inside the larger UV spot, which was about 80 microns in diameter. Spot alignment was accomplished by observing a magnified image of the crystal surface. Time-resolved measurements were made using a vacuum photodiode, which served merely to detect the amplitudes of the three distinct probe pulses (separated by at least 5 nsec). Time was measured from the delay set by the position of the second partial reflector. We established the position for coincident UV and green pulses at the sample by caliper measurement of optical paths, including corrections for group velocities[9] in optical components.

Convenient optical densities (0.2 to 1.0) corresponding to F-center densities up to 10^{17} cm^{-3} were obtained without recourse to amplification of the oscillator output. These densities are well out of the impurity-dominated first coloration stage[10] in our samples of high purity Oak Ridge KCl.[11] Furthermore, the rise time of F-center absorption created as a result of the transfer of electrons onto pre-existing vacancies has been shown to be 0.6 μsec,[12] due to the long lifetime of the first excited state of the F-center. An 0.6 μsec component would not be observable in our experiment. Using a single amplification stage, one 20 psec pulse of fourth harmonic light made the crystal virtually opaque to 5320 Å light on a transient basis.

The samples, typically 0.8 mm thick, were mounted on the cold finger of a liquid helium cryostat for low temperature experiments. The equilibrium sample temperature in this configuration was 10 K ± 2 K. Estimates based on approximate densities of energy deposition and accounting for energy storage in metastable states indicate that the laser pulse used in our quantitative measurements should raise the temperature of the focal region to no higher than 25 K from an initial temperature of 10 K, within the time scale of the experiment. At room temperature the heating is negligible. (Recall that we are operating below dielectric breakdown thresholds.)

The rise of absorption at 5320 Å following two-photon excitation of KCl at an equilibrium temperature of 10 K is shown in Fig. 2. We plot "fractional optical density," meaning the ratio of O.D. measured at the indicated delay to that measured at 10 nsec delay.

"Delay" refers to the time separation between the centers of the damaging and interrogating pulses. The error bars represent one standard deviation above and below the mean of 10 data points.

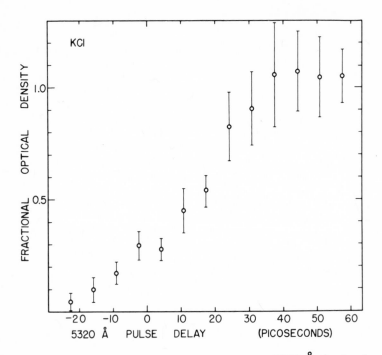

FIGURE 2: The rise of optical absorption at 5320 Å (relative to its value at 10 nsec) following two-photon excitation of pure KCl at an equilibrium temperature of 10 K.

Since our second and fourth harmonic laser pulses were of the order of 20 to 25 psec in duration, much of the apparent rise time results from the convolution of damaging and assessment pulses. In Fig. 3, we have plotted the apparent growth of absorption that would result if the physical defect formation were <u>instantaneous</u> and the laser pulses were of finite duration. The upper part of the figure refers to the case where the damaging and assessment pulses are of equal width (duration). In the lower half of the figure, the assessment pulse is twice as wide as the damaging pulse. Due to competition between squaring effects and differences in group velocities, the actual ratio of pulse widths lies somewhere

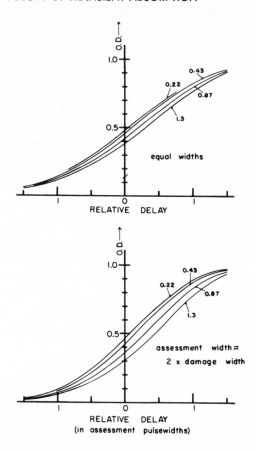

FIGURE 3:   Calculation of apparent absorption measured with damage
and assessment pulses of finite width. The physical rise of
absorption is taken here to be instantaneous. Gaussian pulses
are assumed.

between these limits. These calculations were made for Gaussian
pulses; however, calculations for square pulses and for triangular
pulses yielded very similar quantitative results. In fact these
results are rather insensitive to the detailed shape of any
symmetrical pulse, particularly at zero relative delay. The various
curves are labeled by the final optical density achieved, and are
plotted on normalized scales. The striking feature is that almost
universally (i.e. without much sensitivity to pulse shape (assumed
symmetrical), relative widths, or final O.D. within the indicated
ranges), the apparent optical density has reached about 40% of
its final value when measured by an assessment pulse coincident
with the damaging pulse. Applying this rule to the data in Fig. 2,
we conclude that the physical rise time of absorption is about 10
picoseconds.

At low temperature (25 K > T > 10 K), the absorption does not anneal significantly within the 10 nsec span of observation in the present experiment. Combining these data with electron pulse work at longer times (t > 5 nsec), we infer that the 5320 Å absorption at low temperature is formed in about 10 psec and decays in times on the order of 20 μsec, leaving a 10% residual component. At room temperature we find in the present experiment that the 5320 Å absorption again has a physical rise time faster than the laser probe pulse width, but decay of the absorption in this case occurs in times of the order of 8 nsec due to accelerated recombination of close vacancy-interstitial pairs. A small visually observable residual absorption remains, presumably due to well-separated pairs of vacancies and interstitials as well as impurity stabilization of some of the defects.

The relation of these measurements to proposed mechanisms of Frenkel defect formation will be the subject of a paper to be published elsewhere. Briefly stated, however, the results of the present experiment suggest that the lattice defects are formed out of the higher states of the self-trapped exciton, i.e. those encountered moderately early in the recombination of free carriers. Considerations of the origin of the absorption due to self-trapped excitons in Fig. 1 leads us to speculate that the STE UV absorption may rise even faster than the F-center absorption, while the near-infrared absorption may well rise more slowly. This latter expectation comes in view of the rather large energy separations of the last few levels through which the self-trapped exciton must cascade to reach its lowest states, which give rise to the characteristic red and infrared absorption. Currently we are considering experiments to test these expectations.

REFERENCES

1. R. T. Williams and M. N. Kabler, Phys. Rev. B9, 1897 (1974).

2. R. T. Williams and M. N. Kabler, Solid State Comm. 10, 49 (1972).

3. M. N. Kabler, NATO Advanced Study Institute on Radiation Damage Processes in Materials (Corsica), (1973).

4. D. Pooley, Solid State Comm. 3, 241 (1965); Proc. Phys. Soc. 87, 245 (1966).

5. F. J. Keller and F. W. Patten, Solid State Comm. 7, 1603 (1969).

6. N. Itoh and M. Saidoh, J. Physique 34, C-9, 101 (1973).

7. Y. Toyozawa, Tech. Rept. ISSP (Univ. Tokyo) <u>A648</u> (1974).

8. Y. Toyozawa, International Conference on Color Centers in
    Ionic Crystals (Sendai, Japan), (1974).

9. In a preliminary report on this experiment (International
    Conference on Color Centers in Ionic Crystals, Sendai, Japan,
    1974) only corrections for phase velocities were made, leading
    to an overestimate of the F band rise time.

10. E. Sonder and W. A. Sibley, in <u>Point Defects in Solids</u>, ed. by
    J. H. Crawford, Jr., and L. M. Slifkin (Plenum Press, New York,
    1972).

11. The KCl used in this experiment was grown in the Research
    Materials Program, Solid State Division, Oak Ridge National
    Laboratory. We wish to express special thanks to C. T. Butler
    and W. A. Sibley for the crystals.

12. M. Ueta, Y. Kondo, M. Hirai, and T. Yoshinari, J. Phys. Soc.
    Japan <u>26</u>, 1000 (1969).

Section IV
Glasses

HIGHLY TRANSPARENT GLASSES

Jan Tauc

Division of Engineering and Department of Physics

Brown University, Providence, Rhode Island   02912

Absorption and scattering processes which limit
the transparency of amorphous solids are reviewed
with emphasis on highly transparent glasses for optical
fiber communication. The following topics are briefly
discussed: absorption by electronic transitions (intrinsic, and by defects, impurities and free carriers), absorption by lattice vibrations (intrinsic and by impurities)
and attenuation by scattering (Rayleigh, Mie, Brillouin
and Raman).

Studies of factors limiting the transparency of glasses are of
considerable interest since glasses are the most important optical
materials. In recent years, serious effort was concentrated on the
development of optical fibers for communication applications.
This kind of communication is economically feasible if the losses
in the fiber are below 20 dB/km, that is the absorption constant
$\alpha (=2.3 \times 10^{-6}$ dB/km) must be below 5×10^{-5} cm^{-1} for the frequency
band of interest, which is at least three orders of magnitude smal-
ler than for any other applications. This problem has been solved
with remarkable success, and glass fibers have been produced with
losses as low as 1 dB/km. Methods of preparation and properties
of these fibers have been described in recent review papers by
Maurer[1], Miller et al.[2] and French et al[3].

As shown in Fig. 1, the absorption constant α has a deep minimum between the electronic transitions at high energies and the
vibrational bands at low energies. In this paper I will review
the absorption and scattering processes which limit the transparency of glasses at this minimum, and the present understanding
of them.

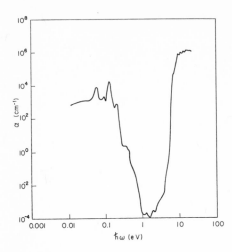

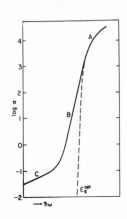

Fig. 1 Absorption constant of a
soda-lime silicate glass (after
Bagley et al.[48,49])

Fig. 2 Absorption edge of amorphous solids. Part B is the
Urbach edge, Part C the low
absorption tail.

1.   Electronic Transitions
1.1   Intrinsic absorption

The absorption edge of many amorphous solids has the shape shown
in Fig. 2. Part A corresponds to band to band transitions which
are broadened (in Part B) by the processes discussed by Dow at this
Conference (Urbach edge). At low absorption levels one often
observes a tail C which is much flatter than the Urbach part. This
part of the absorption edge is very troublesome since it extends far
from the absorption edge and is a limiting factor on the maximum
transparency of the glass in the window between the electronic transitions and vibrational modes. We will discuss the origin of this
tail.

During the early stages of the understanding of the electronic
structure of amorphous solids this tail was associated with the localized states in the gap induced by the disorder[4,5]. Theoretical arguments for the existence of these states proposed by Mott[6], Cohen,
Fritzsche, Ovshinsky[7] and others can be simply understood from a
model of the electronic structure of tetrahedrally bonded monoatomic
semiconductors of Weaire[8,9]. In this approach each atom is surrounded
with four tetrahedrally coordinated neighbors as in the diamond
structure but no assumption is made about the periodicity of the network. Two matrix elements are introduced: V_1 between wavefunctions
associated with the same atoms (but different bonds) and V_2 between
functions associated with the same bond (but different atoms). If
V_1 and V_2 are assumed to be constant throughout the network it can
be shown rigorously that:

$$\text{gap} \geqslant 2\,|V_2| - 4\,|V_1| . \tag{1}$$

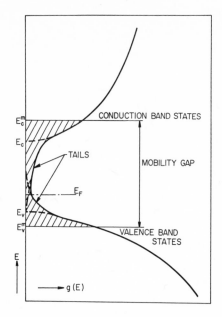

Fig. 3 A model of the density of states g(E) in an amorphous solid.

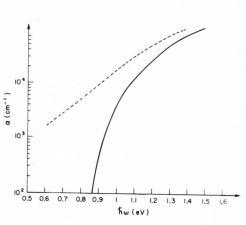

Fig. 4 The change in the absorption edge of an amorphous Ge film produced by annealing (after Theye[11]) Dashed line: as deposited. Solid line: after annealing.

Therefore a gap exists even in a topologically disordered system. However, a network as defined above with fixed V_1 and V_2 cannot exist except in the crystalline structure. A disordered network can be realized if we allow some deviations in angles and interatomic distances throughout the network which lead to distributions of V_1 and V_2 (a quantitative disorder). These distributions would introduce states in the gap as shown in Fig. 3. If we estimate the spread of V_1 and V_2 from the X-ray diffraction data we obtain a high density of states in the gap, which would lead to a significant broadening of the edge and a strong tail C.

Experimental data has not confirmed the expectation of this kind of "intrinsic" tail due to disorder in an amorphous solid in which all chemical bonds are satisfied. Donovan et al.[10] showed in their photoemission experiments on amorphous Ge (a-Ge) that it is possible to prepare amorphous films with sharp optical edges which indicate a low concentration of states in the gap (below the sensitivity of their measurements, $10^{17} cm^{-3}$). In Fig. 4 we show the results of Theye[11] on a-Ge films[12]. One film had a rather broad edge after deposition, but it sharpened significantly after annealing. Experiences with other amorphous materials are similar. Vaško[13] observed in a-Se that the edge is sharpened when the samples are purer until the slope corresponding to the Urbach tail is reached; then a low

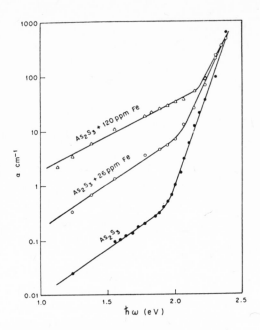

Fig. 5 Absorption edge of amorphous
As_2S_3 nominally pure, and doped with
Fe (after Tauc et al.[14]).

absorption tail is observed (Part C) which decreases if the sample
is further purified. Tauc et al.[14] reported similar observations
on chalcogenide glasses (Fig. 5), and the experience with a-SiO_2
and some related oxide glasses also appears to indicate the absence
of an _intrinsic_ low absorption tail.

These observations are surprising from the point of view of the
theoretical considerations discussed above. There must be some deep
reasons why the above reasoning fails. From the point of view of
the solid state physics of crystals one is used to considering the
structure as rigid (allowing only for vibrations about the equilibrium sites). Usually we do not understand in detail why a particular crystallographic structure is realized but once we know it we
can proceed from this point with the calculation of the electronic
properties. In the case of glasses, the atomic arrangements are
formed during the quenching of the glass. The energies of various
configurations are determined by the electronic states, which in
turn depend on these configurations. One would expect that certain
lowest energy configurations are realized during this process,
extending over regions in which effective interaction is possible.
In other words, in calculating the electronic properties of glasses
one cannot consider the system of electrons and atoms independently
of each other.

Various aspects of this problem have been discussed. Phillips[15] suggested that in annealed films of a-Ge a self-consistent arrangement of atoms maximizes bond energies by ejecting states which would otherwise fall in the gap. He estimates that the number of atoms involved in this process is of the order 10^2. Van Vechten[16] presents thermodynamical arguments to show that fluctuations of the atomic potentials do not produce fluctuations in energies. Anderson[17] noted that the deep impurity states in insulators exhibit large Frank-Condon effects since the lattice nearby displaces considerably when the center is occupied by an electron. He suggested that a similar effect would appreciably shift the energies of localized electron states in amorphous semiconductors and showed that the energies will be shifted towards the band edges. This effect will thus clear the gap of states. Recently, Anderson[18] suggested a model for the electronic structure of amorphous semiconductors. Electrons are localized in bonds and two electrons in the same bond with opposite spins have a strong attractive interaction. The one electron states (optically observable) have a gap. The two electron states (optically non-observable) have a continuous distribution of energies and fix the Fermi level in the middle of the gap. This model explains very well many experimental facts, such as the pinning of the Fermi level, the absence of paramegnetism and optical absorption, and the equality of the optical and electrical gaps.

1.2 Defects and Impurities

If we accept the above results, we must ascribe absorption in tail C (Fig. 2) to defects (unsatisfied chemical bonds) and impurities. This is simple enough in a-Ge. It appears to be well established that the defect states are dangling bonds at the internal surfaces of voids[19]. They introduce both optical absorption and ESR. By annealing the films, the concentration of voids is reduced, ESR decreases and the optical gap sharpens (Fig. 4). Even in this relatively simple case it is not possible to consider the total absorption as a superposition of intrinsic and defect absorption since the disappearance of voids changes somewhat the atomic arrangement in the whole sample.

Thin films of amorphous Mg_3Bi_2 are semiconducting with a gap of 0.15 eV. An excess of Mg or Bi increases the conductivity. This observation was interpreted by Ferrier and Herrell[20] in terms of a "rigid" band model. In this model, states in the gap are produced by intrinsic disorder and are independent of composition. They are filled with electrons according to the composition and the Fermi level moves from one side of the gap to the other. Recently, Sutton[21] studied the absorption edge of this system and has shown that his results are inconsistent with the rigid band model. The data are best described by a gradual filling of the gap with states associated with the excess constituent; the Fermi level is fixed in the gap. Another support for the idea that defects and impurities are responsible for the states in the gap are the results of Hauser[22] on doped amorphous Ge films.

Knights[23] has recently reported a tail in bulk a-As which he believes cannot be due to impurities (in the region from 0.5 to 1.1 eV α varied from 7 to 100 cm^{-1}). a-As can be prepared in the amorphous form chemically, not by quenching the liquid. It appears that this method of preparation introduces some non-satisfied chemical bonds in the amorphous solid.

However, in bulk glasses prepared by quenching the liquid (the "real" glasses), the search for unsatisfied chemical bonds ("dangling" bonds) has led to negative results[14,24,25]. The states in the gap of real glasses appear to be associated with impurities. In this case, the electronic transitions are of two kinds: internal transitions and charge transfer transitions. The former occur within the impurity atom and are characterized by relatively sharp lines and small oscillator strengths. Charge transfer transitions involve neighboring atoms, produce broad bands and have large oscillator strengths (cf. the paper by Blasey et al. at this Conference). It is this kind of transition which is troublesome in high transparency glasses. For example, in a-As_2S_3 Fe produces extended tails as shown in Fig. 5; the oscillator strength is of the order 10^{-1} (as compared with 10^{-4} to 10^{-3} for internal transitions). Taking into account the results of ESR and magnetic susceptibility measurements, the absorption was associated with the $Fe^{3+} \rightarrow Fe^{2+} + e$ transition[24]. One may tentatively describe it as a transition which involves the impurity and the conduction band[26]. However, this kind of transition has been little studied in glasses. The two-photon absorption processes in glasses reported at this Conference by Stolen and Lin may be a very useful tool for the identification of the impurities since the two-photon spectra appear to be much sharper than the one-photon spectra.

Metallic ions such as Cu^{2+}, Fe^{2+}, Ni^{2+}, V^{3+}, Cr^{3+}, Mn^{3+} are the dominant cause of absorption by electron transitions in oxide glasses[1,2]. For example, Miller et al.[2] report that a concentration of 1 part in 10^9 (by weight) of Fe^{2+} or Cr^{3+} limits the transparency of glass (at the peak loss) to 1 dB/km (2×10^{-6} cm^{-1}). To achieve a high level of transparency in fibers for long-distance communications, it is necessary to produce them with the same requirements on impurity content as in semiconductor technology. In fact, the lowest absorption was obtained in SiO_2 fibers (1 dB/km at 1.06 μm[27]) since they could be prepared from extremely pure Si by processes which contaminated the material very little.

At this point, opinions differ as to whether this limit could be obtained in multi-component glasses if one is able to prepare them with the same purity. So far the purification has always

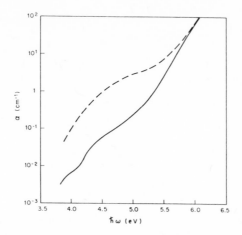

Fig. 6 Absorption edge of a soda-
lime-silicate glass (after Bagley
et al.[48,49]). The solid line
corresponds to a purer glass than
the dashed line.

improved the transparency, as is seen e.g. in Fig. 6. Besides
the obvious technological advantages, the transparency of multicomponent glasses may be less sensitive to some contaminants than
pure SiO_2. The absorption depends on the chemical state of the
metal which is determined by the chemical environment. For
example, it has been observed[28] that Fe atoms produce an order of
magnitude smaller absorption in boron silicate glasses (B_2O_3 -
SiO_2 - Na_2O - Tl_2O) then in soda-lime-silica glasses (Na_2O - CaO -
SiO_2). In some applications, the relatively smaller resistance
of the transparency of multicomponent glasses against high energy
irradiation may be a limiting factor[29].

1.3 Free Carrier Absorption

Part C of the absorption edge (Fig. 2) in a-As_2S_3 becomes indis-
tinguishable at high temperature[30]. However, in As_2Se_3 the absorption in Part C increases with temperature with the same activation
energy as the d.c. conductivity[31]. A similar observation has been
made on the relatively highly conducting $Tl_2SeAs_2Te_3$ glass[32]. It
was suggested[32] that this absorption is due to thermally excited
free carriers. The absence of this absorption in As_2S_3 is explicable
by the large gap of this material.

There are some difficulties with the detailed understanding of
this absorption process. The conductivity calculated from the
absorption constant is 20 to 40 times larger than the d.c. conductivity. The observed frequency dependence of the absorption
constant is different from the case of free-carrier absorption in
crystalline semiconductors, and the theoretical attempts to explain

it have not been successful. At this point, the absorption by
free carriers in glasses is not understood and the experimental
results are scarce. This problem is of considerable practical
interest for optical windows or filters which are exposed to high
temperatures.

2. Absorption by Lattice Vibrations

The present status of the absorption by lattice vibrations has
been recently reviewed[33] and we shall add here a few comments which
are relevant for the high transparency of glasses. In materials
which in the crystalline state have no i.r. one-phonon absorption
bands (such as Ge) the disorder introduces absorption bands.
These reflect the density of one-phonon states which appears in
the first approximation to be similar to the density of phonon
states of the crystalline phase with the same short-range order.
This observation is explained by the relaxation of k-vector
conservation by disorder. If, however, the crystal already has
one-phonon absorption bands, in the amorphous state one usually
observes broad structureless absorption bands at approximately the
same frequencies. However, these bands cannot in general be obtained
by broadening the crystalline bands, since there are important
topological differences between the structure of the crystal and
the glass. From a heuristic point of view one can say that if a
vibration in the crystal is infrared active, the corresponding vibration in the glass is usually much stronger than the disorder
induced absorption bands[34]. This disorder induced lattice vibration absorption is very important in the far infrared region[35]
but does not seem to be troublesome close to the edge.

An important question is the shape of the absorption bands of
glasses since the tails may extend towards the absorption edge and
intrinsically limit the transparency of the glass. One may ask how
the shape of the i.r. bands is effected by disorder, in particular
whether the multicomponent glasses have broader bands than simple
glasses. The questions have not been studied in sufficient detail.
Taylor et al.[36] have observed that the i.r. one-phonon absorption
band in a-As_2S_3 has the Gaussian shape (rather than Lorentzian);
this is attributed to a distribution of Lorentzian oscillators.
The multiphonon absorption in chalcogenide glasses is discussed
at this Conference by Treacy et al., and Howard et al.

Impurities are troublesome not only because of the electronic
transitions as discussed above, but also because of the absorption
caused by their vibrations. A well known example is water in SiO_2
which is present in the form of the ion OH^-. Its fundamental
absorption band near 2.7 µm is the reason why the presence of water
in SiO_2 limits its i.r. transparency. The second and third harmonics of this vibration are observed at 0.95 and 0.75 µm, which
is in the region of interest for communication applications. Mil-

ler et al.[2] estimates that 1 part in 10^6 (by weight) of OH^- ions in oxide glasses limits the transparency at 0.75 µm to 1 dB/km. It is therefore necessary to produce and process the fibers in such a way that contamination by water is avoided.

3. Scattering

Light scattering by fluctuations in the dielectric constant ε is a well known limitation of the transparency of glasses. The invention of technological processes for reducing the scattering of glasses in the last century has been an essential step which has made possible the mass production of optical glass and the development of the optical industry. In the case of fibers for long-distance communications the requirements are of course much more stringent. It is natural to ask what is the lowest scattering limit of a glass. The attenuation constant α_{sc} corresponding to light scattered by fluctuations with a correlation length much shorter than the wavelength of light (Rayleigh scattering)[37,38] is

$$\alpha_{sc} = \frac{8\pi^3}{3} \frac{1}{\lambda^4} V \left\langle (\Delta\varepsilon)^2 \right\rangle \tag{2}$$

where V is the scattering volume, λ the wavelength of light, and $\langle \ \rangle$ denotes the average over volume V. The smallest possible value of α_{sc} corresponds to density fluctuations in thermal equilibrium. It is given by

$$\alpha_{sc} = \frac{8\pi^3}{3} \frac{1}{\lambda^4} \left(\frac{1}{3} n^4 (p_{11} + 2p_{12}) \right)^2 kT\beta \tag{3}$$

where k is the Boltzman factor, β the isothermal compressibility of the material, n the index of refraction, and p_{11} and p_{12} the photoelastic constants. For fused silica at 300 K, $p_{11} = 0.121$, $p_{12} = 0.270$ (from Dixon[39]) and we obtain from Eq. (3) $\alpha_{sc} = 4 \times 10^{-8}$ cm^{-1} (0.02 dB/km). However, a glass is certainly not in thermodynamical equilibrium since it is prepared by quenching. It has been suggested that the proper value of temperature to be used in Eq. (3) is a fictive temperature corresponding approximately to the glass-transition temperature T_g (for fused silica this is about 1700 K). The idea is that density fluctuations corresponding to T_g get frozen during the quenching process. Pinnow et al.[40] suggested this explanation for the higher value of α_{sc} obtained in their measurements on fused silica. They found at 1 µm that $\alpha_{sc} = 1.8 \times 10^{-6}$ cm^{-1} (0.8 dB/km).

Multicomponent glasses have lower T_g's and it follows from the above reasoning that the "frozen-in" density fluctuations should be less troublesome. However, there is an additional contribution to scattering associated with composition fluctuations. Pinnow et al.[40] estimated this contribution using a formula deduced from thermodynamical equations. However, glasses are not in thermodynamical equilibrium and the scattering may be lower for reasons

which are discussed in the paper by Mohr and Macedo at this Conference. In fact, Pinnow et al.[41] have recently reported that in
soda-alumina-silicate glasses the scattering losses were lower
than in fused silica under the same conditions (e.g. only 23% for
the composition $(SiO_2)_{48}$ $(Na_2O)_{40}$ $(Al_2O_3)_{12}$).

Eq. (2) does not hold if the linear dimensions of clusters are
not negligible compared to λ. If they exceed about $\lambda/20$ the spatial
dispersion effects cannot be neglected (Mie scattering). In particular, there is a change in the angular distribution of the scattered
light which, for independent scatterers with dimensions much smaller
than λ, is $1 + \cos^2\theta$ (θ is measured relative to the direction of the
incident light). For scattering by large clusters scattering in
the forward direction becomes increasingly more important. Scatterers of large dimensions have been observed in phase-separated
glasses[42,43].

Everybody working with glasses knows that it is easy to make
glasses which scatter light intensely and that special care must
be taken during the preparation of glasses to make the fluctuation
of the index of refraction small. On the other hand, from the
experience with fibers it appears that scattering is not an insurmountable problem, even in multicomponent glasses.

Brillouin and Raman scattering losses would become non-negligible
if we were able to approach the absolute minimum of Rayleigh scattering determined from Eq. (3). Brillouin losses are determined
by the equation[44]

$$\alpha_{sc} = \frac{8\pi^3}{3} \frac{1}{\lambda^4} (n^4 p_{12})^2 \beta kT \tag{4}$$

which differs from Eq. (3) by a factor of order unity
$(3p_{12}/(p_{11}+2p_{12}))^2$, which is about 1.5 for a-SiO_2 (the difference
between the adiabatic and isothermal β is neglected). Stimulated
Brillouin and Raman scattering[45] become extremely important when
the light intensity in the fiber exceeds a certain critical value.
In fact, they are one of the non-linear effects limiting the maximum power which can be transmitted through the fiber.

4. Glasses for Optical Fibers

In this section we show a few examples of glasses which may have
practical applications as optical fibers in communication systems.
In Fig. 7 the absorption spectrum of very pure fused silica is
shown as reported by Keck et al.[46].

The electronic transitions limit the transmission below 0.7 μm;
their long wavelength tail is due to impurities. According to the
analysis by Keck et al.[46], the vibrational states seen in the figure
are due to the overtones of OH^- vibrations combined with the funda-

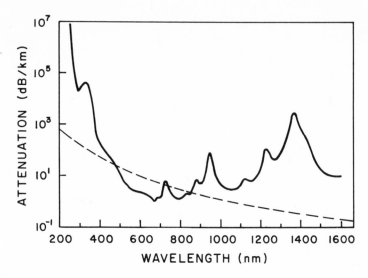

Fig. 7 Absorption (solid line) and scattering (dashed line) in fused silica fibers (after Keck et al.[46]).

mental vibration of SiO_4 units and its overtones. Scattering losses which must be added to the absorption losses are also shown in this figure.

From the practical point of view, one is interested in the windows located at frequencies for which suitable lasers are available. Presently two kinds of lasers are considered to be most promising, the semiconductor $GaAs-Al_xGa_{1-x}As$ lasers which can operate in the wavelength range 0.7 to 0.9 μm, and the Nd:YAG (Neodymium doped Yttrium-Aluminum-Garnet) lasers which operate at 1.06 μm. The longer wavelength is preferable from the point of view of scattering losses, and there is a window at 1.06 μm in the absorption curve shown in Fig. 7. Let us note that the absorption is associated with impurities, and that therefore the transparency could be improved. In fact, more recent development has led to laboratory samples of fused silica fibers which have lower values of absorption at 1.06 μm. In laboratory samples, total losses were made as small as 1 dB/km[3,27,41].

Fibers of multicomponent glasses prepared so far have considerably higher losses. There are reasons to believe that this is

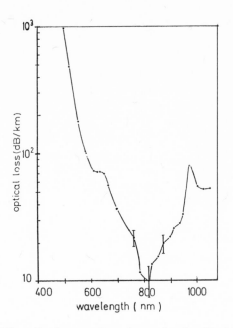

Fig. 8 absorption in a lead-alkali-
silicate glass (after Faulstich et al.[47]).

primarily due to impurities since these glasses were not prepared
under the same stringent conditions which were possible for SiO_2
fibers. As we mentioned in Section 3, in some glasses the scattering losses are surprisingly low, and in some glasses iron ions
seem to introduce less absorption than in SiO_2 (Section 1.2).

As an example of absorption in a multicomponent glass we show
in Fig. 8 the results recently reported by Faulstich et al.[47] on
a new flint-glass (lead-alcali-silicate glass, the composition was
not specified). We note that at 0.81 µm optical loss as small as
10 dB/km was achieved.

The low absorption levels obtained in optical fibers significantly has exceeded original expectations. When the research
started in 1968, the goal was 20 dB/km since at this absorption
level the optical communication systems appeared to be economically feasible. It appears that there is still room for improvement in multicomponent glasses which may have some significant
advantages compared to pure SiO_2 (or SiO_2-GeO_2) fibers. Solution
of the problem of how to produce low-loss materials is only a
part of a more complex problem, since the fibers must be cladded
and the index of refraction must have a certain radial distribution to minimize the pulse dispersion. This imposes additional

requirements on the materials. However, the results achieved
so far indicate that there are no insurmountable materials problems, and that optical communication systems will be experimentally
operated in the near future.

Acknowledgment

During the preparation of this paper, I profited from the
discussions with B. G. Bagley, W. G. French, A. D. Pearson,
D. A. Pinnow and R. H. Stolen. My thanks are also due to
P. W. Anderson, J. C. Knights and C. M. Sutton for the information about their recent results prior to publication.

This work was supported in part by a grant from the National
Science Foundation. It has also benefited from the general support
of Materials Science at Brown University by the NSF.

References

1. R. D. Maurer, Proc. IEEE 61, 452 (1973).

2. S. E. Miller, E. A. J. Marcatili and T. Li, Proc. IEEE 61,
    1703 (1973).

3. W. C. French, J. P. MacChesney and A. D. Pearson, to be
    published.

4. J. Tauc, in Optical Properties of Solids (ed. F. Abelès),
    North-Holland, Amsterdam (1972), p. 277.

5. J. Tauc in Amorphous and Liquid Semiconductors (ed. J. Tauc),
    Plenum Press, 1974, p. 159.

6. N. F. Mott and E. A. Davis, Electronic Processes in Non-Crys-
    talline Materials, Clarendon Press, Oxford, 1971.

7. M. H. Cohen, H. Fritzsche and S. R. Ovshinsky, Phys. Rev.
    Lett. 22, 1065 (1970).

8. D. Weaire, Phys. Rev. Lett. 26, 1541 (1971).

9. D. Weaire, M. F. Thorpe and V. Heine, J. Non-Crystalline
    Solids 8-10, 128 (1972).

10. T. M. Donovan, W. E. Spicer and J. M. Bennett, Phys. Rev.
    Lett. 22, 1058 (1969).

11. M. L. Theye, in Physics of Structurally Disordered Solids
    (ed. S. Mitra), Plenum Press (to be published).

12. Properties of tetrahedrally bonded amorphous semiconductors
    were reviewed by W. Paul, G. A. N. Connell and R. J. Temkin,
    Adv. in Physics 22, 531 (1973).

13. A. Vaško in The Physics of Selenium and Tellurium (ed. W. C. Cooper), Pergamon Press, 1969, p. 241.

14. J. Tauc, F. J. DiSalvo, G. E. Peterson and D. L. Wood, in Amorphous Magnetism (ed. H. O. Hooper and A. M. deGraaf), Plenum Press 1973, p. 119.

15. J. C. Phillips, Comments on Solid State Physics 4, 9 (1971).

16. J. A. Van Vechten, Solid State Comm. 11, 7 (1972).

17. P. W. Anderson, Nature Physical Sci., 235, 163 (1972).

18. P. W. Anderson, to be published.

19. M. H. Brodsky and R. S. Title, Phys. Rev. Lett 23, 581 (1969).

20. R. P. Ferrier and D. J. Herrell, Phil. Mag. 19, 853 (1969).

21. C. M. Sutton, Solid State Comm. 16, 327 (1975).

22. J. J. Hauser, Solid State Comm. 13, 1451 (1973).

23. J. C. Knights, Solid State Comm. 1975, in print.

24. F. J. DiSalvo, B. G. Bagley, J. Tauc and J. V. Waszcak, in Amorphous and Liquid Semiconductors (ed. J. Stuke and W. Brenig), Taylor and Francis, London 1974, p. 1043.

25. The quoted work has shown that the concentration of singly occupied electron states in very pure glasses was below the sensitivity of the methods (about 10^{16} cm^{-3}). More sensitive methods such as the photoluminescence studies reported at this Conference by Bishop and Strom may reveal the existence of non-satisfied chemical bonds at very low concentration levels.

26. D. L. Wood and J. Tauc, Phys. Rev. B5, 3144 (1972).

27. M. D. Rigtering, in Abstracts of the Meeting on Optical Fiber Transmission, Optical Soc. of America, Williamsburg, Virginia, 1975.

28. Y. Ikeda, M. Yoshiyagama and Y. Furuse, Proc. Tenth Int. Congress on Glass, The Ceramic Soc. Japan, 1974, p. 6-82.

29. G. H. Sigel, B. D. Evans, in Ref. 28, p. 6-23.

30. J. Tauc. A. Menth and D. L. Wood, Phys. Rev. Lett. 25, 749
     (1970).

31. J. T. Edmond, Br. J. Appl. Phys. 17, 979 (1966).

32. D. L. Mitchell, P. C. Taylor and S. G. Bishop, J. Non-Crystal-
     line Solids, 9, 1833 (1971).

33. J. Tauc, in Ref. 11.

34. S. S. Mitra, D. K. Paul and Y. F. Tsay, in Tetrahedrally
     Bonded Amorphous Semiconductors (ed. M. H. Brodski and S. Kirk-
     patrick), Am. Inst. Physics, New York 1974, p. 284; cf. also
     the paper by Tsay et al. at this Conference.

35. U. Strom, J. R. Hendrickson, R. J. Wagner and P. C. Taylor,
     Solid State Comm. 15, 1871 (1974).

36. P. C. Taylor, S. G. Bishop, D. L. Mitchell and D. Treacy, in
     Ref. 24, p. 1267.

37. L. D. Landau and E. M. Lifshitz, Electrodynamics of Continuous
     Media, Pergamon Press, 1960.

38. J. Tauc, in Ref. 11.

39. R. W. Dixon, J. Appl. Phys. 38, 5149 (1967).

40. D. A. Pinnow, T. C. Rich, F. W. Ostermayer and M. DiDomenico,
     Jr., Appl. Phys. Lett. 22, 527 (1973).

41. D. A. Pinnow, L. G. VanVitern, T. C. Rich, F. W. Ostermayer
     and W. H. Grodkiewicz, in Ref. 27.

42. B. G. Bagley, in Ref. 5, p. 1.

43. E. G. Rawson, Appl. Opt. 11, 2477 (1969).

44. T. C. Rich and D. A. Pinnow, Appl. Phys. Lett. 20, 264 (1972).

45. R. H. Stolen, E. P. Ippen and A. R. Tynes, Appl. Phys. Lett.
     20, 62 (1972); E. P. Ippen and R. H. Stolen, Appl. Phys. Lett.
     21, 539 (1972); R. H. Stolen and E. P. Ippen, Appl. Phys.
     Lett. 22, 276 (1973).

46. D. B. Keck, R. D. Maurer and P. C. Schultz, Appl. Phys. Lett.
     22, 307 (1973).

47. M. Faulstich, D. Krause, N. Neuroth and F. Reitmayer, in
     Ref. 27.

48. B. G. Bagley and W. G. French, Amer. Ceramic Soc. Bull. 52,
     701 (1973).

49. B. G. Bagley, E. M. Vogel, W. G. French, G. A. Pasteur, J. Gan
     and J. Tauc, to be published.

MULTIPHONON ABSORPTION IN THE CHALCOGENIDE GLASSES As_2S_3 AND GeS_2

D. Treacy
U.S. Naval Academy
Annapolis, Maryland 21402

and

P. C. Taylor
Naval Research Laboratory
Washington, D.C. 20375

Semiconducting chalcogenide glasses generally exhibit a series of well defined multiphonon absorption peaks in the frequency region between about 400 and 4000 cm^{-1}. Measurements of the frequency and temperature dependences of the one-phonon and multiphonon spectra in crystalline and glassy As_2S_3 and glassy GeS_2 are compared with the predictions of a model calculation in which the anharmonic effects are approximated by a Morse potential. These calculations indicate that anharmonic contributions caused by non-linearities in the dipole moment are significant in some chalcogenide glasses.

INTRODUCTION

The chalcogenide glasses are, in general, highly absorbing throughout most of the infrared and visible spectral regions. In the infrared the absorption is dominated by strong vibrational peaks, and in the visible the absorption is due to interband electronic transitions. The frequency dependence of the room temperature absorption in glassy As_2S_3 is shown in Figure 1. The spectrum of Figure 1, which is typical of that found in most chalcogenide glasses, can be divided into five distinct spectral regions.

The first region of Figure 1 extends from microwave frequencies
(0.1 cm^{-1}) up to about 100 cm^{-1}. In this region the absorption is
temperature independent and varies approximately as the frequency
squared. Details of the absorption in this region are available
elsewhere.[1] In the second region, which extends from about 100 to
500 cm^{-1} in most chalcogenide glasses, the absorption is dominated
by strong one-phonon peaks. These peaks are also temperature
independent and have Gaussion lineshapes.[2] At frequencies between
about 500 and 2000 cm^{-1} a series of multiphonon peaks is generally
observed. This third region is the subject of the present paper.
At still higher frequencies in Figure 1 there is a region of relative transparency (region 4 from about 2000 to 10000 cm^{-1}) followed
by a sharply rising absorption at the electronic band edge (region 5).

A tail on the band edge absorption has been observed in
As$_2$S$_3$.[3] The extrapolation of this tail into the gap is shown by
the curved, dashed line in Figure 1. As this absorption tail is
probably not intrinsic but may in fact be related to residual
impurities in the glass, we shall not consider this contribution
further. It is apparent from Figure 1 that the absorption in
As$_2$S$_3$ at wave numbers near 5000 cm^{-1} will be quite small ($< 10^{-4}$)
provided that there is no tail on the band edge absorption. Thus
the use of semiconducting chalcogenide glasses as highly transmitting
infrared windows may be limited by impurity absorption or scattering
from inhomogeneities as is the case for the alkali halides.

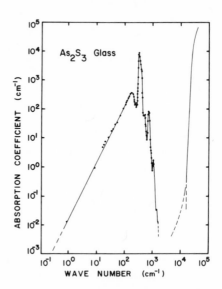

Fig. 1   Room temperature absorption coefficient in glassy As$_2$S$_3$
as a function of frequency. Dashed lines are explained in the text.

EXPERIMENTAL TECHNIQUES

Where the absorption coefficient α in Figure 1 is greater than about 100 cm^{-1}, the data were obtained by a Kramers–Kronig decomposition of the reflectivity spectrum. The data for $\alpha < 100$ cm^{-1} were obtained from transmission measurements performed as a function of sample thickness. All measurements were made at discrete frequencies using a single beam technique. Details of the experimental techniques and data reduction procedures are available elsewhere.[2]

High temperature measurements were performed using a double-beam technique. The samples were inserted in a tubular oven, and a matching oven was inserted in the reference beam to eliminate any extraneous heating effects. Temperatures were monitored with a thermocouple glued to the sample outside of the beam. Double-beam, low temperature measurements were performed by placing the samples on a transmission cold finger in a dewar. KRS-5 windows were employed on the dewar. Low temperature spectra were taken with and without the samples on the cold finger to correct for effects due to the presence of the dewar in the sample beam.

RESULTS AND DISCUSSION

At frequencies greater than the one–phonon frequencies, the optical absorption coefficient for an N phonon summation process α_N is, in general, a complicated sum of virtual excitations of the fundamental mode which decay into N final–state phonons. If the shape of the anharmonic potential is temperature independent and all of the final–state–phonon energies are equal, then the absorption coefficient for an N phonon process at the frequency $\Omega = N\omega$ takes the following simplified form[4]

$$\alpha_N(T,N\omega) = \alpha_N(0,N\omega) \frac{n(\omega)^N}{n(N\omega)} \tag{1}$$

where

$$n(\omega) = [\exp(\hbar\omega/kT) - 1]^{-1}.$$

At low temperatures ($\hbar\omega \gg kT$) this expression approaches the constant $\alpha_N(0,N\omega)$, while at high temperatures the absorption is proportional to T^{N-1} and ω^{1-N}.

Over the range from about 500 to 1000 cm^{-1} in Figure 1, the multiphonon spectrum of glassy As$_2$S$_3$ exhibits a well resolved structure. In this spectral region we emphasize the temperature dependence of α in contrast to the structural detail by plotting the ratio of α at 300K to α at 80K as a function of frequency. These results are shown in Figure 2 where the open circles are points taken on glassy As$_2$S$_3$ and the triangles are points from crystalline As$_2$S$_3$. The dashed line which runs roughly through the data points represents the expected frequency dependence of this ratio obtained from Eq. (1) for a two phonon process (N = 2).

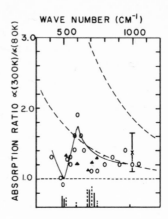

Fig. 2   Ratio of the absorption coefficient at 300K to the
absorption coefficient at 80K as a function of frequency in glassy
and crystalline As_2S_3. Open circles and X represent data for glassy
As_2S_3. Triangles represent data for crystalline As_2S_3. Dashed lines
are theoretical curves for the data as explained in the text. The
vertical solid and dashed markers at the bottom of the figure denote
the positions of peaks in the glass and crystal spectra, respectively,
and the height of the markers their relative intensity.

The upper dashed curve represents the behavior for N = 3. The solid
and dashed lines at the base of the figure indicate the positions
and relative intensities of the major peaks in glassy and crystalline
As_2S_3, respectively. The strong peak at 500 cm^{-1} in glassy As_2S_3
is due to an impurity mode and gives rise to the highly dispersive
character of the $\alpha(300K)/\alpha(80K)$ ratio in this frequency region (as
indicated by the solid line through the data points).[5]

Although there is considerable scatter in the data at the
lower frequencies, Figure 2 clearly indicates that the dominant
contribution to the absorption in glassy As_2S_3 from 500 to 1000 cm^{-1}
is due to two phonon processes. It should be noted that the two
phonon behavior of the absorption near 1000 cm^{-1} is probably not
intrinsic but rather due to an overtone band of the impurity mode
at 500 cm^{-1}. This two-phonon behavior persists between 80K and
300K even in samples of greater purity (the point indicated by
the X in Figure 2).[6] However, at higher temperatures there is
evidence in the purer samples for three-phonon behavior.[7]

At frequencies greater than about 750 cm^{-1} where there is less
scatter in the data, the temperature dependence of the absorption
can be investigated over a wider range. Results are shown in
Figure 3 where the actual absorption coefficient α is plotted as
a function of temperature for several frequencies. The solid lines
are fits of Eq. (1) to the data with N = 2 using $\alpha(300)$ as the

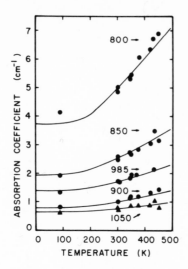

Fig. 3 Temperature dependence of the absorption coefficient in glassy As$_2$S$_3$ at frequencies from 800 to 1050 cm^{-1}. Solid lines are theoretical fits assuming two phonon behavior as described in the text.

fitting point for Eq. (1). Although the frequencies plotted represent peaks and valleys in the multiphonon spectrum (see the solid curve of Figure 4), there is no evidence of any departure from two-phonon behavior in these As$_2$S$_3$ samples over this frequency range.

We now consider the frequency dependence of the multiphonon absorption in glassy As$_2$S$_3$. Figure 4 shows the data of Figure 1 replotted on a semi-log scale (solid line). The dashed line indicates the broadened average absorption coefficient for the layered, crystalline form of As$_2$S$_3$ where only in-plane contributions are averaged.[8] It should be noted that much detailed structure is washed out in the averaging procedure which employs a single Gaussian broadening function[9] with $\sigma = 15$ cm^{-1}. The dotted curve near 400 cm^{-1} indicates the small contribution of the out-of-plane crystalline vibrations to the one-phonon, average spectrum. The out-of-plane contribution to the multiphonon spectrum is not available due to the highly selective cleavage of this layered crystal, but this contribution should be small judging from the one-phonon behavior. At frequencies greater than about 850 cm^{-1} the crystalline data are unreliable because of scattering in the natural crystals employed.

With the exception of the small region dominated by the impurity mode (500 cm^{-1}) in glassy As$_2$S$_3$ and the region dominated by scattering losses in crystalline As$_2$S$_3$, the agreement between

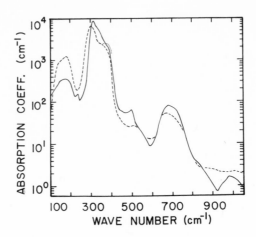

Fig. 4   The solid line is the room temperature absorption
coefficient in glassy As_2S_3 and the dashed line is a broadened
average of crystalline As_2S_3.

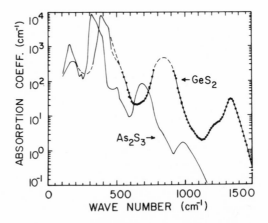

Fig. 5   Room temperature absorption coefficients in glassy As_2S_3
and GeS_2 as functions of frequency.

the glassy and average crystalline spectra is remarkable. This agreement indicates either that the glass retains remnants of the local layer symmetry of the crystal or conversely that the crystal properties are dominated to a surprising degree by the local molecular symmetry.

The multiphonon spectra of the semiconducting chalcogenide glasses are much richer in structural detail than the characteristic exponential fall off observed in alkali halide crystals. These resolved multiphonon peaks enable one to make less ambiguous comparisons with theoretical models than is presently the case for the alkali halides. In the spirit of the models invoked to describe the multiphonon behavior in the alkali halides, we take a molecular approach and assume that the anharmonicity can be described by a Morse potential of the form

$$V(r) = D[1 - e^{-a(r - r_o)}]^2 \qquad (2)$$

where D is the value of the potential at large r (i.e., the dissociation energy). The parameter r_o is the equilibrium separation and the parameter a defines the steepness of the potential for small r.

With this potential, the ratio of the probability for one-phonon transitions to that for two-phonon transitions at low temperature is equal to $8D/\hbar\omega_o$ where $\omega_o = a\frac{2D}{\mu}$ is essentially the one-phonon oscillator frequency.[10] In glassy As₂S₃ where the peak near 700 cm^{-1} can be unambiguously associated with the one-phonon feature centered near 350 cm^{-1}, this ratio is somewhere between 30 and 100 (see Figure 4). In this fashion a value of D of between 0.2 and 0.5 eV is estimated for As-S bonds in As₂S₃. This estimate, although low, does compare favorably with the value of D = 2 eV which is calculated from the electronegativities of As and S.[11] It is reasonable then that the primary contribution to the multiphonon absorption in As₂S₃ comes from anharmonicity in the potential.

The situation in the multiphonon region in glassy GeS₂, which is a three-dimensional, random-network glass similar to SiO₂, is quite different as shown in Figure 5. The oscillator strengths of the strong one-phonon peaks in GeS₂ and As₂S₃ are virtually identical, as one might expect since the ionicity of the Ge-S and As-S bonds are quite similar. However, the multiphonon peaks in GeS₂ are considerably stronger than the corresponding peaks in As₂S₃. In fact, a calculation of D similar to that outlined for As₂S₃ above yields a value of ~0.08 eV from the one-phonon to two-phonon peak ratio, while a value of ~2.4 eV is obtained from ionicity considerations. These two estimates differ by a factor of 30. It is apparent that the Morse potential, with physically reasonable values of the dissociation energy, is incapable of explaining the magnitude of the observed multiphonon absorption in GeS_2. Other forms for the anharmonic potential should yield similar results and will probably be incapable of explaining consistently the results for both GeS_2 and As_2S_3. There are two possible anharmonic contributions to the multiphonon absorption: (1) the anharmonicity of the potential and (2) the anharmonicity of the dipole moment. Since the Morse potential does not adequately explain the multiphonon absorption in GeS_2, the non-linearity of the dipole moment must be considered as a contributing factor.

There is thus strong evidence for the influence of non-linear contributions to the dipole moment in the multiphonon spectra of some semiconducting chalcogenide glasses. Since these glasses not only exhibit well resolved multiphonon spectra but are based on crystals which have molecular character, the molecular approach to anharmonicity, as applied to the alkali halide crystals, may provide more definitive results when applied to the chalcogenide glasses.

REFERENCES

1.   U. Strom, J. R. Hendrickson, R. J. Wagner, and P. C. Taylor, Solid State Commun. 15, 1871 (1974).

2.   P. C. Taylor, S. G. Bishop, and D. L. Mitchell, Solid State Commun. 8, 1783 (1970).

3.   D. L. Wood and J. Tauc, Phys. Rev. B5, 3144 (1972).

4.   J. R. Hardy and B. S. Agrawal, Appl. Phys. Lett. 22, 236 (1973).

5.   The width of the impurity mode near 500 cm^{-1} narrows up at 80K, but the total oscillator strength remains constant. The net effect of superimposing this behavior on a slowly varying two-phonon contribution, as given by the lower dashed curve of Figure 2, is to generate the curve given by the solid line of Figure 2.

6.   This sample was kindly supplied by C. T. Moynihan.

7.   R. E. Howard and C. T. Moynihan, private communication.

8.   P. C. Taylor, S. G. Bishop, D. L. Mitchell, and D. Treacy, Proc. 5th Int. Conf. on Liquid and Amorphous Semicond. (Taylor and Francis, London, 1974), p. 1267.

9.   Crystalline As_2S_3 is a highly anisotropic compound whose local
      structural order consists of AsS_3 pyramidal units linked
      together in a layered configuration. X-ray scattering
      results unambiguously confirm the existence of AsS_3 pyramids
      in the glass but are unable to confirm or deny definitely the
      existence of layer segments in the glass. See the conflicting
      interpretations in: S. Tsuchihashi and Y. Kawamoto, J. Non-
      Cryst. Solids 5, 286 (1971); A. J. Apling, Electronic and
      Structural Properties of Amorphous Semiconductors (P. G.
      LeComber and J. Mort, eds. Academic Press, London, 1973),
      p. 243; A. L. Renninger and B. L. Averbach, Phys. Rev. B4,
      1507 (1973).

10.   H. B. Rosenstock, Phys. Rev. B9, 1963 (1974); L. L. Boyer,
      J. A. Harrington, M. Hass, and H. B. Rosenstock, Phys. Rev. B,
      in press.

11.   See for example L. Pauling, The Nature of the Chemical Bond
      (Cornell Univ. Press, 1960).

MULTIPHONON ABSORPTION IN CHALCOGENIDE GLASSES

R.E. Howard, P.S. Danielson, M.S. Maklad,
R.K. Mohr, P.B. Macedo and C.T. Moynihan

Vitreous State Laboratory, Catholic University
of America, Washington, DC 20064

ABSTRACT

The"molecular model" of Lucovsky and coworkers for vibrational properties of chalcogenide glasses such as As_2S_3, As_2Se_3, GeS_2
and $GeSe_2$ suggests that multiphonon absorption in these materials should be analogous
to overtone and combination vibrational bands
in isolated molecules. A variety of experiments have been carried out whose results are
in reasonable accord with this prediction.
These include Raman spectra of As_2S_3 glass,
measurement·of the frequency dependence of infrared absorption in the multiphonon region
for As_2S_3, As_2Se_3, and mixed As_2S_3-As_2Se_3 and
As_2Se_3-$GeSe_2$ glasses, and measurement of the
temperature dependence of absorption coefficients in the multiphonon region for As_2Se_3
glass.

INTRODUCTION

Chalcogenide glasses such as As_2S_3,As_2Se_3, GeS_2,
and $GeSe_2$ possess open network structures of the types
shown in Fig. 1. For example, As_2Y_3 glass (Y = S or Se)
consists of pyramidal AsY_3 groups bridged by bent As-Y-
As groups, while GeY_2 glass consists of tetrahedral GeY_4
groups bridged by bent Ge-Y-Ge groups. Lucovsky and his
coworkers[1-4] have proposed a "molecular model" for the

vibrational properties of glasses of this sort whose
central feature is the presumption that the high coordination centers, e.g., the AsY_3 pyramidal groups or
the GeY_4 tetrahedra, are coupled vibrationally to one
another only very loosely by the bridging chalcogenide
atoms. As a consequence their infrared and Raman spectra
are to a first approximation expected to correspond to
those of isolated AsY_3 or GeY_4 molecules superimposed
on less intense spectra due to the bridging As-Y-As or
Ge-Y-Ge groups. Raman and infrared spectroscopy studies
of chalcogenide glasses in the fundamental region are
in reasonable accord with the predictions of this model[1-6]
model[1-6].
model

In solids multiphonon absorption takes place when
a high energy photon couples weakly with a transverse
optical mode of the material; this TO mode then decays
into two or more lower energy phonons of frequencies corresponding to fundamental vibrational modes. For a high
energy photon of a given frequency the most probable
multiphonon absorption process involves production of
the minimum number of final state phonons (see papers of
Pohl and coworkers[7,8] and references cited therein). For
crystalline materials such as the alkali halides the
one phonon density of vibrational states is sizeable
over a broad range of frequencies. This may be shown[7-9]
to lead to a predicted absorption spectrum in the multiphonon region which is comparatively featureless at ambient temperature and above, in agreement with experiment.[9,10] For materials such as the chalcogenide glasses, however, the one phonon density of states is predicted by the molecular model to consist of a collection
of discrete vibrational modes broadened only slightly by
the small variations in local structure inherent to the
amorphous state. Multiphonon absorption in these materials should then be restricted to relatively discrete
frequencies $\bar{\nu}$ which satisfy the condition

$$\bar{\nu} = \sum_{i=1}^{n} \bar{\nu}_i \qquad\qquad\qquad (1)$$

where $\bar{\nu}_i$ are frequencies of the n fundamental modes into
which the photon decays[11]. To put it another way, the
molecular model predicts that multiphonon absorption
processes in chalcogenide glasses should be analogous to
combination and overtone bands in isolated molecules.

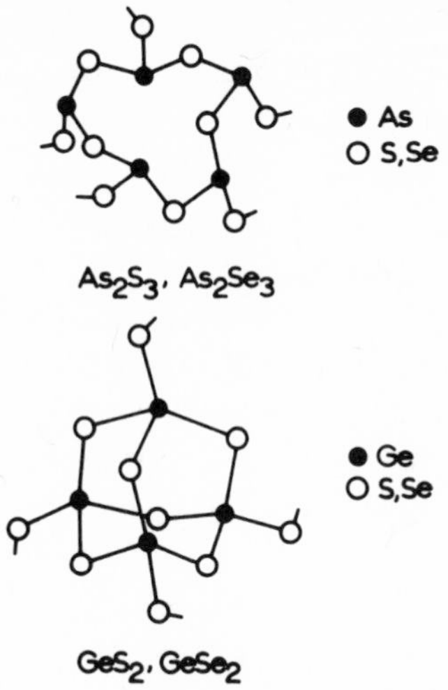

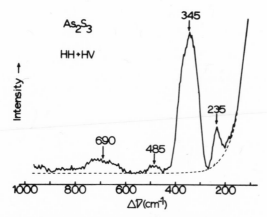

Figure 1. Local structure in chalcogenide glasses.

Figure 2. Raman spectrum of As_2S_3 glass at 63 K.

Further, because the probability of a given multiphonon process at a given frequency falls off rapidly with increasing number of final state phonons, n, multiphonon absorption in chalcogenide glasses should be dominated by combinations and overtones of the highest frequency fundamentals.

In the present paper we report the results of a number of experiments designed to test this hypothesis for multiphonon absorption in chalcogenide glasses. These include Raman spectroscopy of As_2S_3 glass, measurements of the frequency dependence of infrared absorption in the multiphonon region for As_2S_3, As_2Se_3, and mixed As_2S_3 - As_2Se_3 and As_2Se_3 - $GeSe_2$ glasses, and measurement of the temperature dependence of absorption coefficients in the multiphonon region for As_2Se_3 glass.

EXPERIMENTAL SECTION

Chalcogenide glasses were synthesized by reacting the elements (99.9999% purity) in evacuated Vycor tubes. Other preparation details are reported elsewhere[11,12]. Glass densities were determined from the masses and dimensions of the cylindrical specimens used for IR absorption measurements.

At most frequencies infrared absorption coefficients α were obtained from IR spectra of the glasses determined with a Perkin-Elmer Model 467 spectrometer[12]. IR spectra above ambient temperature were measured with the samples thermostatted in a small aluminum heating block; corrections to a spectrum for black-body emission from the hot sample were made using a blank spectrum run at the same temperature with the light beam to the sample blocked.

Absorption coefficient measurements for As_2Se_3 glass in the CO_2 laser wavelength region (920-1090 cm^{-1}) were carried out calorimetrically with a Molectron Corp. Model C250 tunable CO_2 laser and a CRL Model 201 power meter using previously described techniques[13,14]. For measurements above ambient temperature the sample was held in position at the ends of three Teflon-tipped screws and thermostatted in a cylindrical brass oven with baffles to eliminate spurious air currents and scattered radiation.

Raman scattering experiments on As_2S_3 glass were carried out using a Coherent Radiation Model 52 Ar ion laser (wavelength 514.5 nm), a Spex 1401 double monochromator (slit width 20 cm^{-1}), and an ITT FW130 photomultiplier as detector for the photon counting equipment. A 514.5 nm spike filter was placed between the laser beam and the sample to screen out any additional emission lines from the laser. The laser power was about 70 mw focused on a 0.8 mm^2 spot on the sample. To prevent excessive heating by the laser beam the sample was fashioned into a thin plate (0.16 mm thickness) and attached to an aluminum plate on the end of a cold finger with thermal paste and bonding resin. The angle between the incident laser beam and the normal to the sample surface was set equal to Brewster's angle (71.5° for As_2S_3); the scattered laser light was observed in a direction normal to the sample surface.

RAMAN SCATTERING FROM As_2S_3 GLASS

In Fig. 2 is shown the Stokes Raman spectrum of As_2S_3 glass at 63 K. The incident laser light was polarized in the scattering plane; the scattered laser light was unanalyzed. The laser wavelength (514.5nm) lies well inside the electronic absorption region of As_2S_3 glass, so that the intensity of scattered light has been increased considerably by resonance enhancement relative to the scattering intensity expected for incident laser light of wave length well outside the electronic absorption edge[15].

The spectrum of Fig. 2 is similar to those reported previously for As_2S_3 glass below 500 cm^{-1} [5,15,16]. The most pronounced feature is the large peak at 345 cm^{-1}, which has been attributed to stretching vibrations of AsS_3 pyramidal groups [1,2,5]. It presumably contains contributions from both the symmetric (344 cm^{-1}) and antisymmetric (310 cm^{-1}) stretching modes predicted by Lucovsky's molecular model treatment [1,2,5]. The 235 cm^{-1} peak of Fig. 2 is more intense than that in Ward's [16] or Kobliska and Solin's [5,15] spectra; its assignment is uncertain, but it may be due to a stretching vibration of a bent As-S-As group[1,5]. The small peak at 485 cm^{-1} has

also been observed by Kobliska and Solin[5,15] and by
Ward[16]; its assignment is likewise somewhat uncertain,
but it may also be due to a stretch of the bent As-S-As
groups[1,3,5]. A S-S stretching vibration at 475 cm^{-1} is
observed in the Raman spectra of Ward [16] for S-rich As-S
glasses, but this appears to be distinct from the 485cm^{-1}
As$_2$S$_3$ band. A similar weak 485 cm^{-1} band was also observed in GeS$_2$ glasses by Lucovsky et al., who likewise concluded that it was not due to S-S stretching vibrations[3].

The final prominent feature of Fig. 2, the weak band
centered at 690 cm^{-1}, lies at a much higher frequency
than any predicted fundamental modes of As$_2$S$_3$ glass[1,2,5].
Since it occurs at just twice the frequency of the strong
345 cm^{-1} band, it is reasonable to assign it via Eq. (1)
to a 2-phonon scattering process, i.e., to the first harmonic of the AsS$_3$ pyramidal group stretching frequencies.

MULTIPHONON IR ABSORPTION IN As$_2$S$_3$ AND As$_2$Se$_3$ GLASSES

In Fig. 3 is shown a plot of IR absorption coeffi-
cient α versus wavenumber $\bar{\nu}$ for As$_2$S$_3$ and As$_2$Se$_3$ glasses
in the multiphonon region at ambient temperature taken
from our previous paper[11]. The two high frequency bands
in Fig. 2 at 345 and 690 cm^{-1} are also evident in the
multiphonon IR spectrum of As$_2$S$_3$ glass.

We have indicated in Fig. 3 the overtones and combinations of the two highest frequency fundamentals of
As$_2$S$_3$ in Fig. 2, the 345 cm^{-1} band (designated $\bar{\nu}_A$) and
the 485 cm^{-1} band (designated $\bar{\nu}_B$). In line with our molecular model hypothesis for multiphonon absorption in
chalcogenide glasses, these overtones and combinations
correspond well to the prominent features of the As$_2$S$_3$
spectrum (maximum at 690 cm^{-1}, change in slope around
800 cm^{-1}, maximum at 980 cm^{-1}, shoulder at 1050 cm^{-1},etc.)

The similarity of the spectra of As$_2$S$_3$ and As$_2$Se$_3$
glasses in Fig. 3 is a striking demonstration of the iso-
structural character of these two glasses, as suggested
in Fig. 1. The dashed line in Fig. 3 is the spectrum of
As$_2$Se$_3$ scaled to that of As$_2$S$_3$ using a frequency scaling
factor $\bar{\nu}_{As_2Se_3}/\bar{\nu}_{As_2S_3}$ of 0.70 and an amplitude scaling
factor $\alpha_{As_2Se_3}/\alpha_{As_2S_3}$ of 0.63 [11]. The frequency scaling
factor is the same as that found by Zallen et al.[17] to
relate the Raman and IR spectra peak frequencies of both

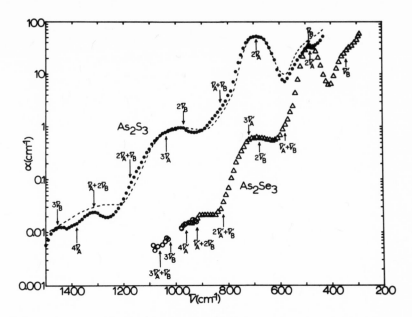

<u>Figure 3.</u>   IR absorption coefficient versus wave-
number for As_2S_3 and As_2Se_3 glasses at
ambient temperature.

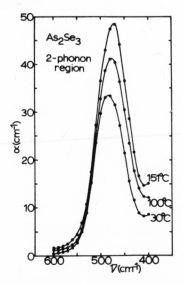

<u>Figure 4.</u> IR absorption coefficient versus wavenum-
ber plots for As_2Se_3 glass as a function
of temperature in the 2-phonon region.

amorphous and crystalline As_2S_3 and As_2Se_3 in the fundamental region.

From the above frequency scaling factor the two highest frequency peaks in the fundamental region for As_2Se_3 glass are predicted to lie at $\bar{\nu}_A{}' = 240$ cm^{-1} (close to the intense Raman peak observed at 227 cm^{-1} [18]) and at $\bar{\nu}_B{}' = 340$ cm^{-1} (observable as a shoulder in Fig. 3). Overtones and combinations of these frequencies are indicated in the As_2Se_3 spectrum of Fig. 3 and correspond well to prominent features in the spectrum.

TEMPERATURE DEPENDENCE OF ABSORPTION COEFFICIENTS

In Fig. 4 are shown plots at different temperatures of α versus $\bar{\nu}$ for As_2Se_3 glass in the vicinity of the 2-phonon absorption peak at 480 cm^{-1}. The absorption coefficient increases in magnitude and the maximum shifts slightly to lower frequencies with increasing temperature. α versus $\bar{\nu}$ plots in the vicinity of the 700 cm^{-1} maximum of Fig. 3 for As_2Se_3 show similar behavior as a function of temperature.

The shift in the absorption peaks to lower frequency with increasing temperature may be understood if we assume that the fundamental vibrational modes (the $AsSe_3$ and As-Se-As group stretching vibrations) are anharmonic (e.g., described by a Morse potential), so that the spacing of the vibrational energy levels narrows with increasing vibrational quantum number. Increasing temperature increases the population of the levels of higher quantum number and hence increases the relative number of mulitphonon absorption processes originating in the higher vibrational states, leading in turn to a decrease in the average frequency for a given multiphonon process. In the temperature range covered in our experiments on As_2Se_3 glass, however, the changes in multiphonon absorption maxima frequency are so small that they may be neglected in our discussion below of the temperature dependence of α.

For a multiphonon absorption process in which a photon of wavenumber $\bar{\nu}$ is absorbed and produces n phonons of the same fundamental wavenumber $\bar{\nu}_f$ the ratio of the absorption coefficient at temperature T to that at 0 K is predicted to be

$$\alpha(T)/\alpha(0) = [1-\exp(-nhc\bar{\nu}_f/kT)]/[1-\exp(-hc\bar{\nu}_f/kT)]^n \qquad (2)$$

where h is Planck's constant, c is the velocity of light, and k is the Boltzmann constant (cf. refs. 7-9,19 and papers cited therein). In Fig. 5 the absorption coefficients of As_2Se_3 glass at 480 cm^{-1} (Fig. 4) are plotted versus temperature. The line for the 480 cm^{-1} data in Fig. 5 is calculated from Eq. (2) using n = 2 and an $\alpha(0)$ value selected to cause the calculated curve to agree with the experimental data at 75° C. The temperature dependence of α at 480 cm^{-1} is in good agreement with that predicted for a 2-phonon absorption process, in accord with our molecular model analysis of the data of Fig. 3. The temperature dependence of α at 700 cm^{-1} shown in Fig. 5 is intermediate between that predicted for n = 2 and that for n = 3, similarly in accord with the molecular model analysis summarized in Fig. 3, although the 3-phonon process appears to dominate absorption at this frequency. In the CO_2 laser region (920-1090 cm^{-1}) the molecular model predicts a variety of 3- and 4-phonon absorption processes for As_2Se_3 as shown in Fig. 3. The temperature dependence of α at 943 and 1026 cm^{-1} shown in Fig. 6 is again in good agreement with this prediction.

MULTIPHONON ABSORPTION IN MIXED CHALCOGENIDE GLASSES

In this section we will discuss multiphonon absorption in the mixed chalcogenide glasses X As_2S_3 - (1-X) As_2Se_3 and X As_2Se_3 - (1-X) $GeSe_2$, where X is the mole fraction of the first component of each pair. Since, as is evident from Fig. 3, a large number of multiphonon processes become possible at high frequencies even for one component glasses, we shall confine our remarks on the mixed glasses for the most part to the low frequency 2-phonon region in which absorption is due to the pyramidal AsY_3 and tetrahedral GeY_4 group stretching modes.

In Fig. 7 are shown α versus $\bar{\nu}$ plots taken from our previous paper[11] for two mixed X As_2S_3 - (1-X)As_2Se_3 glasses. The solid curves are the absorption coefficients predicted on the basis of additivity of the α values of the end member compositions, As_2S_3 (= component 1) and As_2Se_3 (=component 2), at each frequency:

$$\alpha = f\alpha_1 + (1-f)\alpha_2 \qquad (3)$$

f is the volume fraction of component 1:

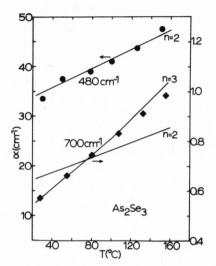

Figure 5. IR absorption coefficient versus temperature at 480 and 700 cm^{-1} for As_2Se_3 glass.
Solid lines calculated from Eq. (2) using
n values shown in the figure and $\alpha(0)$
values selected to give agreement with
the data at 75° C.

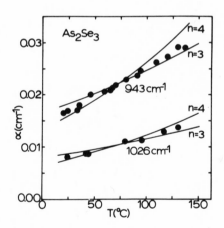

Figure 6. IR absorption coefficient versus temperature at 943 and 1026 cm^{-1} for As_2Se_3 glass.
Solid lines calculated from Eq. (2) using
n values shown in the figure and $\alpha(0)$
values selected to give agreement with the
data at 75° C.

$$f = (XM_1/\rho_1)/[(XM_1/\rho_1) + ((1-X)M_2/\rho_2)]$$

where X, M_1, and ρ_1 are the mole fraction, formula weight, and density of component 1 and $(1-X)$, M_2, and ρ_2 the corresponding quantities for component 2. The glass densities are given in our previous paper[11]. For $X = 0.602$ the experimental and calculated additive α values agree within about 20% except in the region around 585 cm^{-1}. For $X = 0.074$ the agreement is somewhat worse, but still fairly close, except again in the region around 585 cm^{-1}.

Felty and coworkers [20] have observed two reststrahlen bands in reflectivity measurements in the fundamental region of mixed As_2S_3-As_2Se_3 glasses; the two reststrahlen frequencies corresponded closely to those for the pure glasses. Lucovsky[21] suggested that this behavior could be accounted for either by assuming that the mixed glass reflectance spectra were a superposition of the spectra of different AsY_3 polymeric entities in the glass or that the mixed glasses exhibited the two mode reststrahlen behavior found in solid solutions such as CdS_zSe_{1-z} in which the component atoms had large mass differences. In terms of the molecular model for the vibrational characteristics of As_2S_3 and As_2Se_3 the former explanation is to be preferred.

In the mixed As_2S_3 - As_2Se_3 glasses one would expect a structure of the type shown in Fig. 1 with the two types of chalcogen atoms distributed over the As-Y-As bridging groups, so that a substantial portion of the AsY_3 pyramids should consist of mixed AsS_2Se and $AsSSe_2$ groups. The lowered symmetry of these mixed groups would lead to splitting of the degenerate AsY_3 stretching modes. However, because of the large mass difference between S and Se, the stretching modes for the AsS_2Se and $AsSSe_2$ should be close in frequency to those for AsS_3 and $AsSe_3$, i.e., should be approximately equal to $\bar{\nu}_A$ and $\bar{\nu}_A'$ of Fig. 3, in agreement with the observations of Felty and coworkers[20] in the fundamental region of the mixed glasses. In the 2-phonon absorption region ascribable to the AsY_3 stretching modes one would expect to observe combination and overtone bands at $2\bar{\nu}_A = 690$ cm^{-1}, at $2\bar{\nu}_A' = 480$ cm^{-1}, and at $\bar{\nu}_A + \bar{\nu}_A' = 585$ cm^{-1}. The first two of these predicted 2-phonon bands (690 and 480 cm^{-1}) are quite apparent in Fig. 7, but the predicted 585 cm^{-1} band is

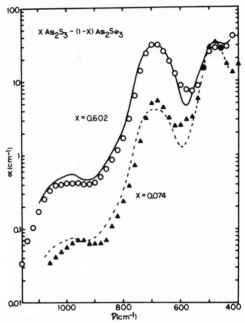

Figure 7. IR absorption coefficient versus wavenumber for X As$_2$S$_3$-(1-X)As$_2$Se$_3$ glasses.
Solid and dashed curves are calculated
from Eq. (3).

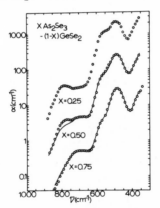

Figure 8. IR absorption coefficient versus wavenumber for X As$_2$Se$_3$-(1-X)GeSe$_2$ glasses.
The α scale is correct for the X = 0.75
glass; for clarity the spectra have been
displaced upward by respective factors of
10 and 100 in α for the X = 0.50 and X =
0.25 glasses. Solid curves are calculated
from Eq. (3).

discernible only in terms of the large deviations from
additivity at this frequency and as a weak shoulder at
about 580 cm^{-1} in the X = 0.074 glass. The relative
weakness of predicted 585 cm^{-1} bands which should occur
only in the mixed As_2S_3 - As_2Se_3 glasses may mean that
the distribution of the S and Se atoms among the bridging groups may be highly non-random[11].

In Fig. 8 are shown α versus $\bar{\nu}$ plots in the multiphonon region for mixed X As_2Se_3 - (1-X) $GeSe_2$ glasses.
We were unable to obtain the end member composition
$GeSe_2$ in the glassy state by rapid quenching of bulk
samples, contrary to the report of Tronc and coworkers[22]
but in agreement with earlier reports of the glass-form-
ing regions in the Ge-As-Se system[23], so that our studies
cover only the composition range X = 0.25 to 1.00. The
peaks at 780 cm^{-1} in Fig. 8 for the compositions X = 0.25
and 0.50 are due to oxide impurity[24], but below 700 cm^{-1}
the spectra of all the glasses are due to intrinsic absorption processes.

The ambient temperature densities of the X As_2Se_3-
(1-X) $GeSe_2$ glasses were [X, $\rho(g/cm^3)$]: 0.25, 4.34; 0.50,
4.40; 0.75, 4.53; 1.00, 4.61. Within experimental error
the molar volumes were additive.

The solid curves in Fig. 8 are the absorption coefficients calculated from Eq. (3) on the assumption of
additivity of the α values of the end member composition
As_2Se_3 (= component 1) and $GeSe_2$ (= component 2). Since
the pure $GeSe_2$ glass could not be prepared, the α_2 values
were calculated from Eq. (3) using the experimental α values for the X = 0.25 compositions. These calculated α_2
values were then used to calculate the solid curves of
Fig. 8 for the X = 0.50 and 0.75 composition.   The a-
greement between the experimental and calculated additive α versus $\bar{\nu}$ curves in the intrinsic region below
700 cm^{-1} is within experimental error.

From Fig. 1 the structure of the mixed As_2Se_3 -
$GeSe_2$ glasses is expected to consist of $AsSe_3$ pyramids
and $GeSe_4$ tetrahedra linked by Se atom bridges. The
molecular model predicts that the vibrations of neighboring $AsSe_3$ and $GeSe_4$ groups should be only very
loosely coupled, so that in the multiphonon absorption
region one expects in turn to see no combination bands of

$AsSe_3$ and $GeSe_4$ fundamental frequencies. The 400 to 600 cm^{-1} region of Fig. 8 is the frequency range in which 2-phonon processes involving the $AsSe_3$ and $GeSe_4$ stretching modes are predicted to occur, and hence the agreement of the spectra of the mixed glasses in this region with those predicted from Eq. (3) on the assumption of additivity of the end member spectra are in complete agreement with this hypothesis.

In summary, then, for mixed chalcogenide glasses the molecular model predicts additivity of absorption coefficients in the 2-phonon region when the high coordination center atoms are mixed, as in the As_2Se_3-$GeSe_2$ glasses, but deviations from additivity when the bridging chalcogen atoms are mixed, as in the As_2S_3-As_2Se_3 glasses. These predictions are in accord with the experimental results reported here.

ACKNOWLEDGMENTS

This research was supported by contracts from the Advanced Research Projects Agency and from the Office of Naval Research. The authors wish to thank Mr. P. Pureza and Mr. J. Lowans for assistance in sample preparation and characterization.

REFERENCES

1. G. Lucovsky and R.M. Martin, J. Non-Cryst. Solids, 8-10, 185, (1972).
2. G. Lucovsky, Phys. Rev. B, 6, 1480 (1972).
3. G. Lucovsky, J.P. deNeufville, and F.L. Galeener, Phys. Rev. B, 9, 1591 (1974).
4. G. Lucovsky, F.L. Galeener, R.C. Keezer, R.H. Geils, and H.A. Six, Phys. Rev. B, 10, 5134 (1974).
5. R.J. Kobliska and S.A. Solin, Phys. Rev. B, 8, 756 (1973).
6. G.N. Papatheodorou and S.A. Solin, Solid State Commun., 16, 5 (1975).
7. D.W. Pohl and P.F. Meier, Phys. Rev. Lett., 32, 58 (1974).
8. T.F. McNeely and D.W. Pohl, Phys. Rev. Lett., 32, 1305 (1974).
9. L.L. Boyer, J.A. Harrington, M. Hass, and H.B. Rosenstock, Phys. Rev. B, in press.

10. T.F. Deutsch, J. Phys. Chem. Solids, 34, 2091 (1973).
11. M.S. Maklad, R.K. Mohr, R.E. Howard, P.B. Macedo, and C.T. Moynihan, Solid State Commun., 15, 855 (1974).
12. C.T. Moynihan, P.B. Macedo, M.S. Maklad, R.K. Mohr, and R.E. Howard, J. Non-Cryst. Solids, in press.
13. D.A. Pinnow and T.C. Rich, Appl. Opt., 12, 984 (1973).
14. F.A. Horrigan and T.G. Deutsch, "Research in Optical Materials and Structures for High Power Lasers," Quarterly Technical Report No. 2, Raytheon Research Div., Waltham, MA , Contract DAAH01-72-C-0194, 1972.
15. R.J. Kobliska and S.A. Solin, Solid State Commun., 10, 231 (1972).
16. A.T. Ward, J. Phys. Chem., 72 4133 (1968).
17. R. Zallen, M.L. Slade, and A.T. Ward, Phys. Rev. B, 3, 4257 (1971).
18. J. Schottmiller, M. Tabak, G. Lucovsky, and A. Ward, J. Non-Cryst. Solids, 4,80 (1970).
19. H. B. Rosenstock, Phys. Rev. B, 9, 1963 (1974).
20. E.J. Felty, G. Lucovsky, and M.B. Myers, Solid State Commun., 5, 555 (1967).
21. G. Lucovsky, Mat. Res. Bull., 4, 505 (1969).
22. P. Tronc, M. Bensoussan, A. Brenac, and C. Sebenne, Phys. Rev. B, 8, 5947 (1973).
23. R.W. Haisty and H. Krebs, J. Non-Cryst. Solids, 1 427 (1969).
24. J. A. Savage and S. Nielsen, Infrared Physics, 5, 195 (1965); Phys. Chem. Glasses, 6, 90 (1965).

THEORY OF MULTIPHONON ABSORPTION IN THE TRANSPARENT

REGIME OF AMORPHOUS SOLIDS

Yet-Ful Tsay* and Bernard Bendow

Solid State Sciences Laboratory, Air Force Cambridge
Research Laboratories (AFSC), Hanscom AFB, MA 01731

Stanford P. Yukon+

Parke Mathematical Labs., Carlisle, MA 01741 and
Dept. of Physics, Brandeis University, Waltham, MA 02154

We predict the multiphonon absorption α in amorphous solids, utilizing the statistical theory of Mitra et al, which relates optical properties of amorphous solids to variations in their local density. We find that if the ratio η of the mean density of an amorphous solid to that of its crystalline counterpart is greater than unity, then α decreases more slowly in the amorphous solid; if $\eta<1$, then either an increase or decrease is possible, depending on the width of the density distribution function. As an application of the method, we perform calculations for III-V semiconductors, utilizing an exactly soluble single-particle model which accounts for both anharmonicity and nonlinear moments. Analysis of their fundamental lattice resonance demonstrates that $\eta<1$ for these solids. Broadening of α and suppression of the temperature dependence is predicted to occur to varying extents for the III-V's investigated. This suggests that various amorphous solids may not be as attractive as their crystalline counterparts for infrared applications requiring high transparency.

*NRC Resident Research Associate.
+Supported by Air Force Cambridge Research Laboratories (AFSC)
under Contract F19628-71-C-0142.

INTRODUCTION

Multiphonon absorption is the principal loss mechanism limiting the transparency of infrared optical materials.[1] Recently, a wide variety of theoretical and experimental studies of multiphonon absorption have been carried out.[2] For certain applications, especially those requiring coatings or a thin-film geometry, the use of amorphous materials has been contemplated.[3] The purpose of the present paper is to investigate the properties of the multiphonon absorption coefficient $\alpha(\omega)$ in amorphous solids, and to contrast them with predictions for the crystalline case.

We here employ the statistical theory of Mitra et al,[4] where the influence of amorphousness is accounted for through fluctuations in local density in the solid. Such an approach proves most useful for "nearly crystalline" solids, which we define below. For this case it is possible to deduce the properties of the amorphous solid from a knowledge of those of the crystalline counterpart under hydrostatic pressure. Our calculations will be applied to an important class of potential infrared materials, namely, the III-V amorphous semiconductors.

To calculate α, we will utilize the single-particle (sp) model[5] which, although rather artificial, allows an exact treatment accounting for both anharmonicity and nonlinear moments. Although the absolute values predicted for α may not be reliable, trends in the frequency and temperature dependence, and trends in the ratio of the amorphous to the crystalline value for α, are expected to be at least qualitatively correct.

LATTICE RESPONSE IN AMORPHOUS SOLIDS

We here outline the principal features of a statistical approach to optical response in amorphous solids, which is described in more detail in Ref. 4. Essentially, the solid is parametrized in terms of two types of atomic configurations: those preserving both short- and long-range order (SL), and those preserving just short-range order (S). Physically, the SL-portion accounts for k-conserving processes, while the S-portion accounts for k-conservation breakdown. Denoting the configurations by n and their probabilities by p_n, then a property θ is given for an amorphous solid as

$$\langle\theta\rangle = \langle\theta\rangle^{SL} + \langle\theta\rangle^{S}$$

$$= \rho^{SL} \sum_n p_n^{SL}\theta^{SL}(n) + \rho^S \sum_{n'} p_{n'}^S \theta^S(n') \tag{1}$$

$$\sum_n p_n^{SL} = \sum_{n'} p_{n'}^S = 1, \quad \rho^{SL} + \rho^S = 1$$

where a convenient normalization has been chosen for the p's. The quantity ρ^S is a measure of the amorphousness of the solid or, equivalently, of the breakdown of $\tilde{k}$-conservation. We will class a solid as "nearly crystalline" if $\rho^S \gtrsim 0.2$; highly disordered solids are characterized by $\rho^S \sim 1$.

The above approach has been applied to the case of tetrahedrally bonded amorphous semiconductors (TBAS). The SL configurations are associated with variations in local density stemming from variations in atomic spacings and bond angles. To calculate the lattice response one requires the dependence of the phonon spectrum on local density; this is approximated by the dependence on (uniform) density in the crystalline case. Calculations were performed assuming a Lorentzian absorption in the crystalline case and a Gaussian distribution for the local density. Comparison of the results with experiment suggest that $\rho^S \sim 0.1$ for typical III-V TBAS; i.e., they are "nearly crystalline". For amorphous Si and Ge, on the other hand, $\rho^S \sim 0.3$.

Our calculations of α for the III-V's will be restricted to the SL portion alone. This is motivated by two considerations: (a) the S terms are small overall in these solids, and (b) the greatest modifications in the density of states appear at the lower frequencies which have a relatively minor effect on multiphonon processes in the transparent regime.[2]

CALCULATION OF $\alpha(\omega)$

The absorption in the sp model is given by[5]

$$\alpha = \frac{4\pi\omega}{\mu c} \, \text{Im}\chi$$

$$\text{Im}\chi = \sum_n |<o|m(r)|n>|^2 \delta(\omega - \omega_{no}) \tag{2}$$

where μ is the refractive index; $|n>$'s are eigenstates and ω_{no}'s eigenergies measured from the ground state of the Hamiltonian

$$H = \frac{p^2}{2M} + \nu(r) \tag{3}$$

where ν is the sp (anharmonic) potential; m is the (nonlinear) moment. Various schemes[6] have been introduced to extract a continuous spectrum mimicking a crystal from the line spectrum of Eq. 2. For the present qualitative purposes it will be sufficient to merely interpolate between peaks to provide a continuous α. The

justification for such procedures stems from more detailed calculations, which indicate substantial smoothing of the multiphonon absorption spectrum at higher frequencies,[2] and from the considerations
described at the end of this section. The Morse potential,

$$\nu = \nu_o (e^{-2\xi r} -2e^{-\xi r})$$ (4)

has been utilized previously in various sp calculations, and will
be employed here as well; the moment will be chosen as[7]

$$m = m_o e^{-2\xi_1 r} .$$ (5)

If $\omega_o(\rho)$ is the harmonic frequency associated with a given
value of density ρ, then α for the amorphous solid is given by

$$\alpha = \int d\rho f(\rho)\alpha(\omega_o(\rho))$$ (6)

where we employ the distribution

$$f(\rho) \propto e^{-A(\rho-\rho_a)^2}$$ (7)

where ρ_a is the mean density in the amorphous solid; A, which
measures the spread in f, is determined by fitting experimental
data on the main-peak absorption.[4] We assume that the range parameters ξ and ξ_1 are the same in the amorphous solid as in the
crystal, and account for changes in ω_o vs ρ through variations in
ν_0. Since we do not account for changes in m_0 (i.e., effective
charge), the absolute magnitude of α will be unreliable, although
the frequency and temperature dependence of α will not be substantially affected.

Before proceeding to the computed results for III-V's, it is
instructive to indicate the general trends anticipated for $\alpha(\omega,T)$.
For purposes of illustration it suffices to utilize an approximate
result for α, valid for a crystal with a linear moment,[1]

$$\alpha(\omega) = \alpha_1 \exp[(\ln(n_o+1)-B) \ \omega/\omega_o -\ln(n(\omega)+1)]$$ (8)

where α_1 and B depend relatively weakly on ω_0, and n_0 is the Bose-
Einstein function for $\omega=\omega_0$; typically, B~4 for zincblende semiconducting crystals.[7] In the vicinity of the crystalline density
ρ_c,

$$\omega_0(\rho) \approx \omega_o(\rho_c)(\rho/\rho_c)^{\gamma_0}$$ (9)

where α is the mode Gruneisen parameter. Thus if $\rho_a/\rho_c \equiv \eta < 1$, ω_0
decreases and $\alpha(\omega)$ falls more rapidly, while the reverse is true
if $\rho_a/\rho_c>1$. The averaging procedure, on the other hand, always

tends to broaden $\alpha(\omega)$. While the two effects act in the same direction for $\eta>1$, they oppose each other for $\eta<1$, which is the case for III-V TBAS. The determining factor is the extent of the spread in local density, as measured by the parameter A in $f(\rho)$. A similar situation prevails regarding temperature dependence. For $\eta<1$, for example, the absorption at fixed ω involves more phonons and thus would be more highly temperature dependent; however, this tendency is counteracted by the averaging procedure which tends to reduce the overall dependence.

The parameters used in our computations for III-V's are listed in Table 1 (values of m_0, ξ and ξ_1 are taken from Ref. 7; A and ρ_a are determined from data in papers referenced in Ref. 4). Calculated results for $\alpha(\omega)$ are illustrated in Figs. 1 and 2. In every instance the amorphous spectra display a lower rate of decrease as a function of ω than do the corresponding crystals. Moreover, departures from exponential behavior are more evident than in the crystalline case. While the changes are relatively small in the case of GaAs, they are very substantial for GaSb and InAs.

The effects of amorphousness on temperature dependence are indicated in Fig. 3 and Table 2. As expected, the temperature dependence at fixed ω is suppressed substantially in the highly amorphous limit, as is the enhancement of the T-dependence with increasing ω which is characteristic of the crystal. Note that while amorphousness has a significant effect at quantum temperatures $(T/\omega_c \lesssim 1)$, the effect on α at high T $(T/\omega_c \gg 1)$ becomes relatively minor.

TABLE I. Parameters Employed in Calculating α.

Solid	$\xi(\text{Å}^{-1})$ [a]	$\xi_1(\text{Å}^{-1})$ [a]	$\nu_0(10^{-12}\text{erg})$ [a]	m_0 [a]	$\dfrac{\rho_a}{\rho_c}$	$\dfrac{A}{\rho_a^2}$
GaP	1.1	0.64	6.3	-1.5	0.89	30
GaSb	1.1	0.52	4.2	-2.1	0.87	21
GaAs	1.3	0.54	4.1	-2.4	0.96	86
InAs	[b] 1.2	[b] 0.53	[b] 4.2	[b] -2.3	0.91	51

[a] Crystalline values.   [b] Estimated values.

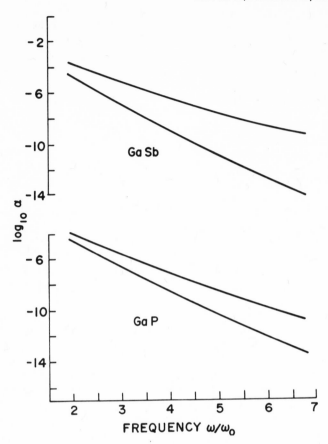

FIGURE 1. Log_{10} absorption coefficient vs. frequency ω/ω_0, where ω_0 is the crystalline harmonic frequency, for GaSb and GaP. The upper curve in each pair is the amorphous result, while the lower is the crystalline.

    A final point worthy of attention is the relation of the in-terpolated α's to frequency-distributed results. Comparison with calculations utilizing a Debye frequency distribution is indicated in Fig. 4. We note that the slopes of each pair of curves, crys-talline and amorphous, are indeed very similar, providing justifi-cation for the use of the interpolated results.

SUMMARY AND CONCLUSIONS

    We have applied the density-fluctuation theory of amorphous solids to predict their multiphonon absorption in the transparent

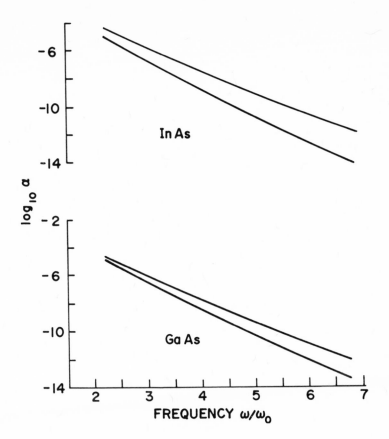

FIGURE 2.   Log$_{10}$ absorption coefficient vs. frequency ω/ω_0, where ω_0 is the crystalline harmonic frequency, for InAs and GaAs.   The upper curve in each pair is the amorphous result, while the lower is the crystalline.

regime. We find that, in general, it is possible for the decrease in α vs ω to be greater or less in the amorphous solid than in the crystalline counterpart, depending on the mean density and the spread in the density distribution. For the amorphous III-V's investigated, $\alpha(\omega)$ was always broadened relative to the crystal, and displayed greater departures from exponential behavior. The temperature dependence of α was found to be suppressed in the amorphous solid, especially in the quantum regime.

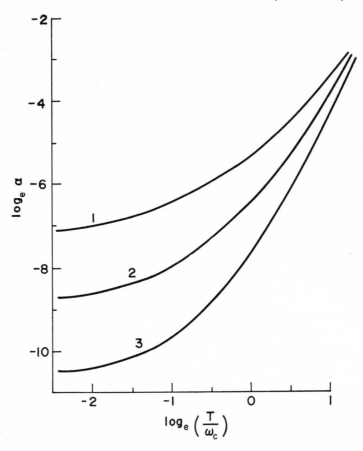

FIGURE 3. $\log_e$ absorption coefficient vs $\log_e(T/\omega_c)$, where ω_c is the crystalline harmonic frequency, at $\omega/\omega_c = 3$. Parameters characteristic of GaSb are employed, except that A is varied. Curve 1: $A=10\rho_a^2$; Curve 2: $A=21\rho_a^2$ (amorphous GaSb); and Curve 3: $A\approx\infty$ (crystalline value).

    The III-V's investigated display a wide range of behavior. The effects of amorphousness are predicted to be relatively minor for GaAs, for example, but substantial for GaSb. Since ρ^s is larger for amorphous Si and Ge than for III-V's, it is reasonable to surmise that α will be even more enhanced and broadened in these solids. Thus various potential infrared materials may, in their amorphous forms, incur inacceptable losses in applications requiring high transparency. Others, such as GaAs, appear to retain adequate transparency in the amorphous state. Some qualification of the

TABLE 2: Temperature Dependence of α(GaSb Parameters)

$\dfrac{A}{\rho a^2}$	$\alpha(\frac{T}{\omega_c}=1.0)/\alpha(\frac{T}{\omega_c}=0.6)$		
	$\frac{\omega}{\omega_c} = 3$	$\frac{\omega}{\omega_c} = 5$	$\frac{\omega}{\omega_c} = 7$
a ∞	3.3	8.0	19.1
b 21	2.0	3.6	4.3
10	1.9	2.1	2.2

aCrystal.   bAmorphous GaSb.

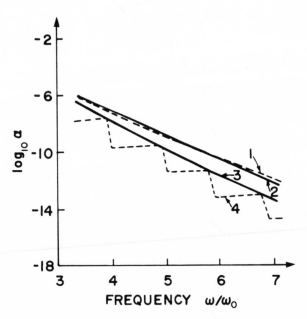

FIGURE 4. Log_{10} absorption coefficient vs frequency ω/ω_0 for GaAs parameters. Curve 1 is the density averaged α, obtained from curve 4, which is the frequency-distributed crystalline α. Curve 3 is the crystalline α, obtained by connecting the δ-function peaks in the sp expression for α, and curve 2 is the density-averaged result obtained from curve 3.

present results is in order: The calculations rely on rather
limited input data, taken on samples whose characterization may
be in doubt (see Refs. 4 and 8). The present techniques for
predicting transparency will be most useful when semiconducting
amorphous films become better characterized, and reliably reproducible.

REFERENCES

1. B. Bendow, J. Elec. Mats. $\underline{3}$, 101 (1974).
2. Some theory papers: B. Bendow, S. C. Ying and S. P. Yukon,
   Phys. Rev. B$\underline{8}$, 1679 (1973); D. L. Mills and A. A. Maradudin,
   ibid. 1617 ($\overline{1}$973); M. Sparks and L. J. Sham, ibid. 3037 (1973);
   K. V. Namjoshi and S. S. Mitra, Phys. Rev. B$\underline{9}$, 815 (1974);
   H. B. Rosenstock, ibid. 1973 (1974); T. C. McGill, R. W.
   Hellwarth, M. Mangir and H. V. Winston, J. Phys. Chem. Sol. $\underline{34}$,
   2105 (1973). Some experimental papers: T. F. Deutsch, J. Phys.
   Chem. Sol. $\underline{34}$, 2091 (1973); J. A. Harrington and M. Hass, Phys.
   Rev. Lett. $\overline{31}$, 710 (1973); L. H. Skolnik, H. G. Lipson and B.
   Bendow, Appl. Phys. Lett. $\underline{25}$, 442 (1974); K. V. Namjoshi, S. S.
   Mitra, B. Bendow, J. A. Harrington and D. L. Stierwalt, ibid.
   $\underline{26}$, 41 (1975).
3. "Proceedings 3rd Conf. on IR Laser Windows", C. A. Pitha et al,
   eds. (AFCRL, Bedford, MA 1974).
4. S. S. Mitra, D. K. Paul, Y. F. Tsay and B. Bendow, in "Tetrahedrally Bonded Amorphous Semiconductors", M. H. Brodsky et al,
   eds. (APS, NY, 1974), and to be published.
5. See, for example, D. L. Mills and A. A. Maradudin, and T. C.
   McGill et al, op cit, Ref. 2; also, B. Bendow and S. P. Yukon,
   this volume.
6. See, for example, H. B. Rosenstock, op cit, Ref. 2, and L. L.
   Boyer et al, this volume.
7. B. Bendow, S. P. Yukon and S. C. Ying, Phys. Rev. B$\underline{10}$, 2286
   (1974).
8. "Tetrahedrally Bonded Amorphous Semiconductors", op cit, Ref. 4.

LIGHT SCATTERING FROM COMPOSITION FLUCTUATIONS IN THE SUPER-SPINODAL REGION OF A PHASE-SEPARATING OXIDE GLASS

R.K. Mohr and P.B. Macedo

Vitreous State Laboratory

The Catholic University of America, Washington, D.C.

Light scattering was used to observe composition fluctuations at temperatures in the single phase region of a phase-separating borosilicate glass. We report the first study of equilibrium fluctuations and the kinetics of their growth by means of light scattering from quenched samples. From the results the spinodal temperature for the glass is obtained. At each temperature observed the relaxation time for the fluctuations was determined. The results are discussed in terms of the classical and mode-mode coupling theories of critical fluctuations.

Light scattering has been used extensively in the last decade to study the critical regions of binary mixtures and simple fluids.[1] The divergent behavior of the appropriate susceptibility, ($\frac{\partial \mu}{\partial x}$) P,T or χ_T (where μ is the chemical potential, x the composition, P the pressure, T the temperature, and κ_T the isothermal compressibility) and the correlation length $\xi(T)$ may be obtained from the intensity and/or linewidth measurements of the Rayleigh scattered light from the material near the critical point. Scaling laws suggest that the dependence of ξ on $\varepsilon \equiv (\frac{T-T_c}{T_c})$ is the source for the divergent behavior of other parameters at the critical point. For densities [2] or compositions[3] different from the critical it has been suggested that the spinodal temperature, T_s, plays the role of the critical temperature.

For simple fluids and mixtures with low viscosities ranging from a fraction of a poise to a few poise typical relaxation times of fluctuations near the critical point range from a fraction of a microsecond to a few seconds. With a digital photon-correlator and good temperature control one can measure

the Rayleigh intensity and the relaxation time characterizing
its linewidth for a liquid in thermodynamic equilibrium in
almost routine fashion. Binary oxide glasses, however, typically have a viscosity 10^7 to 10^{12} times greater than that of
simple liquids near their critical temperatures. Typical relaxation times, τ, of composition fluctuations in glassy systems may range from a few seconds to several years. Clearly
to study fluctuations with large τ, it is impractical to use
a photon correlator technique because of the time required for
measurement ($\gtrsim 100\ \tau$ for large τ). The study of slowly decaying fluctuations, however, is important since it permits a
comparison between approach to equilibrium of a non-equilibrium
system and the decay of thermally driven equilibrium fluctuations.

One of the authors (Macedo et al.)[4] has recently been able
to observe superspinodal T $>$ T_s) composition fluctuations of a
glass in an electron microscope. In this technique composition
fluctuations characteristic of a given temperature are frozen
in by rapidly quenching the sample to room temperature where
standard electron microscopy techniques are used. The long relaxation times of glasses allow the structure to be frozen in.
We report the results of a complementary experiment using light
scattering in which we have observed equilibrium composition
fluctuations and the kinetics of their approach to equilibrium.
Light and x-ray scattering have been used previously[5] to study
the two phase region below the coexistence curve but this, we
believe, is the first study employing light scattering from quenched samples, of the kinetics of composition fluctuations in the
single phase region.

The glass chosen for this study was a commercially available low expansion borosilicate glass. The choice was made
because the existence of a liquid-liquid immiscibility had been
established[6] in a similar glass by other means. The viscosity,
$\approx 10^{10}$ poise,[7] near T_s suggested that a satisfactory quenching
could be acheived. Finally, the low expansion coefficient
makes it possible to rapidly quench a sample large enough for
light scattering experiments without creating excessive thermal
stresses in the sample.

The samples were cut from a single piece of glass to a
size 1/2 cm x 1/2 cm x 1 cm, optically polished, and annealed
in a furnace preheated to the heat treatment temperature and
controlled to $\pm$ 0.5 K. The furnace temperature was recorded
by a Pt, PtRh thermocouple in close proximity to the sample which
reached the furnace temperature within 15 minutes. The samples
were aged for appropriate times, and then quenched to room temperature by dropping in sand. The initial cooling rate is estimated to be approximately 10 K per second. The samples were

repolished if necessary and the Rayleigh scattering was measured. Some of the samples were then returned to the furnace for further heat treatment. This process was repeated until the Rayleigh intensity showed no change with further heat treatment.

In a Rayleigh scattering experiment [8,9] one analyzes the spectrum of the light scattered in a given direction (scattering angle Θ) which corresponds to scattering from fluctuations the magnitudes of whose wavevectors are given by

$$K = 2nk_o \sin (\Theta/2) \qquad (1)$$

(where n is the average refractive index of the sample, $k_o = \frac{2\pi}{\lambda_o}$, λ_o being the incident light wavelength in vacuum). For example, in our experiment light scattered at $90°$ ($K = 2\times10^5$ cm^{-1}) was frequency analyzed by a flat plate Fabry-Perot interferometer. One observes three lines, namely, the Rayleigh and Brillouin components. The Rayleigh component is due primarily to composition fluctuations frozen in at the heat treatment temperature and contains a small contribution from frozen in density fluctuations[9]. The Brillouin components are due to the room temperature phonons and are thus not affected by the heat treatment process. The intensity of the Brillouin lines, 2 I_B, can be used to scale the Rayleigh intensity, I_R, which is then given in terms of the dimensionless Landau-Placzek ratio, R_{Lp}:

$$R_{Lp} \equiv \frac{I_R}{2I_B} \propto I_R \qquad (2)$$

The expression for the intensity of light scattered by equilibrium fluctuations in the limit of small K according to the classical theory[10] of critical fluctuations, is given by

$$\lim_{\substack{k\to 0 \\ T\to T_s}} I_\infty \propto \frac{T}{(T-T_s)^\gamma} \quad ; \quad \gamma = 1 \qquad (3)$$

where we explicitly state the assumption that T_c may be replaced by T_s for non-critical compositions. Experiments[11] indicate that γ is closer to $5/4$ than to 1 for many critical systems. Our present data is insufficient for an accurate determination of γ so for simplicity we plot in Fig. 1 $T/(I_\infty)$ versus temperature for four temperature above T_s. Here I_∞ is the observed Rayleigh intensity after subtraction of a small contribution, R_ρ, from the density fluctuations. R_ρ can be estimated using the results of measurements on a glass similar in composition to ours. R_ρ in our relative intensity units is given by [9]

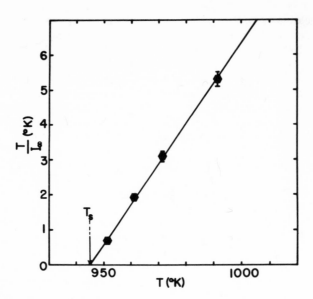

Figure 1 - T/I_∞ versus T for a sodium borosilicate glass.
The temperature intercept of the line drawn through the data
indicates T_S assuming γ in eqn. 3 = 1.

$$R_\rho = \frac{T_f}{T} \left[\left(\frac{V_B(T)}{V_o(T_f)} \right)^2 - 1 \right] \qquad (4)$$

where $V_B(T)$ is the longitudinal sound velocity determined by
light scattering at temperature T; $V_o(T_f)$ is the zero frequency
sound velocity at T_f determined ultrasonically;[12] and T_f is
the fictive temperature for density fluctuations, i.e., the
temperature at which the density fluctuations are frozen in
when a glass is cooled from a higher temperature $> T_f$ to room
temperature. Usually T_f is taken to be the temperature at which
the glass viscosity is 10^{13} poise.[13] The estimated value of R_ρ
is 18.5 which is less than 10% of the observed equilibrium
Rayleigh intensity and thus its accuracy should have little
effect on our conclusions.

The straight line drawn through the data in Fig. 1 inter-
sects the temperature axis at 945 K and is identified as T_S.
The identification follows directly from Eq. (3). If a value
of 5/4 for γ is assumed, the value of T_S obtained is lowered
to approximately 941 K. To determine more accurately the values
of T_S and the exponent γ will require additional and more precise

data.

For each of four temperatures we have made a semilog plot of I_∞ - I_R as a function of heat treatment time as shown in Figure 2. The slopes of the lines drawn through the data give the relaxation times τ if an exponential approach to equilibrium is assumed. The observed relaxation times range from 33 hours at 953 K to .9 hours at 993 K.

We note that the observed relaxation times are orders of magnitude greater than one would expect from a simple relaxation in the shear viscosity in the absence of critical effects. Simply stated, the average shear relaxation time[14] is given by

$$\tau_s = \frac{\eta_s}{G_\infty} \tag{5}$$

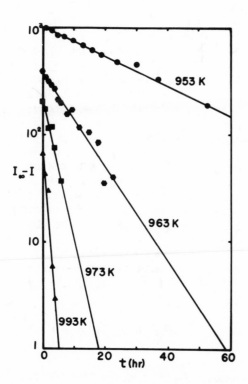

Figure 2 - Log $(I_\infty - I)$ versus t as a function of heat treatment temperature. The relaxation times obtained from the slope of the lines drawn through the data are: τ = 33 hr. @ 953 K; 8.5 hr. @ 963K; 3.5 hr. @ 973K; .9 hr. @ 993 K.

where η_s is the shear viscosity which for our glass is
$\approx 3.2 \times 10^9$ poise at 973 K[7] and G_∞ is the infinite frequency
shear modulus, which has been measured [14] for a similar glass and
is $\approx 2 \times 10^{11}$ dyne/cm^2. This gives $\tau_s \approx 1.6 \times 10^{-2}$ sec. which
is much less than the observed value of 3.5 hr. @ 973 K. The re-
sults are similar for the other temperatures observed.

As a first assumption we assume that the non equilibrium
relaxation we observe is the same as that of an equilibrium
fluctuation (the observed relaxation time is 1/2 that obtained
from equilibrium Rayleigh linewidth measurements). In that
case both the classical theory and the more recent mode-mode
coupling theories[15] of critical fluctuations predict an exponential approach to equilibrium and agree with our data.

Kawasaki[15] has shown using mode-mode coupling that the
relaxation time for composition fluctuations in a super-critical
binary mixture is given by

$$\frac{1}{\tau} = \frac{k_B T}{6 \pi \eta_s \xi} K^2 \tag{6}$$

where k_B is Boltzmann's constant. If we use measured values of
τ, T, K, and η_s to calculate ξ we obtain the unphysically low
result that $\xi \approx 10^{-4}$ A. In a further attempt to see a
positive indication that mode-mode coupling is playing a significant role in the observed relaxation the following experiment
was performed. The relaxation times for two samples with different initial thermal histories and mode populations were
measured as shown in Figure 3. One sample had the thermal history
of the unheatreated or quenched samples and the other had
reached equilibrium at 953 K. Both samples were aged at 973 K
and found to have equal relaxation times, within experimental
error. We thus failed to observed any evidence of mode-mode
coupling either in this experiment or from the observed magnitude
of the relaxation times.

Unfortunately sufficient data is not available for our
glass to allow an estimate of the magnitude of the classical relaxation time[16] τ_{class}

$$\frac{1}{\tau_{class}} = -M K^2 \left[a(T-T_s) + b K^2 \right] \tag{7}$$

where M is a diffusional mobility whose temperature dependence
is that of $1/\eta_s$, a is a constant and b is a coefficient for the
composition gradient energy term in an expansion of the free
energy and which may be temperature dependent. M, a and b
are not available from independent measurements so we can only

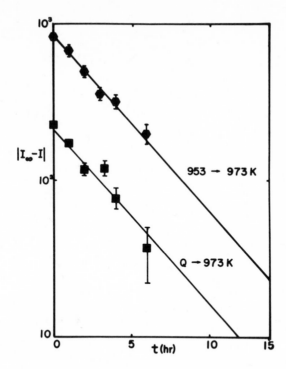

Figure 3 - Log |I∞ - I| versus t for the heat treatment at
973 K of a quenched sample (Q) and one which was previously
equilibrated at 953 K. τ for each is 3.5 hr.

predict the temperature dependence of $\tau_{class} \propto (T-T_s)^{-1}$ assuming
the gradient energy term is negligible   or a constant as a
function of temperature.

        In figure 4 we plot the critical part of the relaxation time
τ/η (dividing out the temperature dependence contained in the
viscosity) versus $T - T_s$. The slope of the line drawn through
the data is ⁻.8 which unfortunately agrees with neither the
classical value of ⁻1 or the nonclassical value of ⁻2/3. The
discrepency is probably outside the uncertainty in the data. Thus
the proper model for describing the critical behaviour of glass
mixtures remains to be determined. More experimentation is required to determine the nature of the apparent differences in
the critical behaviour of low and high viscosity liquid mixtures.

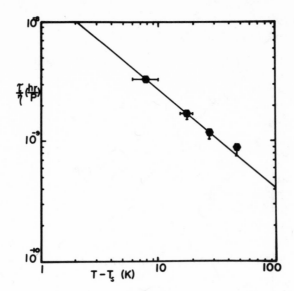

Figure 4 - $\mathcal{T}/\eta$, the critical part of the relaxation time,
versus T - T_s using T_s determined assuming γ = 1 .

In this communication we have discussed an approach to studying the single phase region of a binary glass which exhibits phase
separation. The techniques should find use in glassy systems
with a viscosity in the superspinodal region high enough to make
the system sufficiently slow that a quench through the glass
transition temperature is capable of freezing in the structure
characteristic of the heat treatment temperature. Such a viscosity would be approximately 10^8 to 10^{12} poise. By this technique
we have been able to observe the temperature dependence of equilibrium composition fluctuations which can be explained either by
the classical or nonclassical theories. The kinetic behavior exhibits several interesting features. At least for the one case studied, the relaxation time for the fluctuations depends only on the
heat treatment temperature and not on the previous thermal history.
The temperature dependence of the relaxation time appears to
depart from the predictions of both classical and nonclassical

theories. This apparent disagreement with theory suggests
a fruitful area for future work.

Research supported by the U.S. Office of Naval Research
Contract no. N00014-68-A-0506-0002.

REFERENCES:

1. B. Chu, Ann Rev. Phys. Chem. 21, 145 (1970) and references
    therein.

2. G.B. Benedek, in Polarisation Matiere et Rayonnement,
    edited by Societe Francais de Physique (Presses Universitaires
    de France, Paris, 1969), p. 49.

3. B. Chu, F.J. Schoens and M.E. Fisher, Phys. Rev. 185, 219 (1969)

4. A. Sarkar, G.R. Srinivasan, P.B. Macedo and V. Volterra,
    Phys. Rev. Letters 29, 631 (1972).

5. See for instance, N.S. Andreev and E.A. Porai-Koshits,
    Discussion Faraday Soc. 50, p. 135 (1970).

6. W. Haller, J.H. Simmons, and A. Napolitano, J. Amer.
    Ceram. Soc. 54, 299 (1971).

7. N. Balitactac, Private Communication.

8. J. Schroeder, R. Mohr, C.J. Montrose, and P.B. Macedo,
    J. Non-Cryst. Solids, 13, 313 (1973).

9. J. Schroeder, R. Mohr, P.B. Macedo, C.J. Montrose, J. Amer.
    Ceram. Soc. 56, 510 (1973).

10. L.D. Landau and E.M. Liftshitz, Statistical Physics
    (Pergamon Press, Ltd., London 1958) p. 265.

11. L.P. Kadanoff, W. Gotze, D. Hamblen, R. Hecht, E.A.S.Lewis,
    V.V. Paleiauskas, M. Rayl, J. Swift, D. Aspnes, J. Kane,
    Rev. Mod. Phys. 39, 395 (1967).

12. J.H. Simmons and P.B. Macedo, J. Non-Cryst. Solids 11,
    357 (1973).

13. W. Kauzmann, Chem. Rev. 43, 219 (1948).

14. J.H. Simmons and P.B. Macedo, J. Chem. Phys. 5 2914 (1970).

15. K. Kawasaki, Ann. Phys, (N.Y.) 61, 1 (1970).

16. J.W. Cahn, Trans. AIME 242, 166 (1968).

TWO-PHOTON AND TWO-STEP ABSORPTION IN GLASS OPTICAL WAVEGUIDE

R. H. Stolen and C. Lin

Bell Telephone Laboratories

Holmdel, New Jersey   07733

We have observed a variety of phenomena in studies of
two photon effects in glass optical waveguides. The
long interaction lengths in fibers provide great
sensitivity to weak effects so that these techniques
are applicable to low-loss fibers. The procedure is
to monitor the absorption of a weak CW probe during
and after a strong pulse at a different wavelength.
In general two-step absorption is observed with decay
times of about 10µs. Simultaneous two-photon absorption, bleaching and saturation effects are also
observed.

I.   INTRODUCTION

Fiber optical waveguides have been used to investigate a
variety of nonlinear phenomena in glasses and liquids.[1]   In this
paper we report the observation of both two-photon and two-step
absorption in glass optical fibers. Here as in the previous
experiments, because of the long interaction length in the guide,
weak effects can be observed with quite modest optical powers.

By two-photon absorption we mean the simultaneous absorption
of two photons as is observed in studies of even parity band gap
transitions in crystals.[2]   Such experiments see transitions which
are forbidden in one-photon spectroscopy. Two-photon studies
should provide information about the energy and symmetry of electronic states in glass such as the states of the singly bonded
oxygen in alkali-silicate glasses.[3]   A strong two-photon peak

307

would be of special interest because this would correspond to a
region of enhanced nonlinear susceptibility and thus higher gain
for a parametric four photon amplifier.[4]   Impurities of the iron
group in glass are believed to contribute fairly weakly to the
linear loss in the visible region because their transitions are
"parity forbidden".[5]   One might expect these transitions to
appear strongly in two-photon absorption. Two photon absorption
is also a possible limiting process for strong short optical pulses
in waveguides.

By two-step absorption we refer to sequential processes in
which an impurity is excited by the first photon and later raised
to a higher level by the second. The ability to vary the energy
of the two photons will provide spectroscopic information about
both the initial and the final states. The time response of the
absorption measures the decay time of the first excited level and
might also be a means for identifying specific impurities. Absorption cross sections could be inferred from a combination of one-
photon loss and the power required to saturate the two-step
absorption.

The results reported here are from experiments to determine
the necessary conditions to observe two-photon and two-step
absorption in glass fibers. We have seen both kinds of absorption
in fibers of reasonably high quality. In addition we have seen
bleaching and saturation effects.

II.   APPARATUS

The experimental arrangement is shown in Fig. 1. The procedure

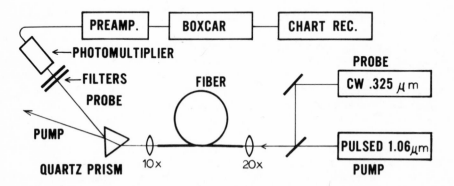

Fig. 1   Experimental arrangement. The cw probe laser is either
         He-Cd at .325μm or He-Ne at .6328μm. The pump is a Q-
         switched Nd:YAG at 1.06μm or .532nm or a tunable flashlamp
         pumped dye laser. The filters are chosen to block the pump.

is to monitor the transmission of a weak cw probe during and after
a strong pump pulse at a different wavelength. Two photon absorption will cause an absorption of the probe only during the pump
pulse while two-step absorption will continue after the pulse for
a time determined by the decay of the first level. The cw lasers
used were He-Cd at .325μm and He-Ne at .6328μm. The pulsed lasers
were a Nd:YAG used at 1.06μm or doubled to 0.532μm and a tunable
flashlamp pumped dye laser. Typical probe powers were about 1 mW
and pump pulses ranged between 5-50 watts in the fiber. Measured
absorption was about 1 percent of the probe power. The fluctuations
in probe power amounted to several percent so that a boxcar amplifier was required to extract the signal. Stray pick up from the
pulsed laser was the major problem and a great deal of effort was
required to shield the detector, preamplifiers, and the pulsed
laser. The cw probe tends to saturate sensitive photomultipliers
and the simple R.C.A. 1P22 has so far been the best detector. Light
is coupled in and out of the fiber with microscope objectives.
Inexpensive uncoated objectives have the least loss at .325μ. For
our experiments the linear absorption of the probe is higher than
that of the pump and maximum signal is obtained for a fiber of
length $1/\alpha$ where α is the absorption coefficient at the probe
wavelength.

III.   TWO-PHOTON ABSORPTION IN GLASS

We have seen what appears to be two photon absorption in a
Soda-Lime-Silicate (SLS) fiber using a 1.06μm pump and a .6328μm
probe. The sum of 3.1 eV falls in the range of ligand field
transitions from iron group metal ions which are one-photon "parity
forbidden".[5] Figure 2 shows the .6328μm absorption as the boxcar
gate was scanned in time. For comparison the 1.06μm pulse was also
measured and this is the solid line on the curve. Within the time
response of the system the absorption appears to be at least as
fast as the pump pulse. The halfwidths of about 300 ns are reasonable when one considers the 100 ns (FWHM) pump pulse, the 100 ns
boxcar gate and the response time of the detector and preamplifiers.
The fact that the pump pulse appears longer than the absorption
signal probably arises from two causes. First, the 1.06μm pump
pulse was measured using a photodiode which had a longer decay time
than the photomultiplier. Second, there are oscillations in the
base line caused by stray pick up in the preamplifiers. The SLS
fiber had a core diameter of 6μm and a loss in the visible of about
100 db/km. The length was 60 cm which was optimal for .325μm
rather than .6328μm.

The most fundamental problem accessible to two-photon studies
is that of the electronic states associated with the singly bonded
oxygens in alkali doped silica glasses. Here one-photon absorption
is observed in the 6-8 eV range [3] (.15-.20μm). We have looked

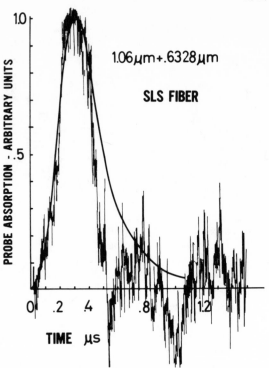

Fig. 2   Scan of .6328μm probe absorption vs. time using a 1.06μm
         pump. For comparison a scan of the 1.06μm pump is also
         included as the solid line. The absorption shows the same
         time response as the pump as expected for a two-photon
         absorption.

for two photon absorption in the same SLS fiber using the .325μm
cw probe and the flashlamp pumped dye laser tuned between .44 and
.60μm which covers the 5.9 - 6.6 eV region. So far any two-photon
absorption is masked by strong two-step absorption.

IV.   TWO-STEP ABSORPTION

        In most experiments we observed an absorption with a decay
time much longer than the pump duration which is characteristic
of a two-step process. Figure 3 shows a scan of such an absorption process. Note that the maximum absorption occurs at 600 ns
which is seen from Fig. 2 to be after the pump pulse. This would
be expected if the decay time of the first level is much longer
than the pulse since the number of excited states is proportional

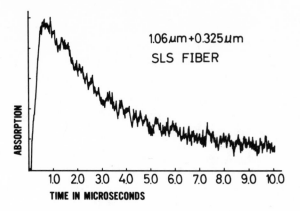

Fig. 3   Two-step absorption characterized by long decay time and
         delay of the absorption maximum to the end of the pump
         pulse.

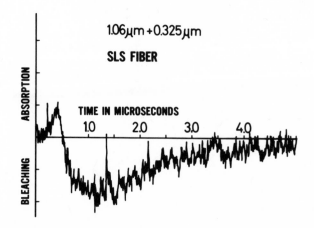

Fig. 4   Scan showing switch from two-step absorption to bleaching.

to the integral of the pump pulse. To our surprise we also saw
similar behavior with an increase rather than a decrease in the
transmission of the probe. Figure 4 is a scan where a switch
from absorption to bleaching was observed.

    We found that for the SLS fiber we could in fact change from
absorption to bleaching by varying the coupling to the fiber.

Absorption could be favored by coupling both probe and pump into
the lowest order guide mode for which most of the energy propagates
in the center of the core. The fiber was near single mode at
1.06μm but supported several modes at .325μm. Bleaching was
observed with the probe in a mode where most energy propagated
near the core-cladding boundary. This indicates that the impurities contributing to absorption and bleaching are at different
distances from the fiber axis.

     We can explain these results using a simplified picture of
absorption and bleaching which is presented in Fig. 5. In both
processes the pump excites the systems from level 1 to level 2.
Level 2 decays with time constant τ. In the absorption process
the increased population of level 2 increases the absorption
between levels 2 and 3. The probe should be weak enough so that
its absorption does not affect τ and the pump should not saturate
level 2. In the bleaching process the decreased population of
level 1 decreases the absorption between levels 1 and 3. The
population of level 2 in both cases is obtained from

$$\overset{\bullet}{n} = n/\tau + f(t) \tag{1}$$

where $f(t)$ is the product of the pump power and the cross section
of the transition from level 1 to 2. The population of level 2 is
then:

$$n(t) = e^{-t/\tau} \int_{-\infty}^{t} f(t')e^{t'/\tau}dt' . \tag{2}$$

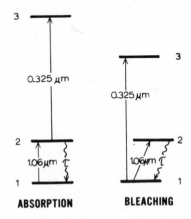

ABSORPTION   BLEACHING

Fig. 5 Simple picture of absorption and bleaching processes.

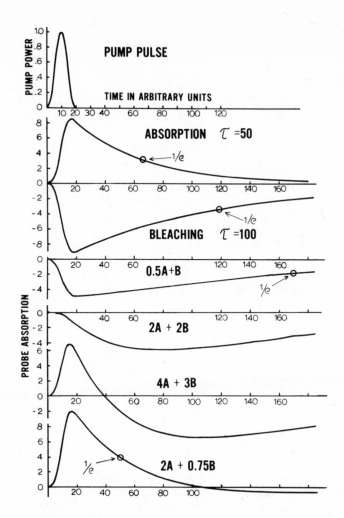

Fig. 6   Absorption (A) and bleaching (B) vs. time are computed for
        the pump pulse pictured. The time constant for A is 50
        in the arbitrary scale units and τ for B is 100. These
        processes are assumed to be independent and are added
        together in various combinations to produce the curves
        shown in the bottom half of the figure.

In Fig. 6 we choose a pump pulse which is plotted on an arbitrary
time scale. The absorption curve is calculated from Eq. 2 using
a time constant of τ = 50 in the same time units. The bleaching
curve is calculated assuming $\Delta n_1 = -n_2$ and choosing τ = 100. If
these processes occur simultaneously some rather interesting
curves appear. In Fig. 6 the absorption and bleaching curves

have been added together in different proportions. Note that the
curve for 4A + 3B looks very much like Fig. 4. At different times
we saw curves similar to all the possibilities pictured in Fig. 6.

The probe absorption should depend linearly on pump power.
Generally this was the case at low pump power but saturation of
the absorption was seen at higher powers. Measurements of the
saturation power and the one-photon absorption coefficient should
permit the determination of both the number of impurities and the
absorption cross section.

A strong two-step absorption was observed with the 1.06μm
pump and .325μm probe in a different fiber with a K doped silica
core. The absorption peak became weaker each time the scan was
repeated and no amount of retuning would restore the original
strong absorption. These results were duplicated with a fresh
piece of the same fiber. This was quite a good fiber: 28 db/km
at 1.06μm and the major impurity was believed to be iron or
copper. One explanation of this decrease in absorption with
repeated scans is that the two-step absorption converted Fe^{++} which
absorbs at 1.06μm to Fe^{3+} which doesn't.

Both bleaching and absorption were observed in an SiO_2 core
fiber using a .532μm pump and a .325μm probe. This fiber had a
loss of 16 db/km in the red and was quite transparent even at
.325μm. In this case 80 meters of fiber were needed to see a
measurable effect.

V. CONCLUSION

We have observed both two-photon and two-step absorption in
glass optical waveguides. We should be able to separate these two
types of processes by using shorter pump pulses to favor two-
photon absorption and longer pump pulses to favor two-step
absorption. These techniques can also be extended to liquid core
fibers.

Two-step absorption was seen in fibers of fairly high quality
(25-100 db/km) which indicates that this technique is sensitive to
small numbers of impurities although we could only guess at their
identity. There is no reason why two-step absorption should not be
observed in fibers with much lower loss since much longer lengths
of guide could then be used. Future experiments using fibers with
known impurity concentrations should provide measurements of
lifetimes and cross sections as well as a more sensitive technique
for identifying specific impurities in optical fibers.

ACKNOWLEDGMENTS

We would like to thank A. Ashkin and J. M. Worlock for many helpful discussions; A. R. Tynes, W. G. French and P. Kaiser for providing the fibers used in the experiments; and W. Pleibel for technical assistance.

REFERENCES

1. E. P. Ippen, Appl. Phys. Lett. $\underline{16}$, 303 (1970); R. H. Stolen, E. P. Ippen, and A. R. Tynes, Appl. Phys. Lett. $\underline{20}$, 62 (1972); R. H. Stolen, J. E. Bjorkholm, and A. Ashkin, Appl. Phys. Lett. $\underline{24}$, 308 (1974); E. P. Ippen, C. V. Shank, and T. K. Gustafson, Appl. Phys. Lett. $\underline{24}$, 190 (1974).
2. J. J. Hopfield, J. M. Worlock and K. Park, Phys. Rev. Lett. $\underline{11}$, 414 (1963); J. M. Worlock in: Laser Handbook (F. T. Arecchi and E. O. Schulz-DuBois, Eds., North-Holland Publ. Co), 1323 (1972).
3. G. H. Sigel, J. Phys. Chem. Solids $\underline{32}$, 2373 (1971).
4. N. Bloembergen, "Nonlinear Optics" (W. A. Benjamin, Inc., 1965); M. D. Levenson, IEEE J. Quantum Electron. $\underline{QE-10}$, 110 (1974).
5. C. R. Kurkjian and G. E. Peterson, "Some Materials Problems in the Design of Glass Fiber Optical Waveguides", 2nd Cairo Solid State Conference, American Univ., Cairo, April 1973; T. Bates, "Modern Aspects of the Vitreous State", Vol. 2, (ed. J. D. Mackenzie, Butterworths, 1962).

OPTICALLY INDUCED EFFECTS IN PHOTOLUMINESCENCE STUDIES OF CHALCOGENIDE GLASSES

S. G. Bishop and U. Strom

Naval Research Laboratory

Washington, D.C. 20375

Enhancement of photoluminescence (PL) by IR light with energy as low as half the band gap is demonstrated to be a restoration of the fatigued PL induced by inter-band excitation radiation. At 6K in glassy As_2Se_3, an increase in the absorption coefficient in the band tail region which accompanies fatiguing of the PL is observed throughout the spectral range of the enhancement band ($\cong 0.6$ - 1.5 eV). Subsequent irradiation by IR light in the 0.6 - 1.5 eV range restores both the IR trans-mission in this range and the PL to their initial cold-dark values.

INTRODUCTION

Photoluminescence (PL) studies of chalcogenide glasses have made substantial contributions to the understanding of the energy distribution of electronic states, interband absorption processes, and radiative and non-radiative recombination mechanisms in these materials. The results of these PL studies can be described con-veniently in terms of several features of the phenomenon which are observed universally in chalcogenide glasses. All chalcogenide glasses exhibit broad PL spectra located well below the forbidden gap energy[1-7] (see Fig. 1). This mid-gap position of the PL has been interpreted either in terms of recombination between long tails of localized states extending deep into the gap below the conduction band edge and above the valence band edge[2] or in terms of recombination between a band of localized states near mid-gap (Mott-Davis model[8]) and one of the band edges.[1,3,4]

PL excitation spectra for chalcogenide glasses have demonstrated that the luminescence is most efficiently excited by light with

317

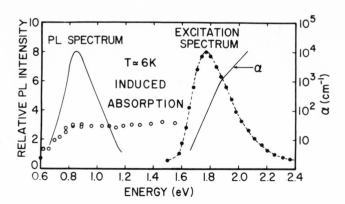

Fig. 1   Photoluminescence spectrum, photoluminescence excitation
spectrum, optical absorption edge, and induced optical absorption
spectrum in vitreous As_2Se_3 at 6K.

penetration depth~100 μm. That is, the excitation of the PL is
predominantly a bulk process and highly absorbed light which penetrates 1 μm or less is quite inefficient in the excitation of PL
(Fig. 1).

Further information concerning the existence and distribution
of localized electronic states in the gap has been obtained through
the study of the dependence of the PL efficiency upon the application of monochromatic light with photon energy less than the forbidden gap energy. In glassy As_2Se_3, $As_2Se_{1.5}Te_{1.5}$, and As_2S_3,
unmodulated monochromatic light with photon energy corresponding
to energies between the PL band and the onset of interband absorption was observed to enhance the modulated PL which is excited by
modulated interband light.[9] These enhancement bands were regarded
as manifestations of optical absorption by localized levels involved
in the recombination process and located at energies corresponding
to high transparency under equilibrium (cold-dark) conditions.

Another feature of the PL which is common to all glasses studied
thus far is the fatiguing or decay of the PL efficiency during continuous excitation by light with wavelength corresponding to the
peak region of the excitation spectrum[3,5,10] (see Fig. 2). Recently
Cernogora et al.[11] have reported an increase in the band tail
absorption of As_2Se_3 which is induced by deeply penetrating PL
excitation light and which accompanies the fatiguing of the PL at
1.6K. In the present work we show that the spectral dependence of
this induced absorption corresponds closely to that of the enhancement band in As_2Se_3 and that the application of near infrared light
within this band restores both the fatigued PL intensity and the

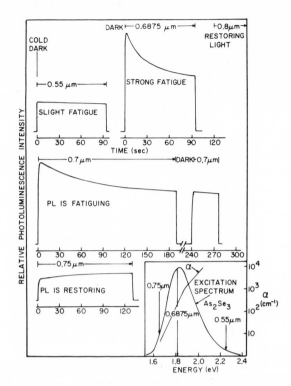

Fig. 2   Time dependence of integrated PL intensity at 6K in
vitreous As$_2$Se$_3$ for various wavelengths of exciting radiation.
The insert shows the location of the exciting wavelengths within
the excitation spectrum.

sample transparency to their cold-dark values. On the basis of
these results it is concluded that the PL enhancement is actually
a restoration of the fatigued PL, and that the observed induced
absorption is the absorption by localized electronic states in
the gap which gives rise to the restoration of the fatigued PL.

FATIGUING AND OPTICAL RESTORATION OF PHOTOLUMINESCENCE

In Fig. 2 the time dependence of the integrated PL intensity
at 6K in glassy As$_2$Se$_3$ is plotted for various wavelengths of monochromatic exciting radiation. The exciting light was chopped at
75 Hz and the modulated PL was detected by a PbS photoconductor
and a lock-in amplifier. Appropriate filters were used to exclude
exciting radiation from the spectral range of the luminescence.
The various curves of Fig. 2 represent the dc output of the lock-
in amplifier as plotted by a strip chart recorder. The wavelength
of exciting light is given for each curve and reference to the
absorption curve shown at the bottom of Fig. 2 provides the penetration depth of the light.

The application of highly absorbed 0.55 μm wavelength light
(top of Fig. 2) produces a weak PL which decays quite slowly. In
contrast exposure to 0.6875 μm exciting light, which penetrates
~100 μm and corresponds to the peak of the excitation spectrum,
produces a rapidly fatiguing PL signal. The extent of the fatiguing
is dependent upon the dose of exciting light. If the incident
photon flux is reduced by a factor of two, the time required to
produce a given percentage fatigue will be doubled.[5,10] The
fatiguing shown in the top portion of Fig. 2 was produced by
0.6875 μm light with intensity of 1 mW cm^{-2} which corresponds to
a photon flux of approximately 3 x 10^{15} cm^{-2} sec^{-1}.

Cernogora et al.[5,10] have reported that the fatigued PL can
be partially restored by thermal annealing at temperatures above
their measuring temperature (1.2K) but below 300K. We find that
exposure to infrared radiation with photon energy corresponding
to the observed enhancement band also restores the PL intensity
to its cold-dark value. After measurement of the fatiguing curve
shown in the top portion of Fig. 2, the sample was exposed to
0.8 μm light for two minutes. This photon energy lies within the
enhancement band observed in glassy As$_2$Se$_3$.[9] Exposure to this
near-infrared light restored the PL so that it again exhibits a
fatiguing effect when PL is excited by 0.7 μm light in the middle
segment of Fig. 2. Also demonstrated in Fig. 2 is the fact that
if the excitation is interrupted for an arbitrary period of time
and subsequently applied again, there is no recovery of PL intensity;
the fatiguing process continues from the point where it was interrupted. Only the application of the near infrared enhancing or
restoring radiation has an effect upon the fatigued PL intensity.

There is a narrow range of weakly absorbed exciting light
energies or wavelengths ($\lambda \cong 0.75$ μm, $\alpha \sim 10$ cm^{-1} in As$_2$Se$_3$) for
which the excitation spectrum overlaps the enhancement or restora-
tion band.[9] Excitation of PL by wavelengths in this region ($\lambda =$
0.75K μm for As$_2$Se$_3$, bottom of Fig. 2) subsequent to fatiguing by
shorter wavelengths produces a PL intensity which grows as a function of time. Light in this excitation regime simultaneously
excites PL and activates the restoration mechanism.

OPTICALLY INDUCED ABSORPTION

Cernogora et al.[11] have reported absorption measurements in
glassy As$_2$Se$_3$ at 1.6K in the range 1.4 - 1.7 eV before and after
irradiation of the sample with 0.6764 μm light from a laser. This
laser wavelength is near the maximum of the PL excitation spectrum
and has a penetration depth of the order of 50 μm. They observed
an increase in the absorption coefficient α at all wavelengths in
this range after the laser irradiation. Typically, a 64 second
exposure to 6 W/cm^2 laser light produced an increase in α of nearly
300 cm^{-1} at 1.5 eV. We have carried out absorption measurements in
glassy As$_2$Se$_3$ at 6K in the range 0.4 - 1.55 eV before and after
irradiation by monochromatic light with wavelength of 0.6875 μm.

This wavelength has a penetration depth of the order of 100 μm and the applied intensity was only 1 mW/cm^2, or a factor of 6000 less than that employed by Cernogora et al.[11]

In Fig. 1 the absorption coefficient induced by a ~100 sec. exposure to the 1 mW/cm^2 0.6875 μm light is plotted in the range 0.6 - 1.55 eV. This induced α maintains a roughly constant value of ~30 cm^{-1} from 1.55 eV to about 0.8 eV. At energies below 0.8 eV the induced α begins to decrease and reaches zero at about 0.55 eV. It should be emphasized that the intensity of the light used to measure the absorption before and after the 0.6875 μm irradiation was reduced by neutral density filters to a level at least two orders of magnitude below the 1 mW/cm^2 intensity which induces the absorption. Hence the measurement does not further perturb the values of α.

The induced absorption remains stable for times at least as long as hours, and it exhibits an excitation dose dependence. That is, several separate exposures to exciting light applied consecutively increase the induced α in proportion to the cumulative dose. As reported by Cernogora et al.,[11] the induced absorption can be "released", i.e. transparency restored, by warming to room temperature. However, we have also observed that transparency can be restored by irradiating the sample with light with wavelength corresponding to the induced absorption band (but below the range of interband absorption) with an intensity roughly equivalent to that of the inducing or exciting radiation. In Fig. 3, the transmission of a sample (which has first been given sufficient dose of 0.6875 μm light to induce a measureable increase in absorption) is shown as a function of increasing exposure to 1 μm light, which lies in the middle of the induced absorption band shown in Fig. 1. This near infrared light had an intensity of ~1 mW/cm^2 and it gradually restores the transparency of the sample, throughout the induced absorption band to its cold-dark value. That is, the 1 μm light is not only effective in restoring transparency for that wavelength, but for the entire 0.55 - 1.55 eV band in which the optically induced increased α was observed. This restoring effect was observed for several wavelengths within the induced absorption band.

Comparison of the rate of restoration of transparency as exhibited by the shape of the recovery curve in Fig. 3 with the rate of restoration of the fatigued PL as shown in the bottom of Fig. 2 indicates that the recovery rates for the two processes are roughly equivalent.

DISCUSSION

The results of the experimental measurements just described can be summarized as follows: Irradiation of glassy As$_2$Se$_3$ at low temperatures with light of wavelength having penetration depths in the range 10 - 1000 μm produces PL which fatigues strongly during

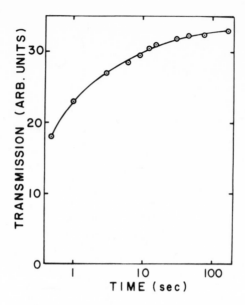

Fig. 3   Recovery of transmission at 1 μm (in arbitrary units) of
As$_2$Se$_3$ at 6K. Sample was previously exposed to 100 sec. of
1 mV/cm^2 0.6875, μm radiation.

continuous irradiation. Accompanying this fatiguing behavior is an
induced optical absorption in the spectral range extending approximately from the band edge to mid-gap. The spectral distribution of
the induced absorption corresponds to that of the PL enhancement
band observed in As$_2$Se$_3$.[9] Subsequent to irradiation by interband
light which causes fatiguing PL and the induced absorption, the
application of near infrared light in the spectral range corresponding to the induced absorption restores both the fatigued PL and the
sample transparency to their cold-dark levels.

It seems obvious that the two processes, fatiguing of the PL
and the increased optical absorption, are closely related. In fact,
the fatiguing or reduction of PL intensity can be explained solely
on the basis of the increased optical absorption. If the increased
optical absorption in the spectral range of the PL excitation spectrum leads to a non-radiative thermalization or recombination path

for the injected electron-hole pairs, there will be a reduction in
the number of excited carriers available for the radiative recombination mechanism and a drop in PL intensity. Changes in the
shape of the PL excitation spectrum which we have observed[12] with
increasing fatigue of the PL are apparent manifestations of just
such an effect. In addition, since the induced optical absorption extends down to energies corresponding to the radiative
recombination spectrum, there is a loss of PL intensity to self-
absorption. These two effects can readily account for the observed
fatiguing of PL efficiency.

Given the close correspondence in energy of the previously
reported PL enhancement band and the optically induced absorption
band, and given the simultaneous restoration of PL intensity and
transparency produced by irradiation in this band of energies, we
may conclude that the observed induced optical absorption is due
to localized states in the gap. These are the same localized states
to which the PL enhancement or restoration effect is attributed.

It is well known that some chalcogenide glasses exhibit a
"photo-darkening" effect upon exposure to highly absorbed interband light.[13-15] This effect is usually described as an optically
induced shift in the absorption edge to longer wavelengths which
can be reversed by thermal annealing. The determination of a
possible relationship between this effect and the optically induced
absorption reported here is made difficult by the widely divergent
experimental conditions employed in the study of the two effects.

Berkes et al.[13] have attributed the photo-darkening effect
to the photodecomposition of the chalcogenide glass. This mechanism
involves the liberation of elemental arsenic which diffuses by a
thermally activated process and forms aggregates or amorphous arsenic
particles. The fact that the absorption edge in amorphous arsenic
occurs at lower energies than in As_2Se_3 is then presumed responsible
for the photo-darkening. There are several reasons why this explanation is not suitable for the induced absorption which accompanies
fatiguing of the PL in As_2Se_3. Berkes et al.[13] found an increasing
induction or delay time in the photo-darkening process with decreasing temperature, amounting to ~10 sec. at 120K. No such induction
time is observed in the time dependence of the optically induced
absorption at 1.6K by Cernogora[11] or at 6K by the present authors.
Furthermore, the shape of the time dependence curves obtained by
Cernogora are very different from the time dependence curves of
Berkes. It seems unlikely that at temperatures as low as 6K photo-
decomposed or liberated arsenic atoms could undergo sufficient
diffusion to form aggregates. It is also difficult to understand
how the near infrared radiation (1 mW/cm^2) which restores the sample
transparency could dissociate and disperse such arsenic particles.
More definite data concerning the connection between photo-darkening
(at 300K) and the low temperature photo effects discussed here is
expected from induced absorption measurements as a function of wave-

length in pure Se as well as other chalcogenide glasses at 300K.

While the photo-darkening effect involves shifts in the absorption edge in the spectral range where the initial $\alpha = 10^2 - 10^4$ cm^{-1}, the induced optical absorption reported here involves an increase from initial $\alpha = 0.1 - 1$ cm^{-1} to an induced α of ~ 30 cm^{-1} in a broad spectral range extending to energies below mid-gap. Absorption increases at such low energies in the normally transparent forbidden gap range of the glass may arise from optical transitions which involve localized electronic states in the gap such as those invoked by Wood and Tauc[16] to explain the weak absorption tails observed in glassy As_2S_3. The interband exciting light renders these states optically active, or alternatively, greatly increases the number of such states. For example, the localized states may be impurity or disorder-induced long-lived traps which can be optically populated and depopulated by light of different photon energies. Evidence for the existence of such long-lived traps in chalcogenide glasses at low temperatures is provided by the very long decay times for low temperature photoconductivity.[17]

In order to estimate the density or concentration of localized states or centers giving rise to the induced absorption, a value for the optical cross section of the absorbing center must be assumed. If an F-center model for an isolated defect is chosen,[18] the optical cross section is assigned a value of order 10^{-16} cm^2 and integration of the 30 cm^{-1} absorption coefficient from 0.8 to 1.55 eV yields a concentration of localized states of approximately 10^{17} cm^{-3}. The exact origin of the optical transitions which give rise to this absorption remains unspecified and it is not possible at this point to determine whether the absorption is attributable to localized levels that are an intrinsic property of the disordered state or to the presence of impurities or defects.

ACKNOWLEDGEMENTS

The authors wish to acknowledge D. L. Mitchell and J. Tauc for helpful discussions.

REFERENCES

1. B. T. Kolomiets, B. T. Mamontova, and A. A. Babaev, J. Non-Cryst. Solids 4, 289 (1970).

2. R. Fischer, U. Heim, F. Stern, and K. Weiser, Phys. Rev. Lett. 26, 1182 (1971).

3. R. A. Street, T. M. Searle, and I. G. Austin, J. Phys. C6, 1830 (1973).

4. S. G. Bishop and C. S. Guenzer, Phys. Rev. Lett. 30, 1309 (1973).

5. J. Cernogora, F. Mollot, and C. Benoit a la Guillaume, Phys. Stat. Sol. (a) 15, 401 (1973).

6. S. G. Bishop and D. L. Mitchell, Phys. Rev. B8, 5696 (1973).

7. R. A. Street, T. M. Searle, and I. G. Austin, Phil. Mag. 29, 1157 (1974).

8. E. A. Davis and N. F. Mott, Phil. Mag. 22, 903 (1970).

9. S. G. Bishop, U. Strom, and C. S. Guenzer, Proc. 5th Intl. Conf. on Amorphous and Liquid Semiconductors, ed. J. Stuke and W. Brenig, (Taylor and Francis, London, 1974), p. 963.

10. F. Mollot, J. Cernogora, and C. Benoit a la Guillaume, Phys. Stat. Sol. (a) 21, 281 (1974).

11. J. Cernogora, F. Mollot, and C. Benoit a la Guillaume, Proc. 12th Intl. Conf. on Physics of Semiconductors, ed. M. H. Pilkuhn (B. G. Teubner, Stuttgart, 1974), p. 1027.

12. S. G. Bishop and U. Strom, to be published.

13. J. S. Berkes, S. W. Ing, Jr., and W. J. Hillegas, J. Appl. Phys. 42, 4908 (1971).

14. A. Hamada, M. Saito, and M. Kikuchi, Solid State Comm. 11, 1409 (1972).

15. Y. Asahara and T. Izumitani, J. Non-Cryst. Solids 16, 407 (1974).

16. D. L. Wood and J. Tauc, Phys. Rev. B5, 3144 (1972).

17. E. A. Fagen and H. Fritzsche, J. Non-Cryst. Solids 4, 480 (1970).

18. F. Stern, Solid State Physics, Vol. 15, ed. F. Seitz and D. Turnbull (Academic Press, New York, 1963), p. 378.

Section V
Multiphoton Processes and Nonlinear Effects

NONLINEAR SPECTROSCOPY IN TRANSPARENT CRYSTALS[†]

N. Bloembergen, M. D. Levenson and R. T. Lynch, Jr.

Division of Engineering and Applied Physics

Harvard University, Cambridge, Massachusetts 02138

1. INTRODUCTION

A review will be given of recent nonlinear optical experiments, which utilize only light beams at frequencies in the transparent region of the sample, and yet yield information about the dispersive properties of crystalline excitations in strongly absorbing regions of the spectrum. Such experiments are based on the dispersive characteristics of nonlinear susceptibilities, which can be measured over a wide, and continually variable, range of frequencies, due to the development of dye lasers. In a typical experiment two tunable laser beams, at frequencies ω_1 and ω_2 respectively, and with wave vectors k_1 and k_2 respectively, are made to overlap inside the specimen. Among several other nonlinear responses, a nonlinear polarization is created at the combination frequency $\omega_3 = 2\omega_1 - \omega_2$. The ith Cartesian component of this polarization may be expressed in terms of a third-order susceptibility tensor,

$$P_i(2\omega_1 - \omega_2, r) = \Sigma \, \chi^{(3)}_{ijk\ell} \, (-\omega_3, \omega_1, \omega_1, -\omega_2)$$

$$E_j(\omega_1) \, E_k(\omega_1) \, E_\ell(-\omega_2) \, \exp\left[(2k_1 - k_2) \cdot r\right] \qquad (1)$$

The theoretical expression for the complex nonlinear susceptibility contains a variety of resonance denominators.[1] Resonances do not only occur when the incident frequencies, ω_1 and ω_2, or the newly

---

[†] Supported by the Joint Services Electronics Program.

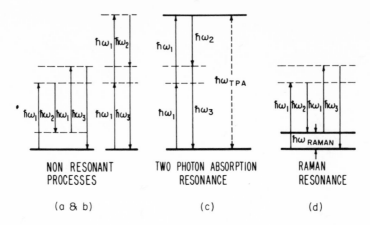

Figure 1: Parametric Processes contributing to three-wave-mixing, in which intensity at $\omega_3 = 2\omega_1 - \omega_2$ is generated by two incident light waves at ω_1 and ω_2 respectively. The resonant contributions from two-photon absorption and Raman processes are shown in c) and d). They interfere with the nonresonant processes a) and b).

created wave at ω_3, corresponds to an energy excitation of the samples, but some terms display resonant behavior when $2\omega_1$ corresponds to an eigenfrequency, for example of an exciton level; or $\omega_1 - \omega_2$ corresponds to a resonance, for example of an optical phonon. It is possible for ω_1, ω_2 and ω_3 all to lie in the visible transparent region of the crystal, while $\omega_1 - \omega_2$ corresponds to an optical phonon or polariton frequency in the infrared. It is also possible for $2\omega_1$ to correspond to UV resonance of an exciton level. The observed intensity at ω_3 is proportional to $|\chi^{(3)}(-\omega_3,\omega_1,\omega_1,-\omega_2)|^2$, and this quantity displays the interference of the possible pathways for this parametric process, which are shown in Figure 1. The real energy levels are chosen such that the material is transparent, in a linear sense, at the frequencies ω_1, ω_2 and ω_3. The momentum conservation conditions for these different channels is depicted in Figure 2. The interference between the non-resonant electronic nonlinearity, and the resonant two-photon absorption or Raman processes,[1] was discussed more than a decade ago in a pioneering paper by Maker and Terhune.[2] Recently detailed experimental data have become available.[3] In Section 2 some salient features of the dispersion of $\chi^{(3)}$ in centrosymmetric media will be briefly reviewed.

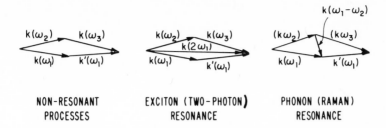

NON-RESONANT   EXCITON (TWO-PHOTON)   PHONON (RAMAN)
PROCESSES   RESONANCE   RESONANCE

Figure 2: The momentum conservation relations between the wave vectors of the incident light waves, $\vec{k}(\omega_1)$ and $\vec{k}(\omega_2)$, and
the wave vector of the generated wave $\vec{k}(\omega_3)$. In the resonant processes there is energy and momentum match with
an intermediate excitation of an exciton or optical phonon.

The reader who wishes more detail should consult Reference 3, and
other papers quoted therein. In Section 3, a few features characteristic for non-centrosymmetric crystals will be discussed.

2. NONLINEAR DISPERSION IN CENTRO-SYMMETRIC CRYSTALS

Diamond shows a strong and narrow Raman resonance. The obser-
ved intensity created at $\omega_3 = 2\omega_1 - \omega_2$ by two incident light beams
at ω_1 and ω_2 is shown as a function of the difference frequency
$\omega_1 - \omega_2$ in Figure 3. The frequency ω_1 is held fixed, corresponding
to a wavelength $\lambda_1 = 545$ nm. At $\omega_1 - \omega_2 = 1332$ cm^{-1}, corresponding
to the Raman-active mode of the diamond lattice, ω_3 may be regarded
as the anti-Stokes frequency. Note the minimum or antiresonance,
where the intensity is eight or nine orders of magnitude smaller
than at the maximum. At the minimum the negative real part of the
Raman susceptibility cancels the electronic nonlinear susceptibility. Since we are many optical phonon line widths away from the
phonon resonance the imaginary part is very small. Curves such as
those shown in Figure 3 make it possible to trace the shape of the
optical phonon line out to about one hundred Lorenzian widths. The
line in diamond remains rather accurately Lorenzian that far out

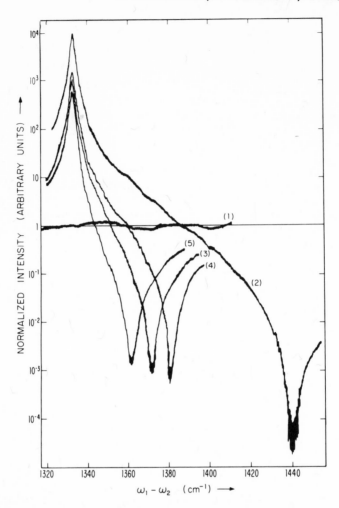

Figure 3: Three-wave-mixing, generating intensity at $\omega_3 = 2\omega_1 - \omega_2$, in diamond as a function of $\omega_1 - \omega_2$. The numbers in parentheses refer to the polarization conditions shown in Figure 4.

into the wings. The data may also be used to calibrate the non-resonant electronic nonlinearity in terms of the known Raman scattering cross section.

The five curves shown in Figure 3 are for the five different geometries, shown in Figure 4, of the directions of the electric field polarizations of the three light beams with respect to the crystallographic axes. There is a very large anisotropy in the pro-

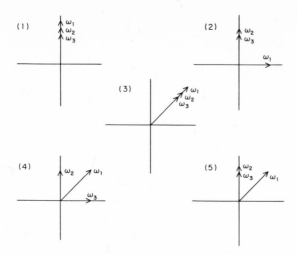

Figure 4: Five polarization geometries used for the data in Figure
3. The arrows show the directions of the incident elec-
tric fields at ω_1 and ω_2 and the polarization at ω_3, with
respect to the cubic crystal axes.

duction of light at $2\omega_1 - \omega_3$. In geometry 1, when all electric
field vectors are parallel to the cubic axis, no Raman resonance
occurs. The optical phonon cannot be excited in this case. The ten-
sor element $\chi_{xxxx}^{(3)}$ $(-\omega_3,\omega_1,\omega_1,-\omega_2)$ has only a non-resonant contribution in the diamond lattice. The two other independent elements
$\chi_{xyyx}^{(3)}$ $(-\omega_3,\omega_1,\omega_1,-\omega_2)$ and $\chi_{xxyy}^{(3)}$ $(-\omega_3,\omega_1,\omega_1,-\omega_2)$ have both resonant
and non-resonant contributions. The three elements for the electro-
nic nonlinearity have been measured with considerable precision,
not only for diamond, but also for the homologous fluoride series,
CaF_2, SrF_2, CdF_2 and BaF_2. It has been found that there is a
significant cubic anisotropy,

$$\sigma = \left[2 \chi_{1122}^{(3)E} + \chi_{1221}^{(3)E} - \chi_{1111}^{(3)E} \right] / \chi_{1111}^{(3)E} \tag{2}$$

and that the ratio $\chi_{1122}^{(3)E} / \chi_{1221}^{(3)E}$ differs significantly from unity.
This value would prevail if Kleinman's symmetry were strictly valid.
In isotropic materials, such as glasses or liquid, σ given by Eq.
(2) vanishes.

The geometry in which the electric field vectors, $\underset{\sim}{E}(\omega_1)$ and
$\underset{\sim}{E}(\omega_2)$, make an angle of $45°$ with respect to each other, is particularly useful in measuring the ratio of the tensor elements.[4]

Kleinman's symmetry can be tested directly by rotating the crystal by 45° so that in the first case $\underset{\sim}{E}(\omega_1)$ is parallel to a cubic axis, and in the second case $\underset{\sim}{E}(\omega_2)$. The direction of the polarization $\underset{\sim}{E}(\omega_3)$ is compared for these two situations. The direction of $\underset{\sim}{E}(\omega_3)$ can also be observed as a function of the difference frequency $\omega_1 - \omega_2$, for a fixed crystallographic orientation. In the vicinity of the Raman resonance the plane of polarization rotates through a full 360° angle. This phenomenon may be called a nonlinear dispersion of the plane of polarization. Comparison of polarization components in the nonlinear process provides an accurate means of measuring ratios of tensor elements of the nonlinear susceptibility.

Finally the variation of $\chi^{(3)}$ as a function of ω_1, or $2\omega_1$, should be mentioned. When $2\omega_1$ is larger than the band gap, two photon processes contribute to the imaginary (and real) part of $\chi^{(3)}$. As a result, the anti-resonance becomes less pronounced, and the minimum of the curves shown in Figure 3 becomes shallower. This effect has been observed in diamond for a wavelength $\lambda_1 = 407$ nm. At the same time the distance Δ between the maximum and the minimum becomes smaller at shorter wavelengths λ_1. This reflects an

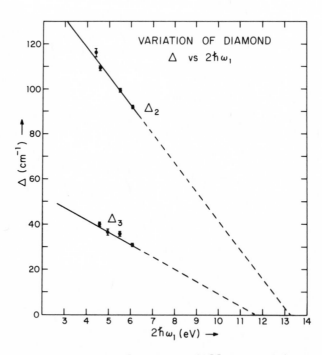

Figure 5: Variation in the frequency difference Δ between the minimum and maximum intensity observed in three wave mixing in diamond as a function of $2\hbar\omega_1$.

increase in relative importance of the electron nonlinearity with respect to the Raman contribution, as shown in Figure 5. Extrapolation of the data suggests that the average band gap, or maximum joint density of states, for electronic states with the same parity, is about 12 - 13 ev.

A few data points have been obtained for a calcite crystal. It is clear that it would be possible to measure the large number of independent tensor elements of the non-resonant nonlinear susceptibility in various crystals of lower symmetry. The interference with more than Raman-active mode may be studied in such crystals. It is conceivable to study the generation at frequency $\omega_4 = \omega_1 + \omega_2 - \omega_3$, by three incident light beams while $\omega_1 - \omega_3$ is resonant with one phonon branch and simultaneously $\omega_2 - \omega_3$ with another. An attempt has been made to detect the resonance from second order Raman scattering resonances. This was not successful in diamond and alkalihalides, but it might succeed in crystals with the perovskite structure.

3. NONLINEAR DISPERSION IN NON-CENTROSYMMETRIC CRYSTALS

The same kinds of phenomena also occur in crystals which lack a center inversion, but the interpretation is more involved because a nonlinear polarization of lower order now also exists,

$$P_i(2\omega_1) = \sum_{j,k} \chi^{(2)}_{ijk} (-2\omega_1, \omega_1, \omega_1) \, E_j(\omega_1) \, E_k(\omega_1) \tag{3a}$$

$$P_i(\omega_1 - \omega_2) = \sum_{j,k} \chi^{(2)}_{ijk} (-\omega_1 + \omega_2, \omega_1, -\omega_2) \, E_j(\omega_1) \, E_k(-\omega_2) \tag{3b}$$

This second-order polarization creates electric fields at the intermediate frequencies $2\omega_1$ and $\omega_1 - \omega_2$. These fields may be represented as the sum of local fields from the second-order polarization in neighboring unit cells, from driven waves by the nonlinear source terms in distant unit cells everywhere in the infinite crystal, and of freely propagating waves which are needed to satisfy the boundary conditions of Fourier components at $2\omega_1$ and $\omega_1 - \omega_2$ at the surface of the crystal. These fields interact, in turn, with the nonlinear polarizability of a unit cell or volume element of the cyrstal to yield a contribution to the polarization at the combination frequency $\omega_3 = 2\omega_1 - \omega_2$.

$$P_i(\omega_3) = \chi^{(2)}_{ijk} (-\omega_3, 2\omega_1, -\omega_2) \, E_j(2\omega_1) \, E_k(-\omega_2) +$$

$$\chi^{(2)}_{ijk} (-\omega_3, \omega_1, \omega_1 - \omega_2) \, E_j(\omega_1) \, E_k(\omega_1 - \omega_2) \tag{4}$$

It is clear that this two-step contribution, resulting from Eqs. (3) and (4) is proportional to $\{\chi^{(2)}\}^2$, and must be added to the direct contribution given by Eq. (1). Complete formal expressions for these effects have been given by Bedeaux.[5] Flytzanis et al.[6,7] have analyzed the nonlinear contributions of infrared Fourier components of ionic and electronic motions in great detail. For the purpose of this review it may suffice to remark that the electric field component at $\omega_1 - \omega_2$ may be lumped with the optical phonon vibration at $\omega_1 - \omega_2$ into a polariton mode. The diagrams on the right hand side in Figures 1 and 2 retain their validity, if the polariton energy is used for $\hbar(\omega_1 - \omega_2)$, with appropriate momentum $\hbar \underset{\sim}{k}(\omega_1 - \omega_2)$ from the polariton dispersion relation. The nonlinear coupling coefficients with the polariton mode involve both the Raman polarizability $\partial \alpha_{ij}/\partial Q_k$ and the electronic nonlinear susceptibility $\chi_{ijk}^{(2)}$. The net result is that in crystals of the class $\bar{4}3$ m, e.g. CuCl, GaP or ZnSe, similar effects are seen as in diamond. The production of intensity at $\omega_3 = 2\omega_1 - \omega_2$ shows an interference between the non-resonant contribution and resonant Raman-type scattering from the polariton excitation. Detailed measurements are in progress and should yield information on polariton dispersion and damping.

The interference of a sharp resonance at $2\omega_1$ and the non-resonant $\chi^{(3)E}(-\omega_3,\omega_1,\omega_1,-\omega_2)$ has been observed and reported[8] for CuCl. This resonance is due to a longitudinal exciton branch. The signal is qualitatively similar to the Raman-interference curves. It should be emphasized that the exciton is generated in the bulk of the crystal, many linear absorption depths away from the nearest boundary, by second harmonic interaction with a near-infrared dye laser beam at ω_1. The observed frequency ω_3 is also in the near-infrared, but it displays characteristics of the strongly absorbing UV excitation. The exciton frequency damping has been studied in detail as a function of temperature by this nonlinear method, as is reported elsewhere in this volume.[9] The investigation of CuCl with two infrared dye lasers permits the investigation of a double resonance interference phenomenon, in which $2\omega_1$ is tuned through the exciton resonance and simultaneously $\omega_1 - \omega_2$ is tuned through the infrared polariton dispersion characteristic.

4. CONCLUSION

Nonlinear parametric optical mixing processes, in which only laser beams with frequencies in the transparent region of the samples are utilized, permit the detailed investigation of dispersion characteristics in strongly absorbing regions of the spectrum. Nonlinear susceptibility tensor elements, describing the nonlinear properties of crystalline excitations, can be measured over a wide range of frequencies with sufficient precision that the terminology "Nonlinear Spectroscopy of Crystals" is justified. This field of endeavor is still in its infancy, but has good prospects for further growth.

REFERENCES

1. N. Bloembergen, Nonlinear Optics, Benjamin, New York, 1965.

2. P. D. Maker and R. W. Terhune, Phys. Rev. 137, A 801 (1965).

3. M. D. Levenson and N. Bloembergen, Phys. Rev. B 10, 4447 (1974).

4. R. T. Lynch, M. D. Levenson and N. Bloembergen, Phys. Letters 50A, 61 (1974).

5. D. Bedeaux and N. Bloembergen, Physics 69, 57 (1973).

6. E. Yablonovitch, Chr. Flytzanis and N. Bloembergen, Phys. Rev. Letters 29, 865 (1972).

7. Chr. Flytzanis, "Infrared Dispersion of Third Order Susceptibilities in Dielectrics", paper presented at the Taormina Conference, September 1972; Chr. Flytzanis and N. Bloembergen, submitted for publication in "Progress in Quantum Electronics".

8. S. D. Kramer, F. G. Parsons and N. Bloembergen, Phys. Rev. B 9, 1853 (1974).

9. S. D. Kramer and N. Bloembergen, Proceedings of the Conference on Highly Transparent Solids.

HF AND CO_2 LASER MEASUREMENTS OF DISPERSION OF THE NONLINEAR

SUSCEPTIBILITY IN ZINC-BLENDE CRYSTALS

J. A. WEISS

NAVAL RESEARCH LABORATORY

WASHINGTON, D. C. 20375

Measurements have been made of the nonlinear susceptibility for second harmonic generation ($\chi^{(2)}$) of several zinc-blende crystals using pulsed HF and CO_2 transverse-excitation lasers. This study was undertaken to determine the existence of dispersion in $\chi^{(2)}$ in regions of crystal transparency. Previous measurements of dispersion in $\chi^{(2)}$ have been restricted to absorbing regions because of a lack of suitable laser sources in the infrared. Analysis of the experimental data indicates increases in the magnitude of $\chi^{(2)}$ from 10.2 μm to 2.87 μm which greatly exceed those predicted by theoretical calculations of the dispersion of the nonlinear susceptibility in crystals with this structure. Single crystals of GaAs, CdTe and ZnTe were used in this experiment with the ZnTe crystal being used as a reference sample. Increases in $\chi^{(2)}$ on the order of 140% for GaAs and 70% for CdTe have been measured.

I. INTRODUCTION

Dispersion measurements of the modulus of the nonlinear optical susceptibility for second harmonic generation $|\chi^{(2)}(2\omega)|$ in zinc-blende crystals have been reported by several groups.[1-4] Although the results of these experiments have differed to some degree there is sufficient evidence to indicate that the pronounced structure measured can be correlated with the presence of critical points in the joint density of states of the lowest conduction and highest valence bands. Contributions from critical points to the structure appear to occur when the critical point energy is approximately equal to either the fundamental or the harmonic photon energy.

All but one of the previously reported studies have used laser sources of sufficiently high photon energy that the harmonic, and often both the fundamental and the harmonic, photons are totally absorbed. This is consistent with the expectation that most, if not all, of the structural features in the $|\chi^{(2)}|$ dispersion spectra would occur at harmonic photon energies greater than the band-gap energy E_o. However, calculations by Bell[5,6] of the dispersion of the nonlinear susceptibility using critical point contributions indicate that some dispersion should also occur at energies of approximately $E_o/2$. Although the magnitude of the expected increase in $|\chi^{(2)}|$ at this energy is less than that reported at higher energies, the effects of this critical point should be measurable for energies at which both the fundamental and harmonic photons are transmitted. Dispersion in this energy region can only be explained by a quantum treatment of the nonlinearity since classical anharmonic oscillator analysis does not predict an increase in $|\chi^{(2)}|$ until the photon energy is approximately equal to E_o.

The fact that there had been no experimental investigation of the nature and magnitude of the predicted $E_o/2$ dispersion prompted us to undertake to measure relative second harmonic generation (SHG) in several zinc-blende crystals using transverse excitation HF and CO_2 lasers. For most zinc-blende materials the CO_2 laser photon energy (.11-.13 ev) is sufficiently small that the measured $|\chi^{(2)}|$ can be considered to be a low frequency limiting value, where low frequency in this case is defined to be above all lattice resonance frequencies but far removed from any electronic resonances. In this limit the dispersion curve of $|\chi^{(2)}|$ is featureless and the magnitude may be estimated with reasonable accuracy using static perturbation theory. The small degree of dispersion in this region agrees well with a simple anharmonic oscillator fit. At the HF photon energy (.44 ev), however, crystals such as CdTe and GaAs with band gaps at about 1.43 ev could be expected to exhibit increases in $|\chi^{(2)}|$ due to contributions at $E_o/2$ from the critical point at E_o.

In the present experiment we report measurements of $|\chi^{(2)}(2\omega)|$ in GaAs, CdTe and ZnTe using HF and CO_2 lasers. The data presented here indicates that the magnitude of $\chi^{(2)}$ in GaAs and CdTe increases at the HF photon energy relative to that measured using the CO_2 laser (> 70% for CdTe, > 140% for GaAs). An increase of this magnitude is considerably greater than that predicted by Bell's calculations and possible explanations are discussed.

II. EXPERIMENTAL

A. Laser sources and detection apparatus

The experimental arrangement for the present study is shown in Figure 1. The CO_2 and HF lasers have been described

previously.[7,8] Each of the nonlinear crystal samples were cut
and polished as wedges with apex angles between 3-5°. The fundamental laser radiation was focused onto the wedge sample with a
25 cm BaF$_2$ lens, a portion being deflected by a BaF$_2$ beam splitter
onto a Ge:Au detector which served to monitor the input laser peak
power. After exiting the nonlinear crystal the second harmonic
and fundamental beams were collimated by a 7.5 cm BaF$_2$ lens and
allowed to spatially separate over a 25 cm path. The harmonic signal was detected by a second Ge:Au detector onto which the beam was
focused by a 2.5 cm BaF$_2$ lens. This second detector could be used
to determine the exit angles of the second harmonic and fundamental
beams in order to measure the indices of refraction at the two wavelengths.

Because of the small wedge angles the beams did not separate
sufficiently over the allowed path to permit detection of the weaker
second harmonic without filtering. This presented no problem when
using the CO$_2$ laser, since a 1/8" sapphire plate completely blocked
the fundamental radiation. The Ge:Au detector is also two orders
of magnitude more sensitive at the 5.1 µm harmonic wavelength of
CO$_2$ than at the fundamental. However, with the HF laser portion
of the experiment no single filter material was readily available.
A dielectric coated shortpass filter provided four orders of magnitude rejection of the fundamental, but this was not sufficient
due to the low signal level of the 1.435 µm harmonic. After investigating several crystals and liquids it was determined that 1 cm
optical cell filled with acetone provided an additional 10^3 rejection of the fundamental while only slightly attenuating the harmonic. The compound filter, consisting of the dielectric filter and
the acetone cell, was used throughout the HF experiments.

As is characteristic of TE lasers, both the HF and CO$_2$ lasers
produced considerable electrical interference which became im-
pressed onto the signal detection electronics. This problem is
considerably more severe with HF than with CO$_2$ lasers since the HF
laser pulse follows the discharge by only a few hundred nanoseconds. The detection electronics were therefore isolated from
the laser noise source by enclosure in a copper Faraday cage with
filtered power line inputs. The Ge:Au detectors remained outside
the cage but signals from the detectors were transmitted to the
electronics along double-shielded coaxial cable enclosed in grounded
copper braid.

The lasers exhibited considerable pulse to pulse jitter as
well as long term (~10 min) drifts in out power due to pressure
variations, line voltage fluctuations, heating, etc. The shot-to-
shot amplitude fluctuations of the HF laser were about ± 5%, while
for the CO$_2$ laser they were on the order of ± 10%. A dual-channel
signal averaging system was used in order to compensate for these
long and short term instabilities. The second harmonic and fundamental detector signals were fed through amplifiers into a dual-

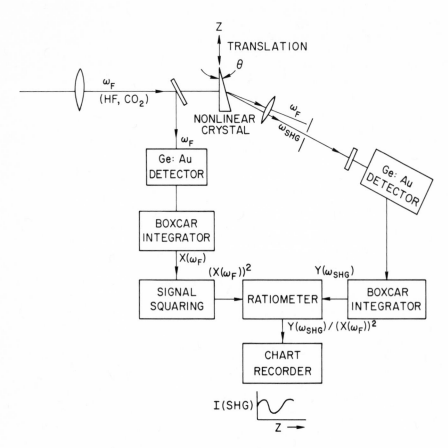

FIGURE 1.   Schematic of experimental arrangement

channel boxcar integrator (Molectron Model LSDS). The integrators of both channels were double-gated to correct for baseline shifts and synchronous noise closely following the signal. The output of the fundamental wavelength monitor was sent into a multiplier unit which squared the signal. Both the second harmonic and squared fundamental signals were then ratioed in a nulling ratiometer (HP 3420B). The output of the integrator-ratiometer system was found to be stable to ± 3% over tens of minutes of operation. More significantly, the output remained constant over changes in laser intensity of more than an order of magnitude, which corresponds to changes in second harmonic signal strength of over two orders of magnitude.

B.   The wedge technique for SHG measurement

The use of wedged samples of nonlinear crystals to avoid multiple reflection effects and simplify spatial beam separation in SHG experiments has become a standard technique since having been introduced by Wynne and Bloembergen[9] in 1969. The usefulness of the technique was extended by Boyd, Kasper and McFee (BKM)[10] by their treatment of the effects of absorption and finite beam size on SHG data. Sherman[11] further refined the BKM calculation by considering together absorption at both fundamental and harmonic wavelengths and beam size corrections.

For a wedge of thickness ℓ_o at a point x_o the thickness ℓ of the wedge at a point $x = x_o + x_1$ is given by $\ell(x) = \ell_o + x_1 \tan\theta$ where θ is the apex angle of the wedge. If the incident field E_1 has a Gaussian distribution with beam radius w_1 then the intensity of the fundamental field at a point (x,y) on the crystal surface is given by

$$S_1(x,y) = \frac{n_1}{2} \sqrt{\frac{\varepsilon_o}{\mu_o}} E_1^2 e^{-2(x^2+y^2)/w_1^2} \tag{1}$$

The power of the fundamental field is given by the integral of S_1 in the (x,y) plane and in terms of S_1 is

$$P_1 = \frac{\pi}{2} w_1^2 S_1 e^{2(x^2+y^2)/w_1^2} \tag{2}$$

From BKM and Sherman we have for the plane wave case that the intensity of the harmonic field in terms of the fundamental intensity in the crystal is given by

$$S_2(x,y) = \pi K S_1^2 \ell_c^2 e^{-\alpha_2 \ell} \left[\frac{(1-e^{-\alpha\ell})^2 + 4e^{-\alpha\ell} \sin^2\left[\frac{\pi}{2}\frac{\ell}{\ell_c}\right]}{\alpha^2 \ell_c^2 + \pi^2} \right] \tag{3}$$

where $K = \sqrt{\frac{\mu_o}{\varepsilon_o}} \cdot \frac{\omega_2^2 d_{eff}^2}{2\pi n_1^2 n_2 c^2}$

and $d = \frac{1}{2}\chi^{(2)}$, $\ell_c = \frac{\pi}{\Delta k}$, $\alpha = \alpha_1 - \frac{\alpha_2}{2}$ (α_1, α_2 being the power losses at the fundamental and harmonic, respectively.) Jerphagnon and Kurtz[12] have shown that (3) is valid for case of Gaussian beams at normal incidence to a nonlinear crystal. We may use (1) and (2) in (3) to write an expression for the harmonic field intensity at (x,y) in terms of the fundamental power.

Thus

$$S_2(x,y) = \frac{4}{\pi w_1^2} K P_1^2 \ell_c^2 e^{-4(x^2+y^2)/w_1^2} \left[\frac{(1-e^{-\alpha\ell})^2+4e^{-\alpha\ell}\sin^2\phi}{\alpha^2\ell_c^2+\pi^2}\right]$$
$$\cdot e^{-\alpha_2\ell}$$

(4)

where $\phi = \frac{\pi}{2\ell_c} \ell(x)$.

The second harmonic power is calculated by integrating (4) over the (x,y) plane with the resulting expression for P_2 being given by[11]

$$P_2 = \frac{K}{w_1^2} \cdot P_1^2 \ell_c^2 e^{-\alpha_2\ell} \left[\frac{(1-e^{-\alpha\ell})^2+2e^{-\alpha\ell}(1-e^{-\pi^2\eta^2/16}\cos2\phi)}{\alpha^2\ell_c^2+\pi^2}\right]$$

(5)

with $\eta = \frac{w_1}{\ell_c} \tan\theta$.

To relate (5) to experimentally observed second harmonic powers, it is necessary to correct P_2 to allow for reflection losses of the fundamental and the harmonic fields. The second harmonic power transmitted out of the crystal in terms of the incident fundamental power is calculated to be

$$P_{2,ext} = \sqrt{\frac{\mu_o}{\epsilon_o}} \frac{w_2^2 d_{eff}^2 \ell_c^2}{2w_1^2\pi} \left(\frac{1}{n_1+1}\right)^2 \left(\frac{1}{n_2+1}\right) F \cdot P_{1,ext}^2$$

(6)

where F is a loss term which includes all corrections for absorption. Letting $\cos2\phi = -1$ in (5) we have that at a maximum of second harmonic power

$$F = e^{-\alpha_2\ell} \left[\frac{(1-e^{-\alpha\ell})^2+2e^{-\alpha\ell}(1+e^{-\pi^2\eta^2/16})}{\alpha^2\ell_c^2+\pi^2}\right]$$

(7)

The relative nonlinear susceptibilities of two crystals A and B in terms of the externally measured second harmonic powers can then be expressed as

$$\frac{d_B}{d_A} = \left(\frac{P_{2,ext}^B}{P_{2,ext}^A}\right)^{\frac{1}{2}} \left(\frac{\ell_c^A}{\ell_c^B}\right) \left(\frac{F_A}{F_B}\right)^{\frac{1}{2}} \left(\frac{n_1^B+1}{n_1^A+1}\right) \left(\frac{n_2^B+1}{n_2^A+1}\right)$$

(8)

This expression reduces to (35) of BKM for the case of $\alpha\ell \ll 1$.

C. Nonlinear crystals

The choice of nonlinear materials for this experiment was dictated by the requirement that the crystals have zinc-blende structure and be transparent at the fundamental and harmonic wavelengths of both the HF and CO_2 lasers. It was necessary that single crystals be available in order to obtain specific sample orientation for determination of the nonlinear susceptibility coefficients. Single crystals of GaAs, CdTe and ZnTe were used as samples. The ZnTe was chosen as the reference material since the large band gap energy (2.25 ev) could be expected to lead to minimal change in the magnitude of the susceptibility between the wavelengths used in this experiment. Measurements of $|\chi^{(2)}|$ in CdTe at HF and CO_2 laser wavelengths had indicated the presence of an increase in magnitude at shorter wavelengths but the size of the increase was difficult to determine accurately due to the lack of a reference material. [13]

The GaAs and CdTe crystals were oriented so that the wavevector of the incident fundamental beam was normal to a {110} crystal plane. The laser was polarized with the electric field vector parallel to a [110] direction lying in the plane of the entrance face. Zinc-blende crystals ($\overline{4}$3m point group symmetry) have only one independent, non-zero element of the susceptibility tensor for second harmonic generation, d_{14}. For the crystal orientation described the nonlinear polarization will then be parallel to a [100] direction in the plane of the entrance face and the effective nonlinearity for harmonic generation is $d_{eff} = d_{14}$.

ZnTe could not be obtained in large single crystals with specific orientation; thus, the orientation of our ZnTe wedge was determined by X-ray diffraction after cutting and polishing. From the X-ray data the effective nonlinear susceptibility was then calculated. With the laser beam propagating in a direction approximately 18° from a [111] direction, d_{eff} was calculated to be $d_{eff} = 0.65d_{14}$.

Data on the absorption and index of refraction of the nonlinear crystals is given in Table I. Absorption measurements at the fundamental wavelengths were made using the lasers and the signal averaging system, correcting for reflection losses. Absorption at the harmonic wavelengths was determined using Cary and Perkin-Elmer spectrophotometers and was then correlated with the data obtained at the fundamental wavelengths. The coherence lengths were measured by translating the crystals normal to the laser beam and measuring the distance between adjacent harmonic minima. Coherence lengths calculated from available index data are also given. For the ZnTe

crystal the strong absorption of the CO_2 laser beam made it impossible to directly measure the coherence length at 10.2 μm.
The length was calculated from the index of refraction data obtained by measuring the exit angles of the fundamental and harmonic beams. The error in determining the coherence length in
this manner is proportional to the error in measuring $\Delta n = n_1 - n_2$,
thus the accuracy of this calculation may exceed the precision of
the index of refraction measurement since systematic errors will
cancel.

The GaAs was obtained from Laser Diode Laboratories, Inc. and
was undoped, with a resistivity of $\rho = .086\Omega$-cm. The CdTe was
bought from II-VI, Inc. and was electrooptic grade with $\rho = >10^8$ Ω
cm. The ZnTe was an undoped low resistivity test sample obtained
from Gould.

III. RESULTS

Using the data from Table I and (8) the following relationships can be calculated:

At 10.2 μm

$$\frac{d_{14}(\text{GaAs})}{d_{14}(\text{ZnTe})} = 1.57 \left(\frac{P_2(\text{GaAs})}{P_2(\text{ZnTe})}\right)^{\frac{1}{2}}$$

$$\frac{d_{14}(\text{CdTe})}{d_{14}(\text{ZnTe})} = 2.48 \left(\frac{P_2(\text{CdTe})}{P_2(\text{ZnTe})}\right)^{\frac{1}{2}} \tag{9a}$$

and at 2.87 μm

$$\frac{d_{14}(\text{GaAs})}{d_{14}(\text{ZnTe})} = 1.09 \left(\frac{P_2(\text{GaAs})}{P_2(\text{ZnTe})}\right)^{\frac{1}{2}}$$

$$\frac{d_{14}(\text{CdTe})}{d_{14}(\text{ZnTe})} = 0.64 \left(\frac{P_2(\text{GaAs})}{P_2(\text{ZnTe})}\right)^{\frac{1}{2}} \tag{9b}$$

Substituting the measured second harmonic powers into (9a)
and (9b) one can calculate the magnitudes of the nonlinear susceptibilities at the experimental wavelengths. These magnitudes are
given in Table II along with estimates of the predicted increases
at the shorter wavelength calculated using Bell's three-band theory
as well as the increase calculated using a classical anharmonic
oscillator analysis. In calculating the ratios of the susceptibilities from the experimentally measured powers we have corrected
for the fact that the ZnTe reference crystal should exhibit approximately a 5% increase in the magnitude of d_{14} at 2.87 μm relative
to its value at 10.2 μm. The overall accuracy of the data at each
wavelength is approximately ± 20% including possible errors in
laser power measurement, coherence length and index of refraction.

TABLE I. Absorption and Index of Refraction Data for the Nonlinear Crystals used in the Present Experiment.

	GaAs	CdTe	ZnTe
Wedge Angle θ	3.2°	4.5°	4.7°
α$_{10.2}$ μm	1.63 cm^{-1}	<.005 cm^{-1}	29.7 cm^{-1}
α$_{5.1}$ μm	0.53 cm^{-1}	<.005 cm^{-1}	27.5 cm^{-1}
α$_{2.87}$ μm	0.53 cm^{-1}	<.005 cm^{-1}	5.65 cm^{-1}
α$_{1.435}$ μm	1.944 cm^{-1}	<.01 cm^{-1}	5.65 cm^{-1}
$\ell_c^{10.2}$ (expt)	96 ± 5 μm	184 ± 10 μm	77 ± 5 μm
$\ell_c^{10.2}$ (calc)		202 μm	
$\ell_c^{2.87}$ (expt)	9.2 ± 5 μm	13.1 ± .5 μm	19.5 ± 1 μm
$\ell_c^{2.87}$ (calc)	10.25 μm	13.8 μm	23.0 μm
n$_{10.2}$	3.27	2.671 (17)	2.69
n$_{5.1}$	3.30	2.684 (17)	2.71
n$_{2.87}$	3.33 (16)	2.696 (17)	2.72 (18)
n$_{1.435}$	3.40 (16)	2.748 (17)	2.74 (18)

TABLE II.   Summary of Results

	Relative d_{14} at 10.2 μm	Relative d_{14} at 2.87 μm*	PERCENTAGE INCREASE		
			Experi- mental	Bell Model	Anharmonic Oscillator Model
GaAs	1.30	3.12	140	~15	7
CdTe	.98	1.70	73	~20	4
ZnTe	1.00	1.05	5	5	3

*Corrected for increase in susceptibility in ZnTe reference at
 this wavelength.

Thus, the accuracy of the measured ratios $d^{2.87}/d^{10.2}$ is approximately ± 40%.

The ratio of the susceptibilities of GaAs and CdTe measured
at 10.2 μm in this experiment is

$$\frac{d_{14}(\text{GaAs})}{d_{14}(\text{CdTe})} = 1.32 \pm 15\% \tag{10}$$

which differs by only two percent from the 1.35 value calculated
by Bell using an anharmonic oscillator model in the low-frequency
limit (6). The magnitude of the nonlinear susceptibility can be calculated with reasonable accuracy at these wavelengths using static
perturbation theory and classical methods. In this low-frequency
limit the differences in results between various methods of calculation tend to diminish. For example, Phillips and Van Vechtan[14]
have calculated a value of unity for (10). The only report of an
experimentally determined value for this ratio was given by Patel[15]
as 0.8. However, the accuracy of this data has been questioned
previously by Sherman since it appeared to yield too large a value
of $d_{14}(\text{CdTe})$.

Examining the ratio of susceptibilities of GaAs and ZnTe at
10.2 μm, the measured value from the present experiment is 1.30
compared with a value of 1.60 calculated by Bell.[6] This 23%
difference is within the combined experimental and theoretical
errors.

IV. DISCUSSION

Bell's three band model[5] of the frequency dependence of $|\chi^{(2)}(2\omega)|$ was originally developed to explain the experimental data of Chang, Ducuing and Bloembergen (CDB).[1] They had measured $|\chi^{(2)}|$ for several zinc-blende crystals in the fundamental absorption region using a series of discrete-frequency laser sources. Later experiments by Parsons and Chang (PC)[2] performed using tunable lasers resulted in significantly different data and necessitated modification of the original calculations.[6] The PC data was explainable by the revised calculations except for the measurements of $\chi^{(2)}$ in GaAs. It was presumed that the PC data was more accurate than that from the CDB experiment since the mode quality of the lasers used in the earlier experiment was uncertain. However, recent experimental results by Lotem et al.[4] are in considerable disagreement with the PC and CDB data specifically for the case of GaAs. These new results have cast doubt on the reliability of all the experimental data to date.

A. GaAs

The principle disagreement between the three sets of experimental data and between all the experimental data and theory is the position of the peak in the $|\chi^{(2)}|$ dispersion curve between 1.2 and 1.6 ev. Such a peak would presumably be due to the E_0 critical point at 1.43 ev. The CDB data placed this peak at approximately 1.6 ev while PC showed it at 1.28 ev and Lotem at about 1.43 ev. Although one might be inclined to accept the latter result since it does agree with the position of the critical point, the conflict between reported data makes this assumption uncertain.

The results of the present experiment yield a relative increase of 140% for $|\chi^{(2)}|$ in GaAs relative to the value at 10.2 μm. Bell's original calculations for this material indicate an increase of about 15% over this wavelength range. It had been suggested[6] that the results obtained for GaAs by PC were unexplainable by present calculations because the almost exact coincidence of the critical point energies E_0 and $E_1/2$ might invalidate the necessary calculational assumption that at every photon energy the behavior of $\chi^{(2)}$ is dominated by only one critical point.

It may be possible that this coincidence of critical point energies leads to an enhancement of the dispersion of $\chi^{(2)}$ due to the critical point at E_0. To resolve the present conflicting experimental results, however, it is clear that a single experiment using a tunable source is necessary. Lotem has noted the effects of sample preparation on experimental results and this has highlighted the difficulty in comparing data from several different experiments.

B. CdTe

The data from the present experiment shows an increase in $|\chi^{(2)}|$ of about 73% at 2.87 μm relative to the measured value at 10.2 μm.   Although the increase is considerably larger than the approximately 20% that would be expected using the Bell calculations the overall uncertainty of ± 40% in the measured ratio $d^{2.87}/d^{10.2}$ makes the discrepancy between theory and experiment considerably less severe than in the case of GaAs. Again, additional experimentation is needed to determine if the measured increase is, in fact, due to a structural feature in the $|\chi^{(2)}|$ dispersion curve at $E_0/2$. Further work would also determine if the behavior of $\chi^{(2)}$ in CdTe is better explainable by theory than that in GaAs. At present only GaAs appears to have anomolous behavior but the conflicting experimental data makes this determination difficult.

V. CONCLUSION

Thus, we have established the presence of dispersion in the lowest order nonlinear susceptibility in GaAs and CdTe. The magnitude of the dispersion in both materials exceeds the theoretically calculated magnitudes and indicates that further experimentation and calculation is needed. Second harmonic measurements using broadly tunable laser sources now available should be able to resolve the present conflicting experimental results.

VI. ACKNOWLEDGEMENTS

The author wishes to thank Dr. Lawrence S. Goldberg of NRL for numerous helpful discussions on this research. This work was submitted in partial fulfillment of the requirements for the Ph.D. in Physics at Harvard University.

REFERENCES:

1. R.K. Chang, J. Ducuing and N. Bloembergen, Physical Review Letters, 15, 415 (1965).
2. F.G. Parsons and R.K. Chang, Optics Communications, 3,173 (1971).
3. J.J. Wynne, Physical Review Letters, 27, 17 (1971).
4. H. Lotem, G. Koren and Y. Yacoby, Physical Review B, 9, 3532 (1974).
5. M.I. Bell, in Electronic Density of States, edited by L.H. Bennett, National Bureau of Standards Special Publication 323 (1971).
6. M.I. Bell, Physical Review B, 6, 516 (1972).
7. J.A. Weiss and L.S. Goldberg, IEEE Journal of Quantum Electronics, QE-8, 757 (1972).

8. J.A. Weiss and L.S. Goldberg, Applied Physics Letters, 24, 389 (1974).
9. J.J. Wynne and N. Bloembergen, Physical Review, 188, 1211 (1969).
10. G.D. Boyd, H. Kasper and J.H. McFee, IEEE Journal of Quantum Electronics, QE-7, 567 (1971).
11. G.H. Sherman, Ph.D. Thesis, University of Illinois, 1972 (unpublished).
12. J. Jerphagnon and S.K. Kurtz, Journal of Applied Physics, 41, 1667 (1970).
13. J.A. Weiss and L.S. Goldberg, presented at the meeting of the Optical Society of America, Denver, Colorado, March, 1973.
14. J.C. Phillips and J.A. vanVechtan, Physical Review, 183, 709 (1969).
15. C.K.N. Patel, Physical Review Letters, 16, 613 (1966).
16. B.O. Seraphin and H.E. Bennett in Semiconductors and Semi-metals, edited by R.K. Willardson and A.C. Beer, Academic Press, 1967.
17. L.S. Ladd, Infrared Physics, 6, 145 (1966).
18. D.T.F. Marple, Journal of the Optical Society of America, 35, 539 (1964).

MULTIPHOTON IONIZATION PROBABILITY AND NONLINEAR

ABSORPTION OF LIGHT BY TRANSPARENT SOLIDS

S. S. Mitra[*], L. M. Narducci[**], R. A. Shatas
Quantum Physics, Physical Sciences, U. S. Army
Missile Command, Redstone Arsenal, AL 35809

and

Y. F. Tsay[+] and A. Vaidyanathan[++]
University of Rhode Island, Kingston, R.I. 02881

ABSTRACT

The multiphoton ionization probability has
been calculated as functions of wavelength and
electric field intensity for a number of semiconductors and insulators using three different theoretical models. Although there is a
large amount of disagreement among available
experimental data, the values of two-photon absorption coefficients calculated by Keldysh
treatment come closest, while the Basov formula
over-estimates and the Braunstein formula underestimates. Surprisingly, Keldysh formula also
predicts the absolute values and the wavelength
dependence of the one-photon absorption coefficient (absorption edge) extremely well for direct gap semiconductors.

[*]Permanent address: Department of Electrical Engineering
University of Rhode Island, Kingston, R. I. 02881.
[**]Permanent address: Physics Department, Worcester Poly-
technical Institute, Worcester, Mass. 01609.
[+]Department of Electrical Engineering; Present address:
NRC Post-doctoral Resident Research Associate at AFCRL
Hanscom AFB, Mass. 01731.
[++]Department of Physics.

I. INTRODUCTION

Because of the availability of high power laser sources, increasing amount of interest is evidenced in the measurement and understanding of multi-photon processes in transparent materials. Fundamental applications of such investigations in solids are many including the determination of non-linear susceptibility, band perturbation under strong illumination (dynamic Franz-Keldysh effect), parity and selection rule assignments, band structure analysis in semiconductors, etc. Among technical applications of non-linear optics reference may be made of uniform pumping in semiconductor lasers (rather than pumping over skin depth) and pulse shaping and optical limiting, i.e., saturable absorption in reverse.

Most measurements in this area have been confined to two-photon absorption coefficient of direct gap semiconductors. This paper concerns itself primarily with such results. We have considered a few commonly used theoretical models for direct transitions in order to correct certain controversial errors, compare the theoretical results with one another and apprise the relative merits of these theories by reviewing a number of available experimental results. A summary of the most frequently quoted theories of two- and multi-photon direct transitions are given in Table I.

II. THEORETICAL EXPRESSIONS

Keldysh Theory

The electronic transition rate per unit volume is given by

$$W(E) = \frac{2}{9\pi} \omega \left(\frac{m^*\omega}{\hbar}\right)^{3/2} \Phi \left[(2\langle x+1\rangle - 2x)^{1/2}\right] \times$$

$$\left(\frac{e^2E^2}{16m^*\omega^2 E_g}\right)^{\langle x+1\rangle} \exp\left[2\langle x+1\rangle\left(1 - \frac{e^2E^2}{4m^*\omega^2 E_g}\right)\right]$$

$$(1)$$

where the Dawson integral Φ is given by

$$\Phi(z) = e^{-z^2} \int_0^z e^{y^2} dy \qquad (2)$$

TABLE I. Theoretical Models for Multi-photon Processes*

	Models	Calculation scheme	Band structure
Keldysh[1]	two bands	transitions between perturbed bands	$E(k) = E_g\left(1 + \dfrac{\hbar^2 k^2}{m^* E_g}\right)^{1/2}$
Braunstein[2]	two bands	second order perturb. theory through higher conduction band	parabolic isotropic $E(k) \propto k^2$
Basov[3]	three bands	second order perturb. theory	parabolic isotropic $E(k) \propto k^2$
Hassan[4]	two bands	second order perturb. theory through higher conduction band	parabolic and anisotropic $E(k) \propto A(k_x^2 + k_y^2) + Bk_z^2$

*References are given at the end of the paper.

and

$$x = \frac{E_g}{\hbar\omega} \left(1 + \frac{e^2 E^2}{4m^* \omega^2 E_g}\right) \tag{3}$$

<x+1> represents the integer part of the argument. In the above eqs., ω is the angular frequency of light, E the electric field envelope, m^* the reduced electron-hole mass of the semiconductor and E_g the direct band gap.

The two-photon absorption coefficient, $\alpha^{(2)}$ is defined by

$$\frac{dI}{d\ell} = -\alpha^{(2)} I^2 \tag{4}$$

where I is the light intensity and ℓ the length traversed in the solid. In terms of the transition rate, $\alpha^{(2)}$ is given by

$$\alpha^{(2)} = \frac{4\hbar\omega (2R_0)^2 W(E)}{\varepsilon_\infty E^4} \tag{5}$$

where ε_∞ is the high frequency dielectric constant of the crystal, R_0 is the vacuum impedance (377Ω), and the factor four in the numerator comes from spin degeneracy and photon multiplicity.

Braunstein Model

Braunstein[2] uses the second order perturbation theory with the intermediate band located above the conduction band. All bands are parabolic, and the scheme is shown in Fig. 1. After some numerical corrections, his expression for $\alpha^{(2)}$ is

$$\alpha^{(2)} = \frac{\sqrt{2} \pi e^4}{c^2 \hbar \omega n^2} \cdot \frac{(\Delta E - E_g) \Delta E}{(\hbar\omega)^2} \cdot \frac{(2\hbar\omega - E_g)^{1/2}}{\sqrt{m^*}}$$

$$\times \left[\Delta E - \hbar\omega + \frac{\alpha_n + \alpha_v}{\alpha_c + \alpha_v} (2\hbar\omega - E_g)\right]^{-2} \tag{6}$$

where n is the index of refraction, ΔE the energy difference between the intermediate (n) and the valence band,

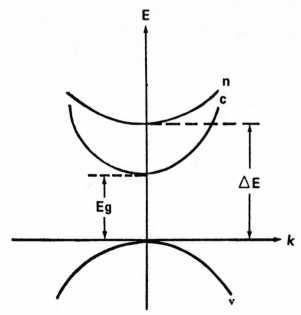

Fig. 1   The energy band model used in Braunstein's calculation.

E_g the band gap energy and $\alpha_i(v,c,n) \equiv \dfrac{m_i^*}{m}$ are the effective mass ratios. It is to be noted that the eq. (6) is in c.g.s. electrostatic units and is specialized to $\omega_1 = \omega_2$, i.e., both the photons having the same frequency. The formula includes the spin and the photon degeneracy factors, and an index of refraction correction has also been made.

Basov Model

Basov gave expressions for $\alpha^{(2)}$ using both one and two valence bands, using parabolic bands. The difference between the calculated values of $\alpha^{(2)}$ using two or one valence band is slight. Here we quote the result specialized to one valence band, and after appropriate corrections:

$$\alpha^{(2)} = \frac{2^8 \sqrt{2} \; \pi e^4 \; (2\hbar\omega - E_g)^{3/2}}{n^2 c^2 \; (\hbar\omega)^5} \cdot \frac{E_g}{\sqrt{m^*}} \tag{7}$$

Since, $\alpha^{(2)}$ is independent of the field intensity, the controversy on whether to use peak or r.m.s. value of field is irrelevant.

III. RESULTS AND DISCUSSION

Two-photon absorption

The relevant semiconductor parameters used in this calculation are given in Table II. The theoretical values of $\alpha^{(2)}$ calculated by the three models are given in Table III. It is interesting to note that the Braunstein formula gives the lowest values whereas the Basov formula the highest - about three orders of magnitude higher. The Keldysh results are intermediate. In Table IV we quote the available experimental results for a number of direct gap semiconductors. It should be noted that in some cases the experimental results vary almost by an order of magnitude. A comparison of Table III and IV reveals that in general the Braunstein formula underestimates, and the Basov formula somewhat over-estimates the values of $\alpha^{(2)}$. This discrepancy is not at all surprising because of the following facts: (i) the experimental results need be much more reliable so that they agree with each other. (ii) the three theoretical models considered here are all approximate in so far as they use parabolic band structure. More realistic models might improve the agreement. (iii) a realistic estimation of the oscillator strengths should be made using the appropriate wave functions of the bands involved. And (iv) the effect of the field on the band gap should be explicitly considered, i.e., a dynamic (high frequency a.c.) field may reduce the band gap thus changing the order of the photon process involved in a transition. All these questions are right now being considered.

TABLE II. Experimental Parameters used in the Calculations

Crystal	E_g (eV)	m_e^*	m_h^*	ε_∞	ΔE (eV)	λ (μm)
GaAs	1.435(300°K) (St)	0.07(300°K) (L)	0.68(0°K) (L)	10.9 (K)	4.2 (G)	1.06 1.318
InSb	0.288(77°K) (F)	0.0145(77°K) (F)	0.39(77°K) (F)	15.68 (K)	3.4 (G)	10.6
InP	1.28(300°K) (P)	0.073(300°K) (L)	0.4 (P)	9.56 (K)	4.2 (G)	1.06
CdS	2.53(300°K) (P)	0.2(0°K) (P)	5(0°K) (L) ∥ to c-axis	5.32 (K)	6.2 (B)	0.694
ZnSe	2.58(300°K) (P)	0.17(300°K) (L)	0.6 (R)	5.90 (K)	6.7 (C)	0.694
CdTe	1.50(300°K) (P)	0.11(300°K) (P)	0.35 (P)	7.21 (K)	5.16 (G)	1.318
CdSe	1.74(300°K) (P)	0.13 (P)	2.5(calcul.) (P) ∥ to c-axis	6.1 (K)	----	1.06 1.318

St: M.D.Sturge, Phys. Rev. 127, 768 (1962).

L: D.Long, Energy bands in semiconductors, Interscience Publ., N.Y.,pp. 196,197.

K: E.Karlheuser in Polarons in Ionic Crystals and Polar Semiconductors, edited by J. T. Devreese, North Holland Publ. (1972).

G: D.L.Greenaway and G.Harbeke, Optical Properties and Band Structure of Semi-conductors, Pergamon Press (1968).

F: H.J. Fossum and D. B. Chang, Phys. Rev. 8B, 2842 (1973).

P: J. I. Pankove, Optical Processes in Semiconductors, Prentice Hall, Inc. (1971).

B: S. Bespalov, L. A. Kulevskii, V. P. Makarov, M. A. Prokhorov, A. A. Tikhonov, Sov. Phys. JETP 28, 77 (1969).

R: H. Rieck, Semiconductor lasers, MacDonald, London (1970).

C: T. Collins, in Polarons in Ionic Crystals and Polar Semiconductors, edited by J. T. Devreese, North Holland Publ. (1972).

TABLE III. Comparison of Theoretical Results: $\alpha^{(2)}[\text{cm/Mw}]$

Crystal	Wavelength in μm	Keldysh	Braunstein	Basov
GaAs	1.06	0.019	0.0067	1.41
GaAs	1.318	0.027	0.0083	1.52
InSb	10.6	2.1	0.72	17.88
InP	1.06	0.026	0.0084	1.84
CdS	0.694	0.0061	0.0022	0.436
ZnSe	0.694	0.0064	0.0022	0.448
CdTe	1.318	0.027	0.0096	1.80
CdSe	1.06	0.018		1.18
CdSe	1.318	0.017		0.99

One-photon absorption in the Keldysh Theory

We have discovered a curious fact, not realized by Keldysh, that his formula for the transition probability holds extremely well for photon energies in the range $\hbar\omega \lessgtr E_g$ as well as that in the one-photon transition region.

The usual perturbation calculation for one-photon absorption, $\alpha^{(1)}$ around the band gap predicts[5] an absorption edge of the form

$$\alpha^{(1)} \sim (\hbar\omega - E_g)^\gamma \tag{8}$$

The values of the exponent γ for various types of transitions are given in Table V.

It may be pointed out that the dependence of $\alpha^{(1)}$ on $(\hbar\omega - E_g)$ is not explicit functionally from Keldysh formula [eqs. (1) and (5)]. However when $\alpha^{(1)}$ versus $(\hbar\omega - E_g)$ is plotted for the semiconductors listed in Table I, one obtains $\gamma = 0.5$ above the band gap. We show the result for GaAs in Fig. 2 where the Keldysh calculation is compared with the experimental results (300K) of Moss and Hawkins[6]. It should be noted that in the calculation the following parameters (300K) were used:

TABLE IV. Selected Experimental Results of Two-photon
Absorption Coefficient

Crystal	$\alpha_{exp}^{(2)}$[cm/mw]	Reference
GaAs(1.06μ)	9	Basov et al, JETP 23, 366 (1966)
	5.6	Jayaraman, APL 20, 392 (1972)
	0.8	Arsen'ev, JETP 29, 413 (1969)
	0.3	Grasyuk, JETP letts. 17, 584 (1973)
	0.2	Oksman, Sov. Phys. Semic. 6, 629 (1972)
	0.36	Lee, APL 20, 18 (1972)
GaAs(1.318μ)	0.033	Kleinman, APL 23, 243 (1973)
	0.04	Ralston, APL 15, 164 (1969)
InSb(10.6μ)	0.256	Fossum, PR B8, 2842 (1973)
InP(1.06μ)	0.26	Lee, APL 28, 18 (1972)
CdS(0.694μ)	0.66	Braunstein, PR 134, A499 (1964)
	0.03	Arsen'ev, JETP 29, 413 (1969)
	0.07	Catalano, PR 9B, 707 (1974)
ZnSe(0.694μ)	0.04	Arsen'ev, JETP 29, 413 (1969)
CdTe(1.318μ)	0.2-0.3	Ralston, APL 15, 164 (1969)
CdSe(1.06μ)	0.95	Basov, J. Phys. Soc. Japan (Suppl.) 21, 277 (1966)
CdSe(1.318μ)	0.2-0.3	Ralston, APL 15, 164 (1969)

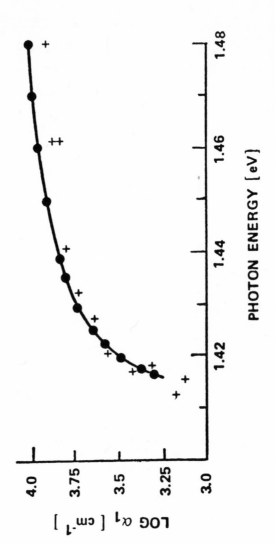

Fig. 2 Comparison of calculated and measured linear absorption constant log α_1[cm^{-1}] at the band edge of GaAs. ●: Keldysh model calculation without adjustable parameters. +: measurements taken from ref. 6.

TABLE V. Exponent γ of eq. (8) for Various Types of Transitions.

Y	Type of transition
0.5	Allowed direct
1.5	Forbidden direct
2.0	Indirect
0.5	Allowed indirect to exciton states

E_g = 1.413 eV and m^*/m = 0.065, and no fitting parameters were used.

As pointed out before that the Keldysh formulas do not explicitly give a band edge of the type given in eq. (8). However, a least square fit of the Keldysh results (Fig. 2) yields an empirical eq. above the band gap of the form

$$\alpha^{(1)} = 4.4 \times 10^4 \ (\hbar\omega - E_g)^\gamma (cm^{-1}) \tag{9}$$

where E_g = 1.413 eV and γ = 0.4992. We have also repeated least square fitting of Keldysh values using parameters recommended for GaAs by Sturge[7], and found

$$\alpha^{(1)} = 4.47 \times 10^4 \ (\hbar\omega - E_g)^\gamma (cm^{-1}) \tag{10}$$

where E_g = 1.435 eV and γ = 0.505.

REFERENCES

1. L. V. Keldysh, Sov. Phys. JETP 20, 1307 (1965).
2. R. Braunstein, Phys. Rev. 125, 475 (1962); R. Braunstein and N. Ockman, Phys. Rev. 134, A499 (1964).
3. N. G. Basov, A. Z. Grasyuk, I. G. Zubarev, V. A. Katulin and O. N. Krokhin, Sov. Phys. JETP 23, 366 (1966).
4. A. R. Hassan, Nuovo Cimento 70B, 21 (1970).

364   S. S. MITRA ET AL.

5.   See for example J. I. Pankove, "Optical Processes in Semiconductors", Prentice Hall, Englewood Cliffs, N. J. (1971), pp. 34-46.
6.   T. S. Moss and T. D. F. Hawkins, Infrared Phys. $\underline{1}$, 111 (1961).
7.   M. D. Sturge, Phys. Rev. $\underline{127}$, 768 (1962).

NONLINEAR SPECTROSCOPY OF EXCITONS IN CuCl[*]

S. D. Kramer and N. Bloembergen

Gordon McKay Laboratory

Harvard University, Cambridge, Massachusetts 02138

1. INTRODUCTION

The sharp Z_3 exciton level in CuCl which at liquid helium temperatures lies about 3.21 eV above the valence band has been investigated extensively by linear (one photon) absorption spectroscopy,[1-3] by second harmonic generation,[4] by two photon absorption spectroscopy,[5,6] and by four wave light mixing.[7] The first two methods require that ultra-violet radiation with an energy near that of the exciton be detected. Since the absorption depth in this wavelength region is about 10^{-5} cm, only excitons in the immediate vicinity of a surface can be probed. In general the surface of a CuCl crystal is apt to be chemically contaminated and physically imperfect. As a result, no reliable and precise data about exciton line shape and damping can be obtained with these methods.

Two photon absorption and four wave light mixing can both be described by various forms of the third order nonlinear optical susceptibility, $\chi^{(3)}$. In fact, the two photon absorption coefficient, $\alpha_2(\omega_1)$, at frequency ω_1, is proportional to the product of the imaginary part of $\chi^{(3)}(-\omega_1,\omega_1,\omega_1,-\omega_1)$ and the intensity at frequency ω_1, $I(\omega_1)$. As the absolute calibration of $I(\omega_1)$ is rather difficult, accurate determinations of the intrinsic two photon absorption parameter, $\alpha_2(\omega_1)/I(\omega_1)$, which depends only on the material

---

[*] Supported by the Joint Services Electronics Program.

being studied, are almost impossible. Attempts to calibrate the
fluorescence due to two photon absorption with the fluorescence
following one photon absorption are hindered by the difficulties
associated with the short absorption length of the one photon process.

The method of four wave light mixing, which has already been
reviewed elsewhere in these Proceedings,[8] provides a practical way
to obtain quantitative data on the variation of $\chi^{(3)}$ as a function
of frequency and temperature. By using incident waves at frequencies ω_1 and ω_2, it can be shown that the output beam at frequency
$\omega_3 = 2\omega_1 - \omega_2$ is proportional to the absolute square of the nonlinear polarization:

$$P(\omega_3) = \chi^{(3)}(-\omega_3,\omega_1,\omega_1,-\omega_2) \ E(\omega_1)E(\omega_1)E^*(\omega_2)$$

where

$$\chi^{(3)}(-\omega_3,\omega_1,\omega_1,-\omega_2) = A \left[1 + \frac{B\hbar^{-2}}{\omega_{ex}^2 - 4\omega_1^2 - 2i\omega_1\Gamma}\right] \tag{1}$$

consists of a non-resonant term and a term resonant when $2\hbar\omega_1$ is
equal to the exciton energy, $\hbar\,\omega_{ex}$. Γ is the damping rate of the
exciton. Theoretical expressions for the coefficients A and B have
been given elsewhere.[9,10] These may be compared with the experimental values for the constants derived from four wave mixing data.
$\chi^{(3)}{}''(-\omega_1,\omega_1,\omega_1,-\omega_1)$ which appears in the expression for the intrinsic two photon absorption parameter is almost the same as the
imaginary part of Eq. (1). A small difference of a few percent, due
to the slight dispersion of the crystal in its transparent region,
can be ignored.

2. EXPERIMENTAL RESULTS

The experimental apparatus used to measure $\chi^{(3)}$ is similar in
concept to that used in our earlier work, but the actual design has
been much improved. As shown in Figure 1, a commercial (Raytheon)
Q-switched ruby laser, with a repetition rate of one pulse per second, provided a beam at ω_2. Approximately 90 percent of the ruby
beam is then used to pump a narrow band dye laser which has a
measured spectral width of about .15 cm^{-1}. The output of the dye
laser at frequency ω_1 was tunable over several hundred wavenumbers
about the wavelength of 760 nm. Both the beam at ω_1 and the beam
at ω_2 were focused and overlapped inside a reference crystal of
NaCl and a sample crystal of CuCl. The internal angles between the
beams were chosen to provide phase matching for the generation of
the output light at $2\omega_1 - \omega_2$. Since the frequencies of all the beams
lie in the near infra-red where the crystals are transparent, the

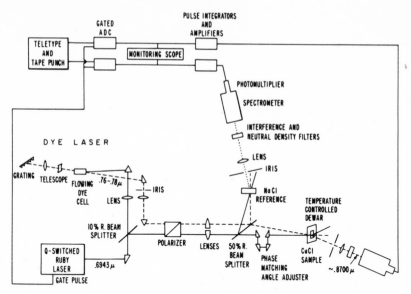

EXPERIMENTAL APPARATUS

Figure 1: Experimental arrangement for four wave mixing experiments in CuCl.

phase matching condition is independent of frequency over the tuning range of the dye laser. The CuCl crystal was mounted in a temperature controlled dewar so that the dispersion of $|\chi^{(3)}(-\omega_3,\omega_1,\omega_1,-\omega_2)|^2$ as a function of ω_1 coud be measured over a temperature range from 4°K to 300°K. A geometry was chosen in which the electric fields in both beams were polarized along a (1,1,0) crystallographic direction while the rays propagated nearly parallel to the (0,0,1) cubic axis. In this orientation only the longitudinal exciton resonance is involved.

Figure 2 shows a plot of the dispersion of $|\chi^{(3)}(-\omega_3,\omega_1,\omega_1,-\omega_2)|^2$, for a crystal of CuCl at 14.9±0.1°K, vs $2\omega_1$ as $2\omega_1$ is varied through the longitudinal exciton resonance. The solid line is a theoretical best fit of $|\chi^{(3)}|^2$ as given by Eq. (1). A constant background term due to scattering from optical imperfections in the crystal has been added. This fit yielded the following values for the constants that appeared in Eq. (1):

$$\hbar\omega_{ex} = 3.2083 \pm 0.0002 \text{ eV}$$
$$B = 0.005 \pm 0.001 \text{ (eV)}^2$$
$$\hbar\Gamma = 1.5 \pm 0.5 \times 10^{-4} \text{ eV}$$

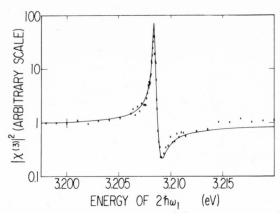

Figure 2: The dispersion curve for $|\chi^{(3)}(-\omega_3,\omega_1,\omega_1,-\omega_2)|^2$ as $2\omega_1$ is tuned through the longitudinal exciton resonance of CuCl at 14.9°K. The crosses are experimental points and the solid line is a best fit using Eq. (1).

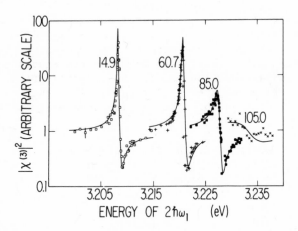

Figure 3: The temperature dependence of the dispersion curve of $|\chi^{(3)}(-\omega_3,\omega_1,\omega_1,-\omega_2)|^2$ in CuCl. Data are shown for four different temperatures in order from left to right, 14.9, 60.7, 85.0 and 105.0°K. The solid lines are a best fit at each temperature using Eq. (1).

Since the instrumental dye laser line width was 0.2×10^{-4} eV, this value for $\hbar\Gamma$ may well represent the true damping constant for our particular sample of CuCl. This damping may be impurity sensitive and may not be the intrinsic value for CuCl. It had been determined previously[7] that $A = 3.3 \pm 2.0 \times 10^{-12}$ cm^3/erg. These measurements give a value for the intrinsic two photon absorption parameter at resonance of $1.6 \pm 1.1 \times 10^{-28}$ (cm)(sec)/eV. The magnitude of the two photon absorption coefficient, consistent with our data, is 0.02 cm^{-1} at resonance for a power flux density of 2×10^7 watts/cm^2. This value is in reasonable agreement with a qualitative determination from direct two photon absorption experiments.[5,6]

The temperature dependence of the observed resonance in $|\chi^{(3)}(-\omega_3,\omega_1,\omega_1,-\omega_2)|^2$ is shown in Figure 3. In the range from 4°K to 105°K the coefficients A and B are constant, to within experimental accuracy. The changes in line position and shape can be accounted for by the variation with temperature of ω_{ex} and Γ which are plotted in Figures 4 and 5 respectively.

Above 20°K the lontitudinal exciton energy is a linear function of temperature with a slope of 2.9×10^{-4} eV/°K. This is close to that found from linear absorption experiments. As Cardona[2] and Brooks[11] have pointed out, the sign of the slope is due to a competition between electron-phonon interactions (explicit temperature effect) and the thermal expansion of the lattice (volume effect).

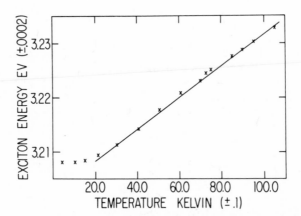

Figure 4: The temperature dependence of the energy difference between the longitudinal Z_3 exciton and the valence band in CuCl.

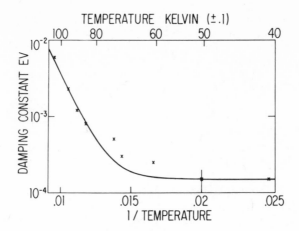

Figure 5: The temperature dependence of the damping constant, $\hbar\Gamma$, of the Z_3 exciton in CuCl. The common logarithm of the damping constant is plotted against the inverse of the temperature. The crosses are experimental points while the solid line is a best fit using Eq. (2).

The temperature dependence of the damping parameter can be fitted by the sum of a constant impurity-dominated part and an exponential term which is indicative of a multi-phonon process. This behavior is shown by the solid line in Figure 5. A temperature variation of this type suggests that the Z_3 exciton can make a transition to the nearest diffuse $Z_{1,2}$ exciton level. These two levels are separated by an almost temperature independent constant of .07 eV.[1] This transition would require the simultaneous absorption of three longitudinal optical phonons which have energies of about .025 eV.[12] The required change in spin configuration can be supplied by the large spin-orbit interaction. In this case, the theoretical temperature dependence would be given by:

$$\Gamma = \Gamma_o + B \left(\frac{1}{e^{\hbar\omega_{ph}/KT} - 1} \right)^3 \tag{2}$$

where the second term is cubic in the phonon density as is required for a three phonon absorption; ω_{ph} is the longitudinal optical phonon frequency and B is a constant that is a measure of the strength of the coupling between the exciton and phonons. For T < 100°K, h $\omega_{ph}/KT \gg 1$ and the exponential term reduces to B $\exp(-3\hbar\omega_{ph}/KT)$. As shown in Figure 5 this expression is in fair agreement with the experimental observation and is quite different from the predictions of Toyozawa.[13] Above 110°K the resonance could no longer be observed due to its large width. Fluorescence data indicates that direct thermal quenching sets in at about 150°K.[1]

In conclusion, it has been shown that four wave mixing experiments using frequencies that lie in the linearly transparent region of a sample afford a powerful method of studying the optical properties of strongly linearly absorbing transitions such as excitons.

REFERENCES

1. M. Ueta and T. Goto, Jour. Phys. Soc. JAP. 20, 401 (1965).

2. M. Cardona, Phys. Rev. 129, 69 (1963).

3. J. Ringeissen, S. Lewonczuk, A. Coret and S. Nikitine, Phys. Rev. Lett. 22, 571 (1966).

4. D. C. Haueisen and H. Mahr, Phys. Rev. Lett. 26, 838 (1971).

5. D. Fröhlich, B. Staginnus and E. Schönherr, Phys. Rev. Lett. 19, 1032 (1967).

6. A. Bivas, C. Marange, J. B. Grun and C. Schwab, Opt. Commun. 6, 142 (1972).

7. S. D. Kramer, F. G. Parsons and N. Bloembergen, Phys. Rev. B 9, 1858 (1974).

8. N. Bloembergen, these Proceedings.

9. E. Yablonovitch, Chr. Flytzanis and N. Bloembergen, Phys. Rev. Lett. 29, 865 (1972).

10. Chr. Flytzanis and N. Bloembergen, paper presented at the Taormina Conference, October, 1972 (to be published).

11. H. Brooks, Advances in Electronics, Edited by L. Marton, vol. 7 (Academic Press, New York, 1957).

12. M. Krauzman, R. M. Pick, H. Poulet, G. Hamel and B. Prevot, Phys. Rev. Lett. 33, 528 (1974).

13. Y. Toyozawa, Prog. in Theor. Phys. 20, 53 (1958).

NONLINEAR WAVE PROPAGATION IN A TRANSPARENT MEDIUM

N. TZOAR AND J. I. GERSTEN
THE CITY COLLEGE
The City University of New York
Convent Avenue at 138th Street
New York, New York 10031

When a sufficiently intense light pulse propagates through a material, it distorts the atomic configuration, which in turn causes the refractive index to acquire a time dependence. The time-varying index alters the phase of the optical wave as it propagates, leading in general to a broadening of the pulse spectrum. This process has been called self-phase-modulation (SPM) and has been observed by many researchers.[1]

There are two processes which limit SPM, namely dispersion and absorption. The dispersion tends to broaden the pulse without changing its energy, thus decreasing the field strength. Dissipation merely decreases the intensity thus reducing the effectiveness of the nonlinear time dependent modulation of the dielectric function.

In this paper we will focus our attention on highly transparent materials and take the dissipation to be zero. We thus concentrate on dispersion effects alone which for a plasma were shown to support an envelope soliton without SPM.[2]

We consider a plane wave propagating through a transparent medium obeying the wave equation:

$$\frac{\partial^2 E}{\partial z^2} - \frac{1}{c^2} \frac{\partial^2 D}{\partial t^2} = \frac{4\pi \omega^2 \epsilon_2}{c^2} \mid E \mid^2 E \tag{1}$$

where

$$D(z,t) = \int \varepsilon(t-t') \, E(z,t') \, dt' \tag{2}$$

Consider next a solution for the electric field E given by:

$$E(z,t) = A(z,t) \, e^{i\omega t - ikz} \tag{3}$$

Using Eqs. (1), (2) and (3) we obtain the wave equation for the envelope A as:

$$\frac{\partial^2 A}{\partial z^2} - 2ik\frac{\partial A}{\partial z} - k^2 A + \int \frac{d\omega'}{2\pi} K_{\omega+\omega'} A_{\omega'}(z) \, e^{i\omega t} =$$

$$\frac{\partial^2 A}{\partial z^2} - 2ik\frac{\partial A}{\partial z} - k^2 A + K_{\omega-i\frac{\partial}{\partial t}} A = \frac{-4\pi\omega^2 \varepsilon_2}{c^2} A^3 \tag{4}$$

where

$$K_{\omega+\omega'} = \left(\frac{\omega+\omega'}{c}\right)^2 \varepsilon_{\omega+\omega'} \tag{5}$$

Our purpose is to find out under what condition Eq. (4) will yield a steady state envelope solution for A without phase modulation; i.e. A is a real function of $\eta = (t-z/v_g)$. We first realize that such a solution exists only if the dispersion effects are weak enough such that only the first and the second order time derivatives are considered in the formal expansion of the operator $K_{\omega-i\frac{\partial}{\partial t}}$. That means a small spread in ω' around ω, the frequency of the carrier. We thus consider experimental situations in which for the weak intensity limit the pulse will propagate at a group velocity proportional to $(\partial K_\omega/\partial\omega^2)$. Considering next A to be real, Eq. (4) can be cast into two simultaneous wave equations given by:

$$2k\frac{\partial A}{\partial z} + K_\omega \frac{\partial A}{\partial t} = 0 \tag{6a}$$

$$\frac{\partial^2 A}{\partial z^2} - k^2 A + K_\omega A - \tfrac{1}{2}K_\omega'' \frac{\partial^2 A}{\partial t^2} = -\frac{4\pi\omega^2 \varepsilon_2}{c^2} A^3 \tag{6b}$$

The first equation shows that A is a function of $(t-z/v_g)$ as required. The solution of Eqs. (6a) and (6b) is given by:

$$A = A_o \, \text{Sech}\left(\frac{t-z/v_g}{\tau}\right) \tag{7}$$

where the phase velocity is $v_p = \omega/k$, the group velocity is $v_g = 2k/(\partial K_\omega/\partial \omega)$, the wave number is given by $k^2 = K_\omega(1+\chi)$ and $\chi = 4\pi\varepsilon_2 A_0^2/\varepsilon_\omega$. It is easy to verify that for a weak intensity, when $\chi \to 0$, the group velocity reduces to $v_g \to \dfrac{1}{\partial k/\partial \omega}$

as expected.

These parameters can be expressed in terms of the dielectric function of the medium and are given by:

$$v_p = \frac{c}{\sqrt{\varepsilon_\omega}\,\sqrt{1+\chi}} \quad ; \quad v_g = \frac{c}{\sqrt{\varepsilon_\omega}}\,\frac{\sqrt{1+\chi}}{(1+\frac{1}{2}\frac{\varepsilon_\omega}{\varepsilon_\omega})}$$

$$k = \frac{\omega}{c}\sqrt{\varepsilon_\omega}\,\sqrt{1+\chi} \quad ; \quad \chi = \frac{4\pi\varepsilon_2 A_0^2}{\varepsilon_\omega} \tag{8}$$

The fundamental ingredient[3] of this particular solution is the relation between the width of the pulse and the nonlinear dielectric function given by:

$$\omega^2\tau^2 = \frac{1}{\chi}\,\frac{(1+\frac{\omega}{2}\frac{\varepsilon_\omega'}{\varepsilon_\omega})^2}{1+\chi} - (1 + \frac{2\omega\varepsilon_\omega'}{\varepsilon_\omega} + \frac{\omega^2}{2}\frac{\varepsilon_\omega''}{\varepsilon_\omega}) \tag{9}$$

Consider next, for example, a solid with dielectric function:

$$\varepsilon_\omega = \varepsilon_\infty(\omega^2 - \omega_\ell^2)/(\omega^2 - \omega_t^2) = \varepsilon_\infty[1-(\omega_\ell^2 - \omega_t^2)/(\omega^2-\omega_t^2)]$$

when $\omega^2 \gg \omega_t^2$. Thus $\varepsilon_\omega \approx \varepsilon_\infty$ and $\chi < 1$.

After simple algebra we obtain for the "intensity-width" relation;

$$\omega^2\tau^2 \approx 3\,\varepsilon_\infty^2\,\frac{\omega_\ell^2 - \omega_t^2}{\omega^2\,4\pi\varepsilon_2 A_0^2} - 1 \tag{10}$$

which indicates an upper limit for the field strength of the pulse. Here the nonlinear dielectric function $4\pi\varepsilon_2 A_0^2/\varepsilon_\infty$ must be smaller than the dispersive part of ε_ω given by $\varepsilon_\infty(\omega_\ell^2 - \omega_t^2)/\omega^2$. The result definitely indicates that the solution depends on the balance between the dispersive term which tends to widen the pulse and the nonlinear term which tends to compress it.

Let us consider a realistic case with $\epsilon_\infty \approx 10$, $\omega_\ell = \sqrt{2}\omega_t$, $\omega = 10\,\omega_t$, $\epsilon_2 \approx 10^{-12}$ and $\omega\tau = 1000$. For these parameters we obtain $A_o \approx 10^3\,\dfrac{SV}{cm}$ which indicates a situation suitable for soliton propagation. For most experimental situations when SPM has been seen we estimated the field strength at the center of the pulse A_o to be of the order of $10^5\,\dfrac{SV}{cm}$, i.e. two order of magnitude larger than what was calculated here.

We proceed now to discuss the propagation of intense laser beam in a transparent medium for which the nonlinearity dominates over the dispersion. Taking the displacement vector $\vec{D}$ to include nonlinearity and absorption, for a wave propagating in the z direction we obtain the equation

$$\frac{\partial^2 E}{\partial z^2} - \frac{\partial^2 D}{\partial t^2} = 0 \tag{11a}$$

where

$$D = \epsilon_o E + \epsilon_2 E^3 - 2f \int_{-\infty}^{0} dt\, E\,(t + \tau)\, e^{\gamma t/2}\, \mathrm{Sin}\,\omega_a \tau \tag{11b}$$

It will be assumed that both E and D are describable by a slowly varying amplitude and phase. Here ω is the radiant frequency, $k = \dfrac{\omega}{c}\sqrt{\epsilon_o}$ is the speed of light of a weak signal in the medium.

Take $E = \mathcal{E} \,\mathrm{Cos}\,\emptyset$, $D = D_1\,\mathrm{Cos}\,\emptyset + D_2\,\mathrm{Sin}\,\emptyset$, $\emptyset = kz - \omega t + \varphi$, (12)

Where $\mathcal{E}$, D_1, D_2 and φ depend on z and t. Use Eqs. (11) and (12) neglecting all second order derivatives and take the nonlinearity and absorption to be small such that only linear terms in $\mathcal{E}$ and φ are retained to obtain the following coupled equation for the envelope $\mathcal{E}$ and the phase φ:

$$\frac{\partial \mathcal{E}}{\partial t} + v\frac{\partial \mathcal{E}}{\partial z} = \frac{\pi\omega t}{2\epsilon_o}\,\mathcal{E}\,\Delta(\omega_a - \omega + \frac{\partial \varphi}{\partial t})$$

$$\frac{\partial \varphi}{\partial t} + v\frac{\partial \varphi}{\partial z} = \frac{3}{8}\,\frac{\omega\epsilon_2}{\epsilon_o}\,\mathcal{E}^2 \tag{13}$$

Here $\Delta(x)$ is a function which peaks at $x = 0$ with unit intensity and hereafter we shall take $\Delta(x)$ to be $\delta(x)$, the Dirac delta function. From Eq. (13) we see that for $f \to 0$ we obtain the usual SPM solution. Here the amplitude is a function of $(t - \frac{z}{v})$ and remains a constant

amplitude for an observer which travels at the speed of light v. The phase, on the other hand, changes with increasing z. Define $\xi = t - \frac{z}{v}$ and $\eta = t + \frac{z}{v}$. The solution for the case where $f \to 0$ is given by:

$$\mathcal{E} = H(\xi)$$

and

$$\varphi = \frac{3}{16} \frac{\omega \epsilon_2}{\epsilon_0} H^2(\xi)(\eta - \xi) \tag{14a}$$

where

$$H(t) = \mathcal{E}(z = 0, t) \tag{14b}$$

In the case with finite absorption we have two regions in the ξ, η plane to be considered, separated by the curve on which the argument of the δ function vanishes. As long as the SPM introduces instantaneous frequencies smaller than ω_a, the solution for the phase is given by Eq. (14). Beyond the resonance curve the argument of the δ function again vanishes so the general solution to Eq. (13) is given by

$$\varphi = F(\xi) + \eta G(\xi) \tag{15}$$

where

$$G(\xi) = \frac{3}{16} \frac{\omega \epsilon_2}{\epsilon_0} \mathcal{E}^2(\xi) \tag{16}$$

Here F and G could in principle be determined by the jump across the curve defined by the vanishing of the argument of the δ function. It was shown elsewhere[4] that this procedure yields two highly nonlinear differential equations for F and G which we shall not discuss here. It is only needed to point out that G has an upper limit which reads:

$$G(\xi) \leq \frac{3}{16} \frac{\omega \epsilon_2}{\epsilon_0} H^2(\xi) e^{-\left| \gamma \frac{d}{d\xi} H^2(\xi) \right|} \tag{17}$$

where

$$\varphi = \frac{3\pi\omega^2 \epsilon_2 f}{64 \epsilon_0}$$

We proceed now to calculate an upper limit on the spectral function whose dominant contribution is given by

$$S(\sigma) = \frac{1}{4} \left| \int_{-\infty}^{+\infty} dt\, \varphi\, e^{i(\sigma t - \omega t + \varphi)}\, \mathcal{E} \right|^2 \tag{18}$$

Using the Schwartz inequality we may formally write:

$$S(\sigma) \leq \frac{1}{4} \left(\int_{-\infty}^{+\infty} dt\, H(\xi) \right) \left(\int_{-\infty}^{+\infty} dt\, \frac{\mathcal{E}^2}{H(\xi)} \right)$$

Using Eqs. (16) and (17) and replacing the exponent in Eq. (17) by its maximum value we arrive at

$$S(\sigma) \leq \left(\int_{-\infty}^{+\infty} dt\, H(\xi) \right)^2 e^{-\left| \gamma \frac{d}{d\xi} H^2(\xi) \right|_{max}} \tag{19}$$

Eq. (19) indicates that the spectral function $S(\sigma)$ will tend to zero with the increasing of f [see Eq. (17)], the coupling to the absorption line.

We thus conclude that the effect of line absorbers on both sides of the carrier frequency on the spectrum of the propagating signal is to limit it predominantly to the region between the two absorbing frequencies.

This effect has been studied experimentally and is in agreement with the theory presented here [4]. It should be noted that the spectrum beyond an absorption line can result not only from SPM but also by parametric conversion of the intense broad band signal produced by SPM between the absorbing lines. This makes comparison between theory and experiments rather difficult.

References:

[1]
M. J. Lighthill, J. Inst. Math. Appl. 1, 269 (1965); V. I. Karpman and E. N. Krushkal, Zh. Eksp. Teor. Fiz 55, 530 (1968) [Sov. Phys. JETP 28, 277 (1969)]; Akira Hasegawa and Fredric Tappart, Appl. Phys. Lett. 23, 142 (1973).

[2]
N. Tzoar and J. I. Gersten, Phys. Rev. Lett. 28, 1203 (1972).

[3]
See References 1 and 2.

[4]
R. R. Alfano, J. I. Gersten, G. A. Zawadzkas and N. Tzoar, Phys. Rev. A 10, 698 (1974).

PICOSECOND LASER-INDUCED DAMAGE IN TRANSPARENT MEDIA[*]

W. L. Smith, J. H. Bechtel and N. Bloembergen

Gordon McKay Laboratory

Harvard University, Cambridge, Massachusetts 02138

1. INTRODUCTION

A high degree of transparency is a material characteristic of great importance for all laser systems. For high-power systems, that characteristic becomes a vital requirement of many solid components.

However, even though a material exhibits utmost transparency to low-level light waves, an optical frequency electric field of sufficient strength is capable of converting the state of the material into one of high absorption and scattering - a plasma. Figure 1 illustrates such an occurrence. In this 1973 experiment by Fradin, Yablonovitch and Bass,[1] a Q-switched laser pulse was focused into an initially transparent NaCl crystal. The energy transmitted through the crystal was monitored, yielding the displayed oscilloscope trace. When the electric field reached the breakdown threshold, plasma buildup rapidly ensued and transmission was promptly reduced. The process leading to the rapid transmission cutoff may be initiated with a single free electron located in the focal volume.[2] Such an electron may be produced by thermal- or photoionization of shallow traps or by multiphoton emission. The electron gains energy from the optical field, meanwhile undergoing collisions on the average roughly every 10^{-14} seconds. The dominant collision mechanism depends on the electron energy. If the electron gains sufficient energy to

---

[*] Supported by Advanced Research Projects Agency and National Aeronautics and Space Administration.

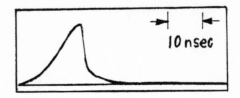

Figure 1: Laser-induced breakdown by a Q-switched, 0.69 μm pulse
in NaCl, from Ref. 1. This tracing of an oscillograph
displays energy, transmitted through the focal volume,
versus time, and illustrates the suddenness with which
avalanche ionization may reduce transparency.

cause by impact ionization a second electron, then those two electrons may produce four, etc., and the avalanche has begun. After
30 to 40 generations, the plasma density has reached roughly 10^{18}
electrons/cc, a sufficient density to produce sharp attenuation
and lattice disruption. The resulting damage constitutes a limiting
factor to the performance of transparent optical components in high-
power laser systems.

In many practical cases the damage threshold is set by the presence of absorbing inclusions, or by the occurrence of self-focusing.
It has, however, been established that an intrinsic damage threshold,
characteristic of the bulk transparent material, can be defined. If
the effects of inclusions have been eliminated and if one has corrected for the effects of self-focusing, the threshold is determined
by the onset of avalanche ionization. Since the conduction electron
density has to build up from the initial density of about 10^8 to
10^{10} electrons per cc to the critical value of about 10^{18} electrons

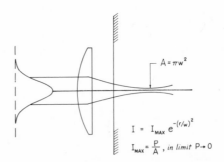

Figure 2: Illustration of the external focusing of a Gaussian laser
pulse into a material.

per cc during the presence of the laser pulse, the damage threshold
has a characteristic dependence on pulse duration. Most available
experimental data were taken with pulses of nanoseconds or longer
duration. In fact, heretofore, there was only one quantitative data
point taken with a picosecond pulse - that of Fradin, Bloembergen
and Letellier in 1973.[3] The purpose of this paper is to provide
systematic picosecond data on the damage threshold in fourteen
transparent materials. The data obtained are characteristic of the
material and are very reproducible.

Consider now a low-power, diffraction limited, Gaussian pulse
with an initial intensity radius ρ. A corrected lens will focus
this pulse to a spot having a radius w equal to $\lambda f/2\pi\rho$, as illustrated in Fig. 2. If the pulse power is P, and we define a focal
area $A = \pi w^2$, then the focal plane intensity distribution will be
given by

$$I = I_{max} \; e^{-(r/w)^2}. \tag{1}$$

I_{max}, in the low power limit, will equal P/A.

However, due to the high pulse powers necessary to reach a
value I_{max} sufficient for breakdown, self-focusing[4] unavoidably
plays a role in this experiment. The self-focusing of a light pulse
has its origin in the intensity dependence of the index of refraction,

$$n = n_o + n_2 \; |E_{RMS}|^2, \tag{2}$$

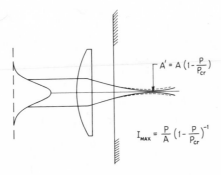

Figure 3: Illustration of the external focusing into a material of
          a Gaussian laser pulse having power P that is nonnegligible
          compared to P_{cr}. Self-focusing reduces the focal area from
          A to A', thereby increasing I_{max}.

where n_0 is the usual linear index, n_2 is the nonlinear refractive index, and E_{RMS} is the RMS electric field.

A quantity characteristic of this self-action is the critical power for self-focusing, P_{cr}, given by

$$P_{cr} = c\lambda^2 / 32 \pi^2 n_2. \tag{3}$$

It represents an equilibrium power at which an <u>unfocused</u>, Gaussian laser pulse, propagating in the medium, will neither diffract nor self-focus - the two tendencies just balance each other. For power P less than P_{cr}, diffraction always dominates; for P exceeding P_{cr}, the pulse will self-focus in the medium.

Now, for the situation in which the pulse is externally focused to begin with, the critical power enters into the experiment as indicated in Fig. 3. If A is the same low-power limit focal area that was introduced above, then for high power pulses where self-focusing tendencies cannot be ignored, that quantity A is reduced to A' by a power-dependent factor:

$$A' = A (1 - P/P_{cr}). \tag{4}$$

The same factor serves to <u>increase</u> I_{max}, according to

$$I_{max} = (P/A) (1 - P/P_{cr})^{-1}. \tag{5}$$

That equation may be rearranged to give

$$P^{-1} = I_{max}^{-1} A^{-1} + P_{cr}^{-1}, \tag{6}$$

an equation of the form y = ax + b. If one plots, therefore, the reciprocal of the measured input power, necessary to produce the value of I_{max} required for breakdown, as a function of reciprocal focal area, one should obtain a straight line. The slope of the straight line would be I_{max}^{-1}, and would yield the breakdown RMS electric field via

$$E_b = \left[I_{max} / n_0 \varepsilon_0 c \right]^{1/2}. \tag{7}$$

The intercept with the vertical axis would yield P_{cr}, thereby providing n_2, using Eq. 3. Additionally, the appropriate element of the third-order nonlinear susceptibility tensor $\chi^{(3)}$ is obtained according to

$$\chi_{xxxx}^{(3)E} (-\omega,\omega,\omega,-\omega) = n_0 n_2 / 12 \pi. \tag{8}$$

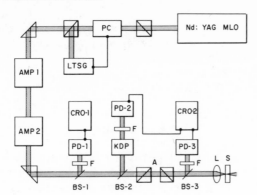

Figure 4: Schematic diagram of the experimental arrangement. MLO,
          mode-locked Nd:YAG oscillator; PC, Pockels cell; LTSG,
          laser-triggered spark gap; AMP, Nd:YAG amplifier; BS,
          beam splitter; F, filter; PD, biplanar photodiode; CRO,
          oscilloscope; KDP, second-harmonic generation crystal;
          A, double Glan prism variable attenuator; L, focusing
          lens; S, dielectric sample.

This method of determining P_{cr} was first suggested by Zverev and
Pashkov[5] in 1970, and was employed by Fradin[6] in 1973 to measure
n_2 in sapphire.

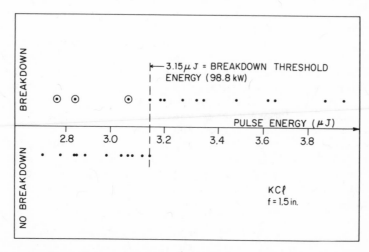

Figure 5: Results of breakdown experiment in KCl for f = 1.5 in.
          lens. The average pulse duration was used to convert energy
          to power. The three circled points resulted from atypically
          short pulses.

In all the experiments that we have performed, the input power never exceeded P_{cr}, and typically it was less than $0.5\ P_{cr}$.

2. EXPERIMENTAL APPARATUS AND PROCEDURE

The experimental apparatus is schematically illustrated in Fig. 4. MLO is a saturable absorber mode-locked TEM_{00} osciallator Single pulse switch-out is accomplished by the Pockels cell PC and the spark gap LTSG. A selected pulse is amplified as needed by 2 YAG amps. The various photodiodes, oscilloscopes, and the KDP crystal are used to measure the pulse energy, duration and area, as will be discussed in detail elsewhere.[7]

Extensive tests were performed to determine accurately the spatial and temporal intensity distribution - the average 1/e radius of intensity at the lens is 1.1 mm ± .1 mm. The average FWHM pulse duration is 30 ± 6 psec.

Three lenses were used; the focal lengths were 0.5, 1.0 and 1.5 in. Careful consideration of all aberration was included in the determination of the proper focal areas.

Breakdown were detected initially with well dark-adapted eyes, and a later microscopic examination verified any visually uncertain points.

The typical results of an experimental run are shown in Fig. 5. Typically 50 shots were used for each lens in each material. Each value of the pulse power was then plotted - above the axis if breakdown occurred, below if not. The sharpness of the threshold is typical of the behavior encountered. The three circled points were associated with pulses somewhat shorter than the average duration of 30 ps, and did cause breakdown even though their energy was below that designated in the figure for pulses with duration closer to the average. The damage energy was determined in this manner in each case. The reproducibility of the data when probing different volume elements indicates that a property characteristic of the bulk material is observed. Only a few isolated cases of inclusions were observed under the microscope, and they were eliminated. A further detailed account of the morphology of the damage will be given elsewhere.

The three such damage power values that we measured for each material were then plotted versus reciprocal focal area as discussed earlier.

An example of such a plot, is shown in Fig. 6. The fitted line yields the desired quantities - the intrinsic breakdown field E_b,

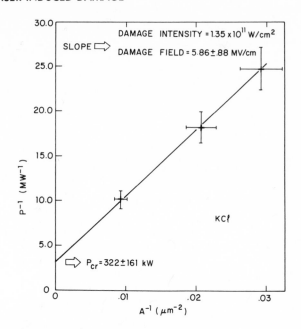

Figure 6: Results of breakdown experiments in KCl for three lenses.
The solid line is a least-squares fit to the data. The
slope equals the reciprocal threshold breakdown intensity
and the vertical-axis intercept equals the reciprocal critical power for self-focusing.

the quantity usually obtained from breakdown experiments, and it
is independent of self-focusing effects. The field values reported
here are all absolute, with an experimental uncertainty of ± 15%.
However, with this method, we also obtain n_2.

3. RESULTS

A. Dielectric Breakdown Threshold Measurements

We now consider the variation of the breakdown threshold among
different materials. Table I displays the results for the materials
which we have tested: 9 alkali-halides and 5 other materials of importance. Also tabulated in parentheses are the thresholds normalized to that of NaCl. The threshold is observed to vary from a minimum of 3.4 MV/cm for RbI to a maximum of 22.28 MV/cm for KDP,
among the materials tested. KDP possesses a threshold so high as
to approach the level necessary to make multiphoton ionization
probable.

TABLE I

RMS BREAKDOWN ELECTRIC FIELD STRENGTHS

(E_b UNCERTAINTY, ±15%; FIELD RATIO UNCERTAINTY, ±10%)

	NaF	NaCl	NaBr	KF	KCl	KBr	KI
E_b (MV/cm)	10.77	7.34	5.67	8.34	5.86	5.33	5.87
E_b/E_b^{NaCl}	1.47	1.00	0.77	1.14	0.80	0.73	0.80

	LiF	RbI	ED-4 Glass	Nd:YAG	Fused SiO$_2$	CaF$_2$	KDP
E_b (MV/cm)	12.24	3.40	9.90	9.82	11.68	14.44	22.28
E_b/E_b^{NaCl}	1.67	0.46	1.35	1.34	1.59	1.97	3.04

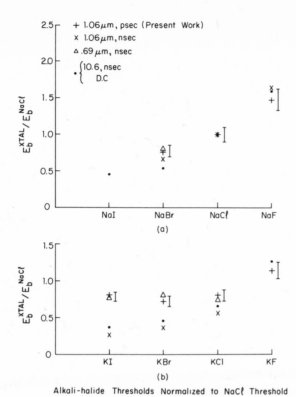

Alkali-halide Thresholds Normalized to NaCℓ Threshold

Figure 7: Experimental trend of the breakdown field through (a) the Na-halides and (b) the K-halides. Shown for comparison is data from previous studies: (x), Ref. 1; (Δ), Ref. 10; (·), DC data from Ref. 8, 10.6 µm data from Ref. 9. In the figure nsec denotes a pulse duration the order of nanoseconds, and psec, picoseconds. Each set of data is normalized to the field for NaCl of that same data set. Error bars for the present work are shown displaced to the right.

Figure 7 displays the trend of our normalized threshold measurements across the periodic table for the Na- and K-halides. Our points are the (+) signs, and the error bars displaced to the right are for our points. For comparison, previous data from DC,[8] and Q-switched CO_2,[9] ruby,[10] and Nd:YAG[1] laser damage studies are presented. Note that all the data for a particular combination of pulse duration and frequency is normalized to the NaCl threshold of that same data set. The systematic variation throughout the alkali-halides is the result of the systematic variation of the parameters

TABLE II.   Results for $\chi^{(3)E}_{1111}(-\omega,\omega,\omega,-\omega)$

| Material | A. Present Work | | | | B. $\chi^{(3)E}_{1111}$ (10^{-14} esu), Other Work | | | | | |
| | P_{cr} (MW) | n_0 | n_2 (10^{-14} esu) | $\chi^{(3)E}_{1111}$ (10^{-14} esu) ± 50% | Experiment(a) | | | | Theory(b) | |
					I ± 300%	II ± 20%	III	Other	I	II
NaBr	0.111	1.62	96.15	4.13					0.8	2.0
NaCl	0.492	1.532	21.69	0.88	1.2	1.7			0.13	0.3
NaF	1.13	1.321	9.45	0.33	0.25				3.1	9.0
KI	0.095	1.638	1↑2.35	4.88					1.3	3.7
KBr	0.075	1.544	142.31	5.83	2.8	3.0			0.9	1.9
KCl	0.322	1.479	33.15	1.30	1.2	1.9			0.28	0.3
LiF	0.438	1.387	24.37	0.90	0.3	0.2	0.34 ± .06		0.4	0.4
CaF$_2$	0.320	1.429	33.35	0.71	0.35		0.43 ± .13		0.4	
Fused SiO$_2$	0.783	1.450	13.63	0.52	0.7	0.7	0.70 ± .12	0.38 ± .03[IV]	1.4	1.1
KDP	0.295	1.494	36.18	1.43						
ED-4	0.503	1.550	21.22	0.87			1.07 ± .08	0.71 ± .05[IV] {1.98 ± .14[IV]		
Nd:YAG*	0.302	1.823	35.34	1.71			2.18 ± .13	1.53 ± .30[V]		0.62

* All experimental values in this row, with one exception, are for pulse propagation along a [111] direction, the usual laser rod configuration. In these cases, the measured susceptibility is $\frac{1}{2}(\chi^{3E}_{1111} + \chi^{3E}_{1122} + 2\chi^{3E}_{1221})$, however. The value in the column designated III is for the single element χ_{1111}, the row pertain to undoped YAG crystal. The present work investigated a Nd:YAG crystal; all other values in the row pertain to undoped YAG crystal.

(a.) I.   Values from third-harmonic generation, Wang and Bardsen (1969). Ref.11
      II. Values from three-wave mixing, Maker and Terhune (1965). Ref.12
      III. Values from three-wave mixing, Levenson and Bloembergen (1974). Ref.13
      IV. Values from intensity-dependent ellipse rotation, Owyoung (1973). Ref.14
      V.   Values from time-resolved interferometry, Bliss et. al.(1974). Ref.15

(b.) I.   Calculated from generalized Miller's, $\chi^{(3)} = (\chi^{(1)})^4 \times 10^{-10}$ esu, Wang (1970). Ref.16
      II. Calculated from Wang's rule, Wang (1970). Ref.16

of the avalanche: the band gap, phonon density of states, and deformation potential

B. Third-Order Nonlinear Susceptibility Measurements

We now turn to the results of our experiments for the third-order susceptibility. As we saw earlier, the vertical-axis intercept of the experimental graphs gives the critical power, from which n_2 and $\chi^{(3)}$ are obtained.

Our results are presented in Section A of Table II. The experimental uncertainty of P_{cr}, and of $\chi^{(3)}$, is $\pm$ 50%. Section B of the table lists values previously obtained by others of $\chi^{(3)}$ for comparison. The smallest $\chi^{(3)}$ value was registered by NaF, a factor of almost 18 less than that of KBr. Agreement with previous values is substantial, given the rather large experimental uncertainties listed. The only serious discrepancy occurs for LiF - a factor of approximately 3. We tested three different samples of LiF, and our results were reproducible. Therefore, we know of no reason to suspect that the error in our value for LiF exceeds the stated uncertainty of $\pm$ 50%.

4. CONCLUSION

In summation, the method of controlled dielectric breakdown has been demonstrated to be a useful way to measure simultaneously the two intrinsic parameters which must be known to enable propagation of maximum picosecond light intensity through a material in a stable, nondestructive manner. Those parameters are the breakdown field strength and the electronic nonlinear refractive index. The measurements were made in a manner allowing elimination of any inclusion effect and measurement of intrinsic processes characteristic of the bulk. The values of $\chi^{(3)}$ presented here are in reasonable agreement with those of other investigators. The measured breakdown field strength have been shown to be those of an avalanche ionization process.

REFERENCES

1. D. W. Fradin, E. Yablonovitch and M. Bass, Appl. Opt. 12, 700 (1973).

2. N. Bloembergen, IEEE, J. Quant. Elect. QE-10, 375 (1974).

3. D. W. Fradin, N. Bloembergen and J. P. Letellier, Appl. Phys. Lett. 22, 635 (1973).

4. S. A. Akhmanov, R. V. Khoklov and A. P. Sukhorukov, in Laser Handbook, edited by F. T. Arecchi and E. O. Schulz-Dubois (North-Holland, Amsterdam, 1972), vol. 2, chap. E3.

5. G. M. Zverev and V. A. Pashkov, Sov. Phys. - JETP 30, 616 (1970).

6. D. W. Fradin, IEEE, J. Quant. Elect. QE-9, 954 (1973). Note that Eq. 2 in this reference should read $I_d = I_o(1 - P/P_{cr})^{-1}$.

7. A comprehensive report of this work has been submitted to the Physical Review.

8. A. von Hippel, J. Appl. Phys. 8, 815 (1937).

9. E. Yablonovitch, Appl. Phys. Lett. 19, 495 (1971).

10. D. W. Fradin and M. Bass, Appl. Phys. Lett. 22, 206 (1973).

11. C. C. Wang and E. L. Bardsen, Phys. Rev. 185, 1079 (1969), and Phys. Rev. B1, 2827 (1970).

12. P. D. Maker and R. W. Terhune, Phys. Rev. A137, 801 (1965).

13. M. D. Levenson, IEEE, J. Quant. Elect. EQ-10, 110 (1974), and M. D. Levenson and N. Bloembergen, Phys. Rev. B10, 4447 (1974).

14. A. Owyoung, IEEE, J. Quant. Elect. QE-9, 1064 (1973).

15. E. S. Bliss, D. R. Speck and W.W. Simmons, Appl. Phys. Lett. 25, 728 (1974).

16. C. C. Wang, Phys. Rev. B2, 2045 (1970).

SECOND-HARMONIC GENERATION OF INTENSE LASER LIGHT

IN TRANSPARENT CENTROSYMMETRIC SOLIDS

Stanisław KIELICH and Roman ZAWODNY

Nonlinear Optics Division, Institute of Physics

A. Mickiewicz University, 60-780 Poznań, POLAND

The authors analyze the feasibility of second-
-harmonic generation (SHG) intensity amplification
in centrosymmetric solids by:
1. Nonlinear spatial dispersion, related with electric and magnetic multipolar transitions;
2. Changes in nonlinear susceptibilities self-induced by strong laser light intensity;
3. Lowering of the intrinsic crystal symmetry e.g.
inversion centre destruction by a DC electric or
magnetic field, or crossed fields;
4. Coupling between self-light-intensity dependent
effects and DC applied-field induced effects.
Supplementing hitherto considered SHG mechanisms,
these new processes are described in terms of 5-th
and 6-th rank polar and axial tensors of electro-
-electric and magneto-electric nonlinear susceptibilities. The nonzero and independent elements of
these new tensors are calculated, thus pinpointing
those classes of centrosymmetric crystals where
SHG can occur with amplified intensity.

INTRODUCTION

Terhune et al [1] observed weak second-harmonic generation (SHG) in the light transmitted by calcite crystal, which
has a centre of symmetry. That earliest experiment was repeated by Bjorkholm and Siegman [2] , who compared the SHG in-

* Supported by Physics Institute of Polish Academy of Sciences

tensities from calcite and ADP crystal. Wang and Dumiński [3]
observed SHG in transmission through thin platelets of glass
and LiF. Bloembergen et al [4] studied SHG in reflection from
a medium with inversion symmetry. This kind of SHG both in
transmitted and reflected light was hitherto attributed to
electric quadrupole and magnetic dipole polarization [5-10].
    The study of SHG in bodies exhibiting a phase transition
from a state without centre of symmetry to a centrosymmetric
state and vice versa e.g. under the influence of a DC electric
field [11] is of the greatest interest [11, 12]. Recently,
Vogt [13] performed such SHG studies on Sodium Nitrite at the
temperature transition from the ferroelectric state, when the
symmetry is m2m, to the paraelectric state, in which the crystal as a whole has a centre of inversion with the symmetry
mmm. Rabin [14] analyzed the conditions for SHG provided by
the introduction of lattice defects into centrosymmetric crystals.
    It is a well known fact that in the process of light
self-focusing the refractive index and thus the susceptibility
of the medium become light-intensity dependent [15]. It is the
aim of this paper to show that in the process of SHG as well,
when using strong laser light, the nonlinear susceptibility
tensors become functions of the light intensity. We suggested
and considered this earlier [16] with regard to SHG by isotropic electrically polarized bodies. Since self-induced nonline-
arities occur in all bodies to a larger or lesser degree, the
nonlinear susceptibilities of materials determined by the method of harmonics generation [17,18] depend on the light intensity itself.
    We shall restrict our discussion to SHG in centrosymme-
tric materials, in which the self-induced light-intensity dependence of the nonlinear magneto-electric dipolar and electric quadrupole susceptibilities are described, respectively,
by a 5-th rank axial tensor of dipolar magneto-electric susceptibility and a 6-th rank polar tensor of electric-electric quadrupole susceptibility. Besides frequency dispersion, we take
into account the nonlinear spatial dispersion derived in a previous theory of multipole transitions [19] , induced in quantal
systems by strong electromagnetic fields. We also discuss amplification in SHG intensity, due to inversion centre destruction by an external DC electric or magnetic field, and able to
exhibit coupling with optically self-induced SHG amplification.

SHG IN THE PRESENCE OF INVERSION CENTRE

Quite generally, the total electric polarization of a medium
at the space-time point $(\underline{r},t)$ is given by the multipole expansion [19] :

$$\underline{P}_e(\underline{r},t) = \sum_{m=1}^{\infty} \frac{(-1)^{m+1}}{(2m-1)!!} \nabla^{m-1} [m-1] \underline{P}_e^{(m)}(\underline{r},t), \qquad (1)$$

$\underline{P}_e^{(m)}(\underline{r},t)$ denoting 2^m-pole electric polarization, and $[m-1]$ an $m-1$-fold contraction between the spatial differential operator ∇^{m-1} and the m-vector $\underline{P}_e^{(m)}$. On the other hand, the electric polarization vector is a function of the electric field:

$$\underline{E}(\underline{r},t) = \underline{E}(\underline{r},t)e^{-i\omega t} = \underline{E}(\omega,\underline{k})\exp[i(\underline{k} \cdot \underline{r} -\omega t)], \qquad (2)$$

and of the similarly defined magnetic field $\underline{B}(\underline{r},t)$ of a light wave, of frequency ω and propagation $\underline{k}$.
SHG is defined by second-order polarization, of the form [19]:

$$\underline{P}_e^{(2)(m)}(\underline{r},t) = \sum_{n=1}^{\infty} \sum_{s=1}^{\infty} \frac{\exp(-i2\omega t)}{(2n-1)!!(2s-1)!!} \cdot$$

$$\left\{ {}_e^{(m)}\underline{\chi}_{ee}^{(n+s)}(2\omega)[n+s][\nabla^{n-1}\underline{E}(\underline{r},\omega)][\nabla^{s-1}\underline{E}(\underline{r},\omega)] + \right.$$

$$\left. + {}_e^{(m)}\underline{\chi}_{em}^{(n+s)}(2\omega)[\nabla^{n-1}\underline{E}(\underline{r},\omega)][\nabla^{s-1}\underline{B}(\underline{r},\omega)] + ... \right\}, \qquad (3)$$

where ${}_e^{(m)}\underline{\chi}_{ee}^{(n+s)}(2\omega)$ is a tensor of 2-nd order electric multipole susceptibility at frequency 2ω taking into account all 2^{n+s}-pole electric transitions. Likewise, ${}_e^{(m)}\underline{\chi}_{em}^{(n+s)}(2\omega)$ is a tensor of electric multipole susceptibility containing contributions from 2^n-pole electric and 2^s-pole magnetic transitions, in conformity with the quantum-mechanical formula:

$${}_e^{(m)}\underline{\chi}_{em}^{(n+s)}(2\omega) = \frac{\varrho}{\hbar^2} S(n,s) \sum_{abcd} \varrho_{ab}$$

$$\left\{ \frac{<a|\underline{M}_e^{(m)}|c><c|\underline{M}_e^{(n)}|d><d|\underline{M}_m^{(s)}|b>}{(\omega_{cb} + 2\omega + i\Gamma_{cb})(\omega_{db} + \omega + i\Gamma_{db})} + \right.$$

$$+ \frac{<a|\underline{M}_e^{(n)}|c><c|\underline{M}_e^{(m)}|d><d|\underline{M}_m^{(s)}|b>}{(\omega_{ca} - \omega - i\Gamma_{ca})(\omega_{db} + \omega + i\Gamma_{db})} +$$

$$\left. + \frac{<a|\underline{M}_e^{(n)}|c><c|\underline{M}_m^{(s)}|d><d|\underline{M}_e^{(m)}|b>}{(\omega_{ca} - \omega - i\Gamma_{ca})(\omega_{da} - 2\omega - i\Gamma_{da})} \right\}, \qquad (4)$$

where ϱ is the number density of the medium, ϱ_{ab} the statistical matrix for the transition a→b with Bohr frequency ω_{ab} and relaxation time Γ_{ab}^{-1}, and $\underline{M}_e^{(m)}$, $\underline{M}_m^{(s)}$ the 2^m-pole electric and 2^s-pole magnetic moment operator of the quantal system. On writing $\underline{M}_e^{(s)}$ instead of $\underline{M}_m^{(s)}$ in (4), we get an expression for the susceptibility tensor ${}_e^{(m)}\underline{\chi}_{ee}^{(n+s)}(2\omega)$.

Eqs (1) and (2) define the influence of spatial dispersion of all orders on SHG. At very high light intensity, the
electric polarization (1) at 2ω contains contributions of
higher even orders besides that of 2-nd order (3) . Thus we
can write the total electric polarization at 2ω , in a satisfactory approximation, as:

$$P_{ei}(2\omega,\underline{r}) = \chi_{ijk}^{eee}(2\omega,\underline{k},I)E_j(\omega,\underline{r})E_k(\omega,\underline{r}) +$$

$$+ \chi_{ijk}^{eem}(2\omega,\underline{k},I)E_j(\omega,\underline{r})B_k(\omega,\underline{r}) + \dots, \qquad (5)$$

Hence, the tensors of nonlinear 2-nd order electric-electric
and electro-magnetic susceptibility, which are the source of
SHG. depend in general not only on frequency dispersion ω but
also on spatial dispersion $\underline{k}$ and on the incident light in-
tensity $I=E(\omega,\underline{r})E^*(\omega,\underline{r})$. For bodies having a centre of inversion we thus get in a satisfactory approximation:

$$\chi_{ijk}^{eee}(2\omega,\underline{k},I) = i\Big\{ \chi_{ijkl}^{eeek}(2\omega) + \chi_{ijklmn}^{eeekkk}(2\omega)k_m k_n +$$

$$+ \chi_{ijklmn}^{eeekee}(2\omega)E_n(\omega,\underline{k})E_m^*(\omega,\underline{k}) + \dots\Big\}k_l, (6)$$

$$\chi_{ijk}^{eem}(2\omega,\underline{k},I) = \chi_{ijk}^{eem}(2\omega) + \chi_{ijklm}^{eemkk}(2\omega)k_l k_m +$$

$$+ \chi_{ijklm}^{eemee}(2\omega)E_l(\omega,\underline{k})E_m^*(\omega,\underline{k}) + \dots, \qquad (7)$$

The expansions (6) and (7) are expressed formally in such a
manner that the individual susceptibility tensors contain the
respective multipole transition contributions resulting from
Eq. (4) ; however, for brevity, we omit the numerical expansion
coefficients. E.g., the polar tensor χ_{ijkl}^{eeek} (1) consists of the
nonlinear electric dipole susceptibility $^e\chi^{(1+2)}$ with electric dipole-quadrupole transition and of the electric quadrupole susceptibility $^{(2)}\chi^{(1+1)}$ with electric dipole-dipole transitions. The polar 6-th rank tensor χ_{ijklmn}^{eeekkk} consists in general of successive multipolar susceptibilities with the appropriate multipolar transitions.

Non-zero and independent elements of the polar tensor
χ_{ijkl}^{eeek} are listed in Table I. The numbers of nonzero and mutu-
ally independent elements of the remaining tensors of Eqs (6)
and (7) for centrosymmetric classes are given in Table II. The
axial tensor elements χ_{ijk}^{eem} and χ_{ijklm}^{eeeem} have been tabulated
previously for all classes [20].

For fields of the form (2) , Maxwell's equations yield:

Table I.
Non-zero and independent elements of the polar tensor χ_{ijkl}^{eeek}
for all centrosymmetric crystallographical classes.

Class	Elements χ_{ijkl}^{eeek}
$\overline{1}$	A≡xxxx,yyyy,zzzz,xxyy,yyxx,xyxy,yxyx,xyyx,yxxy,xxzz,zzxx, xzxz,zxzx,xzzx,zxxz,yyzz,zzyy,yzyz,zyzy,yzzy,zyyz; B≡xxxy,xxyx,xyxx,yxxx,yyyx,yyxy,yxyy,xyyy,xzzy,xzyz,xyzz, yzzx,yzxz,yxzz,zzxy,zxzy,zxyz,zzyx,zyzx,zyxz; C≡xxxz,xxzx,xzxx,zxxx,zzzx,zzxz,zxzz,xzzz,xyyz,xyzy,xzyy, zyyx,zyxy,zxyy,yyxz,yxyz,yxzy,yyzx,yzyx,yzxy,yyyz,yyzy, yzyy,zyyy,zzzy,zzyz,zyzz,yzzz,xxyz,xyxz,xyzx,xxzy,xzxy, xzyx,yzxx,yxzx,yxxz,zyxx,zxyx,zxxy;
2/m	A and B
mmm	A
4/m	D≡xxxx=yyyy,zzzz,xxyy=yyxx,xyxy=yxyx,xyyx=yxxy,xxzz=yyzz, xzxz=yzyz,xzzx=yzzy,zzxx=zzyy,zxzx=zyzy,zxxz=zyyz; E≡zzxy=-zzyx,zxyz=-zyxz,xzzy=-yzzx,xyzz=-yxzz,zxzy=-zyzx, xzyz=-yzxz,xxxy=-yyyx,xxyx=-yyxy,xyxx=-yxyy,yxxx=-xyyy;
4/mmm	D
3	F≡zzzz,xxxx=yyyy=xxyy+xyxy+xyyx,xxyy=yyxx,xyxy=yxyx,xyyx= =yxxy,xxzz=yyzz,xzxz=yzyz,xzzx=yzzy,zzxx=zzyy,zxzx=zyzy zxxz=zyyz; G≡xxxy=-yyyx=-(xxyx+xyxx+yxxx),xxyx=-yyxy,xyxx=-yxyy, yxxx=-xyyy,zzxy=-zzyx,zxyz=-zyxz,xzzy=-yzzx,xyzz=-yxzz, zxzy=-zyzx,xzyz=-yzxz; H≡xxxz=-xyyz=-yxyz=-yyxz,xxzx=-xyzy=-yxzy=-yyzx,xzxx= =-xzyy=-yzxy=-yzyx,zxxx=-zxyy=-zyxy=-zyyx; J≡yyyz=-yxxz=-xyxz=-xxyz,yyzy=-yxzx=-xyzx=-xxzy,yzyy= =-yzxx=-xzyx=-xzxy,zyyy=-zyxx=-zxyx=-zxxy;
3m	F and J
6/m	F and G
6/mmm	F
m3	L≡xxxx=yyyy=zzzz; M≡xxyy=zzxx=yyzz; N≡yyxx=xxzz=zzyy; P≡xyxy=zxzx=yzyz; Q≡yxyx=xzxz=zyzy; R≡xyyx=zxxz=yzzy; S≡yxxy=xzzx=zyyz;
m3m	L, M=N, P=Q and R=S,
Y_h, K_h	L=xxyy+xyxy+xyyx, M=N, P=Q and R=S,

$$B_k(\omega,\underline{r}) = -\left(\frac{c}{\omega}\right)\varepsilon_{klm}E_1(\omega,\underline{r})k_m, \qquad (8)$$

and Eq. (5) can be re-written in the form:

$$P_{ei}(2\omega,\underline{r}) = \chi^T_{ijk}(2\omega,\underline{k},I)E_j(\omega,\underline{r})E_k(\omega,\underline{r}), \quad (9)$$

where we have introduced the tensor of total nonlinear 2-nd order susceptibility:

$$\chi^T_{ijk}(2\omega,\underline{k},I) = \chi^{eee}_{ijk}(2\omega,\underline{k},I) - \left(\frac{c}{\omega}\right)\chi^{eem}_{ijl}(2\omega,\underline{k},I)\varepsilon_{lkm}k_m, \quad (10)$$

with ε_{lkm} —— the Levi-Cività antisymmetric tensor.

REMOVAL OF SYMMETRY CENTRE BY DC FIELDS

An electric field $\underline{E}^O$ or magnetic field $\underline{B}^O$ acts on a body in a way to lower its symmetry, i.a. by removal of its centre of symmetry, thus leading to an enhancement of the SHG power from centrosymmetric crystals [1,2] . The nonlinear suscepti- bility tensors of Eq. (5) now become moreover functions of $\underline{E}^O$ or $\underline{B}^O$, e.g. $\chi^{eee}_{ijk}(2\omega,\underline{k},I,\underline{E}^O)$. If the DC electric field $\underline{E}^O$ app- lied to the centrosymmetric crystal is not excessively strong one can write, besides (6) and (7) , the following expansions:

$$\chi^{eee}_{ijk}(2\omega,\underline{k},I,\underline{E}^O) = \chi^{eeee}_{i(jk)l}(2\omega)E_1^O + \chi^{eeekke}_{ijklmn}(2\omega)k_1k_mE_n^O +$$

$$+ \chi^{eeeeee}_{i(jk)(lmn)}(2\omega)E_1^OE_m^OE_n^O + i\chi^{eeekee}_{ijkl(mn)}(2\omega)k_1E_m^OE_n^O +$$

$$+ \chi^{eeeeee}_{i(jk)lmn}(2\omega)E_1(\omega,\underline{k})E_m^*(\omega\underline{k})E_n^O + \dots, \qquad (11)$$

$$\chi^{eem}_{ijk}(2\omega,\underline{k},\underline{E}^O) = \chi^{eemee}_{ijk(lm)}(2\omega)E_1^OE_m^O + i\chi^{eemke}_{ijklm}(2\omega)k_1E_m^O + \dots, \quad (12)$$

Hitherto, in the interpretation of results of DC electric-field induced SHG studies, only the first term of (11) was used.

If the centrosymmetric body is immersed in a DC magnetic field $\underline{B}^O$, the expansions (6) and (7) have to be supplemented with these:

$$\chi^{eee}_{ijk}(2\omega,\underline{k},\underline{B}^O) = i\chi^{eeekm}_{ijklm}(2\omega)k_1B_m^O + i\chi^{eeekmm}_{ijkl(mn)}(2\omega)k_1B_m^OB_n^O + \dots, \quad (13)$$

$$\chi_{ijk}^{eem}(2\omega,\underline{k},I,\underline{B}^{o}) = \chi_{ijkl}^{eemm}(2\omega)B_{l}^{o} + \chi_{ijk(lm)}^{eemmm}(2\omega)B_{l}^{o}B_{m}^{o} +$$

$$+ \chi_{ijklmn}^{eemmee}(2\omega)B_{l}^{o}E_{m}(\omega,\underline{k})E_{n}^{*}(\omega,\underline{k}) + \chi_{ijklmn}^{eemkkm}(2\omega)k_{l}k_{m}B_{n}^{o} +\ldots, (14)$$

The only available SHG study in the presence of a DC magnetic field is for InSb [21] —— a material of the class $\bar{4}3m$ without centre of symmetry.

SHG observations can also be performed in crossed DC electric and magnetic fields [22]. One now has the nonlinear susceptibilities:

$$\chi_{ijk}^{eee}(2\omega,\underline{k},\underline{E}^{o},\underline{B}^{o}) = \chi_{i(jk)lm}^{eeeem}(2\omega)E_{l}^{o}B_{m}^{o} + \chi_{i(jk)l(mn)}^{eeeemm}(2\omega)E_{l}^{o}B_{m}^{o}B_{n}^{o}, (15)$$

$$\chi_{ijk}^{eem}(2\omega,\underline{k},\underline{E}^{o}_{,}\underline{B}^{o}) = i\chi_{ijklmn}^{eemkem}(2\omega)k_{l}E_{m}^{o}B_{n}^{o} + \chi_{ijkl(mn)}^{eemmee}(2\omega)B_{l}^{o}E_{m}^{o}E_{n}^{o}, (16)$$

The expansions (11) – (16) involve, in addition to the well known tensors χ_{ijkl}^{eeee}, χ_{ijkl}^{eemm}, χ_{ijklm}^{eeeem} and the ones discussed in Section 2, some new polar and axial tensors of ranks 5 and 6 for which the numbers of nonzero and independent tensor elements are given, for centrosymmetric bodies, in Table III.

DISCUSSION AND CONCLUSIONS

By Eqs (5) – (16), the observation and amplification of SHG from centrosymmetric bodies requires that various new mechanisms, related with nonlinear susceptibility tensors of ranks 5 and 6, shall be taken into account. This surely complicates the problem considerably. However, we now have at our disposal the experimental possibilities of determining the values of higher nonlinear susceptibilities [23]. Even when these values are not available, Tables I – III and Eqs (5) – (16) still give us the possibility of ajusting the natural configuration of the crystal and the direction of incidence and polarisation of the light beam so as to achieve the maximal SHG signal.

Existing studies [3,4,24] show that second-harmonic radiation is generated chiefly by the surface layer, where field inhomogeneities are far larger than in the bulk of the crystal. This does not apply to plastic crystals or to ones with structural phase transitions [11 – 13], where conditions favor the recurrence of the SHG processes considered by us. However, in bodies with natural and induced optical inhomogeneities, not only SHG but moreover second-harmonic scattering (SHS) takes place [25 – 28]. In general, both in SHG and SHS, a periodic spatial modulation of the nonlinear susceptibilities intervenes [29].

Table II.
The number of non-zero (N) and independent (I) elements of the nonlinear susceptibility tensors of Eqs. (6) and (7). Symmetricity in indices is denoted by parenthesses (i...).

Class	χ^{eem}_{ijk} N	I	$\chi^{eemkk}_{(il)(jm)k}$ N	I	χ^{eemee}_{ijklm} N	I	$\chi^{eeekee}_{ij(kl)mn}$ N	I	$\chi^{eeekkk}_{(il)(jm)(kn)}$ N	I
$\bar{1}$	27	27	243	108	243	243	729	486	729	216
2/m	13	13	121	52	121	121	365	244	365	112
mmm	6	6	60	24	60	60	183	123	183	60
4/m	13	7	117	26	121	61	363	122	351	56
4/mmm	6	3	56	11	60	30	183	62	183	32
$\bar{3}$	21	9	229	36	233	81	715	162	711	72
3m	10	4	112	16	116	40	359	82	359	40
6/m	13	7	117	20	121	51	363	92	359	40
6/mmm	6	3	56	8	60	25	183	47	183	24
m3	6	2	60	8	60	20	183	41	183	20
m3m	6	1	48	3	60	10	183	21	183	12
Y_h, K_h	6	1	48	1	60	6	183	10	183	6

Table III.
The number of non-zero (N) and independent (I) elements of the nonlinear susceptibility tensors of Eqs. (11) – (16). Symmetricity in indices is denoted by parenthesses (i...).

Class	$\chi^{eeekm}_{ij(kl)m}$ N	I	χ^{eemmee}_{ijklmn} N	I	$\chi^{eeeeee}_{i(jk)lmn}$ N	I	$\chi^{eeeemm}_{i(jk)l(mn)}$ N	I	$\chi^{eeeeee}_{i(jk)(lmn)}$ N	I
$\bar{1}$	243	162	729	729	729	486	729	324	729	180
2/m	121	80	365	365	365	244	365	164	365	92
mmm	60	39	183	183	183	123	183	84	183	48
4/m	119	40	365	183	363	122	361	82	357	46
4/mmm	58	19	183	92	183	62	183	43	183	25
$\bar{3}$	231	54	717	243	715	162	713	108	709	60
3m	114	26	359	122	359	82	359	56	359	32
6/m	119	32	365	143	363	92	361	60	357	32
6/mmm	58	15	183	72	183	47	183	32	183	18
m3	60	13	183	61	183	41	183	28	183	16
m3m	54	6	183	31	183	21	183	15	183	9
Y_h, K_h	54	3	183	16	183	10	183	7	183	4

Surely some progress in the SHG study of centrosymmetric crystals can be expected from the latest observations by Yu and Alfano [30] of double- and triple-photon scattering in diamond crystal. Diamond has the space group (Fd3m) and many other crystals of the sodium chloride (Fm3m) , cesium chloride (Pm3m), cesium fluoride (Fm3m) and perovskite (Pm3m) types have the highest symmetry m3m of all known centrosymmetric crystals and it is to be regretted that, expect LiF [3] , they have not as yet been used for SHG in transmission. Crystals like those, and ones belonging to other centrosymmetric classes especially mmm, 4/mmm, 6/mmm, 3̄m should be tested for SHG in the free fieldless state and for induced SHG in electric and magnetic fields. In the latter case, a particularly interesting situation arises from Eqs (11) and (14) consisting in the simulataneous coupling between the self-induced light intensity dependent effect and DC field effect lowering the crystal symmetry. These optico-electric and optico-magnetic coupling effects are especially large in statistically inhomogeneous bodies, where reorientation of asymmetric microelements can occur. As is seen from Eqs (15) and (16), studies of SHG amplification in appropriately selected centrosymmetric crystals by the method of crossed fields $\underline{E}^o$ and $\underline{B}^o$ can also prove of interest.

Obviously, examples of SHG phase matching conditions would have exceeded the limits of this paper. We primarily hoped to stimulate interest in more intense, experimental studies of the structure of centrosymmetric crystals by the methods of SHG.

REFERENCES

1 R.W.Terhune, P.D.Maker and C.M.Savage, Phys.Rev.Letters,8 /1962/ 404.
2 J.E.Bjorkholm, A.E.Siegman. Phys.Rev.154 /1967/ 851.
3 C.C.Wang and A.N.Dumiński, Phys.Rev.Letters,20 /1968/ 668.
4 N.Bloembergen, R.K.Chang, S.S.Jha and C.H.Lee, Phys.Rev. 174 /1968/ 813.
5 N.Bloembergen and P.S.Pershan, Phys.Rev., 128 /1962/ 606.
6 P.A.Franken and J.F.Ward, Rev.Mod.Phys., 35 /1963/ 23.
7 P.S.Pershan, Phys.Rev., 130 /1963/ 919.
8 E.Adler, Phys.Rev., 134 /1964/ A728.
9 S.S.Jha, Phys.Rev., 140 /1965/ A2020; Phys.Rev., 145 /1966/ 500; S.S.Jha and C.S.Warke, Phys.Rev. 153 /1967/ 751.
10 J.Rudnick and E.A.Stern, Phys.Rev. 4B /1971/ 4274.
11 R.C.Miller, Phys.Rev. 134 /1964/ A1319; R.C.Miller and A. Savage, Appl.Phys.Lett. 9 /1966/ 169; L.S.Goldberg and J. M.Schnur, Radio and Electronic Engineer, 39 /1970/ 279;

J.P. Van Der Ziel and N.Bloembergen, Phys.Rev., 6A, 135 /1964/ A1662; W.A. Nordland, Ferroelectrics, 5 /1973/ 287.

12 V.S.Suvorov and A.S.Sonin, Zh.Eksperim. i Teor.Fiz. 54 /1968/ 1044; I.A.Pleshakov, V.S.Suvorov and A.A.Flimonov, Izv.Akad.Nauk SSSR 35 /1971/ 1856.

13 H.Vogt, Phys.Stat.Solidi /b/ 58 /1973/ 705; Appl.Phys. 5 /1974/ 85.

14 H.Rabin, Int.Conference on Science and Technology of Non-metallic Crystals, New Delhi, India, January 13-17, 1969.

15 S.A.Akhmanov, R.V.Khokhlov and A.P.Sukhorukov, Laser Hand-buch /Ed.F.T.Arecchi and E.O.Schulz-Dubois/ North-Holland, Amsterdam 1972, Vol.2, p.1151.

16 S.Kielich, Opto-electronics 2 /1970/ 5; 13 /1971/ 5.

17 J.G.Bergman,Jr. and S.K.Kurtz, Materials Science and Engi-neering, 5 /1969/70/ 235; R.L.Byer, Ann.Rev.Materials Science, 4 /1974/ 147.

18 S.Kielich, Opto-electronics 2 /1970/ 125; Ferroelectrics 4 /1972/ 257 and references therein.

19 S.Kielich, Proc.Phys.Soc., 86 /1965/ 709; Acta Phys.Polon. 29 /1966/ 875.

20 S.Kielich and R.Zawodny, Acta Phys.Polonica, A43 /1973/ 579; Optica Acta 20 /1973/ 867.

21 H.G.Hafele, R.Grisar, C.Irslinger, H.Wacharering, S.D. Smith, R.B.Denis and B.S.Wherrett, J.Phys. C.4 /1971/ 2637.

22 S.Kielich, Optics Communications, 2 /1970/ 197.

23 S.A.Akhmanov, personal information, Moscov, May 1974.

24 J.M.Chen, J.R.Bower and S.C.Wang, Optics Communications 9 /1973/ 132.

25 I.Freund and L.Kopf, Phys.Rev.Letters, 24 /1970/ 1017.

26 S.Kielich, J.R.Lalanne and F.B.Martin, Phys.Rev.Letters 26 /1971/ 1295; J.Raman Spectroscopy 1 /1973/ 119; S. Kielich and M.Kozierowski, Optics Communications, 4 /1972/ 395.

27 G.Dolino, J.Lajzerowicz and M.Vallade, Phys.Rev. B2 /1970/ 2194; G.Dolino, Ibid, B6 /1972/ 4025.

28 D.Weinmann and H.Vogt, Phys.Stat.Solidi /a/ 23 /1974/ 463.

29 I.Freund, Phys.Rev.Letters, 21 /1968/ 1404; C.L.Tang and P.P.Bey, IEEE J.Quantum Electronics, QE-9 /1973/ 9.

30 W.Yu and R.R.Alfano, private communication, September, 1974.

Section VI

Measurement Techniques

A REVIEW OF TECHNIQUES FOR MEASURING SMALL

OPTICAL LOSSES IN INFRARED TRANSMITTING MATERIALS

Lyn H. Skolnik

Air Force Cambridge Research Laboratories (AFSC)

Hanscom AFB, MA 01731

We review some recent developments in techniques for
measuring small absorption coefficients in the infrared spectral region. This paper considers thermal
methods such as thermocouple, optical, and acoustic
calorimetry as well as some direct loss techniques
including laser differential attenuation measurements
and emittance spectroscopy. The various methods are
compared on the basis of sensitivity and ease of
implementation; advantages and disadvantages of each
technique are discussed. Some consideration is
given to methods for separating surface, bulk, and
scattering losses.

INTRODUCTION

Over the past few years there has been considerable research
into reducing optical absorption losses in nearly transparent infrared transmitting materials. New materials growth techniques
and surface preparation procedures, spurred primarily by the search
for ultra-low loss, high power laser windows, have resulted in
crystals with bulk absorption coefficients, β, below 10^{-4}cm^{-1} at
10.6μm. In addition, near infrared transmitting fiber optic materials now have optical losses approaching the 10^{-6} cm^{-1} range at
1.06μm. Since conventional infrared spectroscopic methods become
insensitive for accurately measuring absorption coefficients much
below the 10^{-2}cm^{-1}, a variety of new techniques have evolved for
determining "residual" losses in the transparent region. It is
the purpose of this paper to review the current status in infrared techniques for measuring bulk absorption losses in primarily

the 2-14μm spectral region, and to compare their advantages and
limitations. The review will be divided into two broad categories;
calorimetric methods and direct loss techniques. The former topic
refers to experiments which depend on the conversion of light to
thermal energy resulting in a temperature increase in the sample
which can then be related to the absorption coefficient at the
frequency of the irradiating source. In the latter section, we
consider techniques which measure optical attenuation directly
including emittance spectroscopy for determining the frequency dependence of β. The problem of separating bulk, surface and absorption losses will also be briefly discussed.

     Before beginning, it is instructive to examine the optical
loss levels thus far achieved in some state-of-the-art infrared
transmitting materials. Table I compiles the lowest absorption coefficients which have been measured in some selected materials of
current interest for fiber or integrated optics and infrared laser
windows. These data indicate that some materials such as CaF_2 and
SrF_2 with measured absorption coefficients at 5.25μm of 5×10^{-4}
and 4×10^{-5} cm^{-1}, respectively, are already close to their intrinsically predicted values of 2×10^{-4} and 2×10^{-5} cm^{-1} at this
wavelength. Other materials, however, such as KBr with a measured
absorption coefficient of 4×10^{-4} cm^{-1} at 10.6μm are over three
orders of magnitude higher than the intrinsic multiphonon value
of $\sim 2 \times 10^{-7}$ cm^{-1}. It remains, then, to determine the nature of
limiting residual loss mechanisms in many of these materials.

CALORIMETRIC TECHNIQUES

Thermocouple Laser Calorimetry

     The most widely adopted method for measuring low level optical
losses in the infrared is that of adiabatic laser calorimetry.[1-6]
Thermocouples are attached to the sample periphery and, for a given
incident laser power, the thermal rise and decay as a function of
laser irradiation time is recorded (see Fig. 1). The magnitude or
rate of thermal rise in the sample is proportional to the absorption coefficient, β, at the irradiating laser frequency which can
be approximated from:[3,4]

$$\beta \approx \frac{mC_p}{LP_T} \frac{2n}{n^2+1} \left[\left(\frac{dT_{gain}}{dt}\right)_{T_1} + \left|\frac{dT_{loss}}{dt}\right|_{T_1} \right], \quad \beta L \ll 1 \qquad (1)$$

where m is the sample mass, C_p is the specific heat at constant
pressure, L is the sample thickness, P_T is the laser power transmitted by the sample, n is the index of refraction at the laser
frequency, and $(dT/dt)_{gain}$ and $|dT/dt|_{loss}$ are the temperature

TABLE I

LOWEST REPORTED ROOM TEMPERATURE ABSORPTION COEFFICIENTS
FOR SOME INFRARED TRANSPARENT SOLIDS

Material	Wavelength (μm)	Absorption Coefficient (cm^{-1})
Supersil W1 (fused Silica)	1.06	5.3×10^{-6} (a)
BaF_2	2.7	4.4×10^{-3} (b)
	3.8	3.7×10^{-3} (b)
	5.25	3×10^{-5} (c)
CaF_2	2.7	1.5×10^{-3} (b)
	3.8	6×10^{-4} (b)
	5.25	5×10^{-4} (c)
SrF_2	2.7	1.5×10^{-3} (b)
	3.8	9.5×10^{-4} (b)
	5.25	4.1×10^{-5} (c)
MgF_2	2.7	8.6×10^{-4} (b)
	3.8	4.7×10^{-4} (b)
	5.25	1.4×10^{-2} (c)
Al_2O_3	2.7	1.5×10^{-3} (b)
	3.8	3.1×10^{-2} (b)
MgO	2.7	4.0×10^{-2} (b)
	3.8	5.4×10^{-3} (b)
$NaCl$	1.06	7×10^{-6} (d)
	10.6	1.3×10^{-3} (c)
KCl	5.25	1.5×10^{-5} (c)
	10.6	7×10^{-5} (c)
KBr	5.25	2.1×10^{-4} (c)
	10.6	4.2×10^{-4} (c)
Si	5.25	5.9×10^{-3} (c)
Ge	5.25	1.8×10^{-3} (c)
	10.6	1.2×10^{-2} (c)

Table continued next page

TABLE I. (continued)

Material	Wavelength (μm)	Absorption Coefficient (cm^{-1})
GaAs	5.0	5×10^{-3} (e)
	10.6	8×10^{-3} (f)
ZnSe	3.0	1.3×10^{-2} (g) (at 373°K)
	5.25	1.6×10^{-3} (c)
	10.6	1.0×10^{-3} (c)
CdTe	5.25	4.9×10^{-4} (c)
	10.6	2.5×10^{-4} (c)
KRS-5	10.6	2.2×10^{-3} (c)

(a) D. A. Pinnow and D. C. Rich, Appl. Optics 12, 984 (1973).
(b) J. A. Harrington, B. Bendow, K. Namjoshi, S. S. Mitra, and
    D. Stierwalt, in "Proc. of the Fourth Conf. on High Power
    Laser Windows", Tucson, Ariz. 18-20 Nov 74 (to be published).
(c) T. F. Deutsch, in "Proc. of the Third Conf. on High Power
    Infrared Laser Window Materials", Vol I. (C. Pitha and B.
    Bendow, eds.) AFCRL-TR-74-0085(I), 14 Feb 74.
(d) M. Hass, J. Davisson, H. Rosenstock, and J. Babiskin, in Ref (b).
(e) L. Skolnik, M. Clark, R. Koch, W. McCann and W. Shields, in
    Ref. (b).
(f) C. P. Christen, R. Joiner, S. Nieh, and W. Steier in Ref. (c).
(g) D. L. Stierwalt in Ref. (c).

gain and loss rates evaluated at the same temperature. Thermal
losses are usually approximated by turning off the laser and
measuring the thermal decay rate back to starting ambient temperature.

Although Eq. (1) is the "work horse" of laser calorimetry, its
validity depends on allowing a sufficient calorimetric rating
period such that each point within the sample experiences the same
rate of thermal rise. In general, the detailed shape of temperature-
time calorimetry curves depends on sample and laser beam geometry
as well as on material and environmental factors including refractive index, specific heat, surface absorption, thermal conductivity,
thermal diffusivity, and heat transfer coefficients. Kahan et al[7]
have calculated the effect of variation of these parameters on
thermal history curves by solution of a three dimensional heat flow
equation for a Gaussian laser beam incident on a cubic sample.

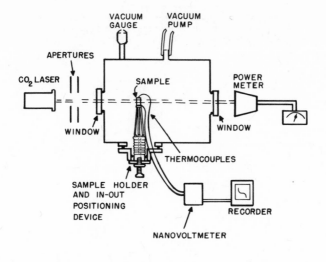

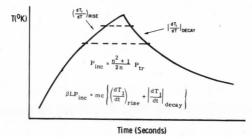

FIGURE 1. Laser calorimeter and typical thermocouple rise and decay curves.

Fig. 2 shows the effect of heat transfer coefficient variation on time-temperature curve recorded for sensor positioned at the center and at the edge of the sample. Note changes in shape of the curves for long time laser irradiations. However if the experiment can be performed rapidly, heat transfer coefficients have relatively little influence on thermal rise curves, especially if the temperature rise is sensed at the irradiating beam center rather than at the sample periphery.[7] We will return to this point later.

One reason that thermocouple calorimetry has been so widely adopted is because of its relative ease of implementation. As shown in Fig. 1, "standard" equipment is required: a laser source and power meter, sensitive nonovoltmeter, thermocouple(s), strip chart recorder, and a calorimeter which can often be simply a plastic box containing a thermal isolation mount for the sample. Two problems, however, are receiving increased attention -- scattering and surface absorption. Direct scattering of incident

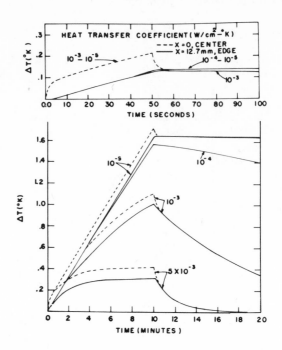

FIGURE 2. Effect of heat transfer coefficient on thermal history curves for a one-inch KCl cube with a bulk absorption coefficient of 1×10^{-3} cm^{-1} and a laser power of 25W. The dashed line is the center temperature rise, while the solid line is the peripheral T-rise. (after Kahan et al[7])

radiation onto the thermocouple appears as an anomalously fast rise in temperature at the moment of laser turn-on followed by an equally sharp decay in thermocouple signal when the laser is switched off.[18] This effect is usually accounted for by subtracting a d. c. level off the thermal rise and decay curves; i.e., a constant scattering contribution is assumed. A more subtle effect, however, involves bulk or surface scattering of incident light with subsequent trapping and absorption thus resulting in indirect heating of the thermocouple. In this case, the absorption path length is no longer simply the sample thickness and, consequently, the calorimetrically measured β is composed of the sum of a single pass absorption coefficient and a scattering absorption loss. A theoretical treatment of scattering and trapping has been given by Winsor.[9] Shiozawa, Roberts and Jost[10] have developed a rather unique calorimetric approach to the scattering and absorption problem. Absorption loss is calorimetrically measured for two (or more) points on a sample. Measurements are repeated after the sample has been painted with a highly absorbing coating so that all scattered

radiation is absorbed and none is allowed to escape. The fraction
of trapped radiation which is absorbed is determined by assuming
that this fraction is constant from point to point on the sample
since there will be fairly uniform redistribution of radiation
throughout the window before absorption occurs. The single pass
absorption coefficient can then be calculated from the difference
between the absorption coefficient measured for the unpainted sample
and the absorption loss determined due to trapped radiation. For
some samples of CdTe it was found that the absorption due to trapping was equal to or exceeded that due to single pass radiation.[10]

Surface absorption is becoming an ever increasing problem as
purification and processing techniques are improved. For example
in reactive atmosphere processed KCl, the calorimetrically
measured absorption loss can be reduced from near 10^{-3} cm^{-1} to below
10^{-4} cm^{-1} if the surfaces are HCl-etched by using a method developed
by Davisson et al at NRL.[11] Yet even at this low level there is convincing evidence that a substantial fraction of this 10^{-4} cm^{-1} value
is still dominated by extrinsic surface loss.[12] The standard calorimetric technique used to separate bulk and surface components is to
measure absorption as a function of sample length;[13] i.e., one
assumes that $\beta L_{meas} = \beta L_{bulk} + 2\sigma$ where σ is the loss per surface.
Thus, by extrapolating a plot of βL_{meas} versus L to zero length, σ
is determined. This technique suffers from (1) the time required to
make such measurements since many lengths are required for accuracy,
and (2) the assumption that all prepared surfaces for different
lengths are identical with respect to absorption loss and scattering
effects. Kahan[7] calculated the influence of surface absorption
on calorimetric thermal history curves for cubic geometries. Recently, Hass et al[14] performed an experiment which separates bulk
and surface absorption in a single measurement without the need for
multiple sample lengths. By using long rod-like geometry and placing
the thermocouple in the central periphery of the z-axis, thermal
diffusion from the cylinder faces can be temporally separated from
diffusion due to bulk heating of the sample. As shown in Fig. 3,
the thermal rise curve will be composed of two distinct portions;
the initial slope is representative of bulk absorption, while after
some time characteristic of thermal diffusion from the faces, an increase in rise rate is noted due to the superposition of bulk and
surface heating. Note, however, that in order to calculate surface loss, the heat transfer coefficient must also be determined.
With extremely careful thermocouple attachment and baffling techniques, Hass et al have measured bulk absorption coefficients as
low as 7 x 10^{-6} cm^{-1} in NaCl at 1.06μm.[14]

Calorimetric techniques have also been employed for measuring
optical losses in fiber optic materials. Pinnow and Rich[15] have
reported attenuation coefficients as low as 2.3 db/km (1 db/km =
2.3 x 10^{-6} cm^{-1}) for Supersil W1 fused silica at 1.06μm. Their

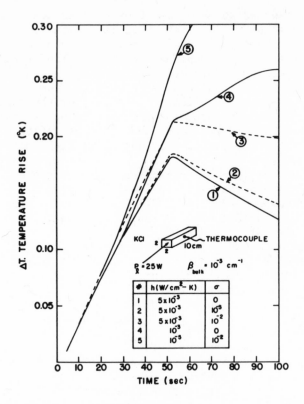

FIGURE 3. Calculated combined effects of surface absorption (σ), and heat transfer coefficient (h) on thermal history curves for long geometry sample of KCl. (A. Kahan, private communication - see also Hass et al[14].)

method consists of inserting the sample within a laser cavity thereby taking advantage of over two orders of magnitude increase in available intracavity circulating laser power. In this way, several tenths of a degree temperature rise can be realized even for very low loss materials.

Laser calorimetric thermal history curves are generated by sources of coherent, highly monochromatic radiation, and therefore interference effects can sometimes be observed.[16,17] As the temperature of the irradiated sample rises, its optical path length, nL, changes which results in sinusoidal variation of transmitted laser beam intensity. Since the power absorbed by the sample is always proportional to the transmitted power,[17] the temperature versus time curve will suffer slope changes corresponding to oscillations in transmitted energy; the slope will be a maximum

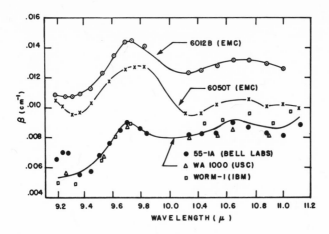

FIGURE 4. Tunable laser calorimetry for three different GaAs samples. (after Christensen et al[19]).

at points of peak transmitted power and a minimum (often going to zero or even negative) when the transmitted power reaches its minimum. This "self-induced thermal" Fabry-Perot effect will, of course, be marked for high index materials where finesse is high and if the temperature increases by more than $\lambda/[2d(n\ell)/dT]$. It has been observed by Weil[16] in GaAs and by Skolnik et al[17] in ZnSe. Wedging the sample to avoid the interference effect introduces a path length uncertainty; for interference situations, it is therefore necessary to calculate the absorbed power point by point from the transmitted laser beam intensity.

In order to assess the mechanism(s) responsible for residual optical losses, both temperature and frequency dependence of the absorption coefficient have been measured calorimetrically for some IR transmitting materials. Harrington and Hass[18] first reported the temperature dependence of β at 10.6μm for NaCl, KCl, and NaF over a temperature range of 300°K to near 1000°K. Christensen et al[19] have calorimetrically measured β versus T for GaAs at 10.6μm while similar experiments were performed on GaAs and ZnSe by Skolnik et al.[20] Calorimetric measurements of temperature variation of β require accurate furnace or dewar temperature control (usually to within a few tenths of a degree) as well as careful attention to scattered laser energy within the furnace. In addition to a good temperature controller, a large thermal mass surrounding the sample and adequate shielding from convection currents are usually necessary to resist sudden temperature fluctuations. Detail must also be given to insuring intimate contact between sample and measuring thermocouple and between reference mass and its thermocouple. The reader is referred to the original literature for more complete experimental discussions.[18-21]

Frequency dependence of the absorption coefficient has been measured calorimetrically for KCl[22,24] and GaAs[19] by using several discrete CO_2 laser lines in the 9.2 to 11.2μm range. Fig. 4 shows some tunable laser calorimetry spectra for three GaAs samples obtained from different sources and grown by different techniques. The similarity of these spectra suggests that high resistivity GaAs may either be approaching its intrinsic limit at 10.6μm or that all melt grown materials share a common impurity or defect. Fig. 5 shows calorimetric temperature dependence data on GaAs which supports the contention that this material is nearly intrinsically dominated at low temperatures but electronically dominated by deep level thermal ionization at elevated temperatures.

Summarizing, then, thermocouple laser calorimetry is the most widely used infrared technique for measuring low optical absorption losses. Its advantages are (1) relative ease of implementation, (2) rapidity of measurement, and, (3) high sensitivity, as high as 10^{-4}-10^{-6} cm^{-1} if vacuum or intracavity calorimetry is used. Disadvantages include (1) extreme care necessary for thermocouple placement and attachment for very low absorbing samples, (2) difficulty in separating bulk and surface scattering effects from true single pass bulk and surface absorption losses, (3) lack of information provided on point-to-point sample homogeneity, and, (4) the limitation of measurements to frequencies where suitable power laser sources are available.

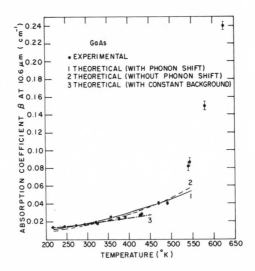

FIGURE 5. Temperature dependence of β measured calorimetrically at 10.6μm for high resistivity GaAs. The departure from theoretical curves is attributed to thermal ionization of free carriers. (after Skolnik et al[20]).

Optical Laser Calorimetry

Some of the aforementioned difficulties with thermocouple calorimetry can be minimized by remotely sensing sample temperature rise. A few optical calorimetric approaches have been devised. Hordvik and Skolnik[23] first employed a Twyman-Green interferometer to monitor temperature rise. By inserting the sample in the interferometric probe beam, fringe movement can be observed due to change in sample optical thickness which results from laser heating. If the expansion coefficient and temperature change of the refractive index are known, or alternatively, if a prior calibration of fringe shift versus temperature rise is performed, then fringe movement can be converted to sample temperature change and a calorimetric temperature versus time curve generated. This technique was refined by Skolnik, Hordvik and Kahan[24] by replacing the conventional Twyman-Green interferometer with a sensitive Hewlett-Packard laser Doppler interferometer (LDI).[25]   A diagram of this set-up is shown in Fig. 6. The interferometer uses a He-Ne laser as a probe beam, while sample heating is accomplished by a laser at whose frequency the absorption coefficient is to be determined. The He-Ne lasing level is Zeeman split into two closely spaced frequencies, f_1 and f_2 (f_1-f_2 ≈ 1.8 Mhz). One line is isolated by polarization filters and the other is sent through the sample and returned. When the sample is irradiated, f_1 undergoes a Doppler shift due to temperature change of sample optical path. The Doppler shift is detected by optically heterodyning and subtracting the two frequencies; the frequency shift is then time integrated to yield optical path length change and hence sample temperature rise.

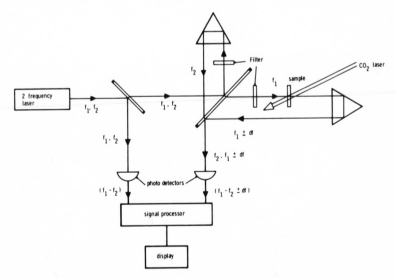

FIGURE 6.   Laser Doppler interferometer (LDI) for measuring small absorption coefficients.   (after Skolnik et al[24]).

That is,

$$dT/dP = [2nL(\frac{n-1}{n}\alpha_L + \delta)]^{-1} \tag{2}$$

where n is the refractive index at the interferometer wavelength λ, L is the sample length, $\alpha_L = (1/L)(dL/dT)$ is the linear expansion coefficient, $\delta = (1/n)(dn/dT)$ is the temperature coefficient of the refractive index, and $P = 2(n-1)L = K\lambda$ is the effective optical path length (K is an integer). By using the LDI, it is possible to measure temperature rise in times short compared to a thermal diffusion time (i.e. in the linear region where $T \propto t$) so that thermal losses can be considered negligible. In this way effects of heat transfer coefficients (Fig. 2) are minimized. Also in the linear time region, focussing the laser beam increases temperature rise so lower power lasers can be used for measuring low losses. Rapid measurement capability is also important for pulsed rather than cw laser irradiation. However, two significant problems may arise with interferometric calorimetry. Firstly, change of index with stress also results in optical path change; this second order effect is difficult to calibrate out since laser heating is usually accomplished with a Gaussian shaped beam. Optical path variations due to stress may be important in those samples containing high residual stresses which may change differently during calibration heating or laser irradiation. In addition, this effect is important in materials with large stress-optic coefficients such as alkali halides. A second problem arises in cases where dn/dT and α_L have opposite signs (again for alkali halides). As shown in Eq. 2, the sensitivity of the technique decreases if the index change and thermal expansion terms are cancelatory. On the other hand, sensitivity can be increased by using longer samples (Eq. 2). Generally, for semiconductor materials, LDI calorimetry can measure absorption losses in the mid 10^{-5} cm^{-1} range, while for alkali halides sensitivity is decreased by an order of magnitude. Bernal has also used a holographic interferometric technique with limited success for these materials.[26]

Nurmikko et al[27] have recently suggested an optical calorimetric technique for measuring losses in semiconductors by monitoring the shift in electronic absorption edge with temperature. By choosing the proper monitoring frequency near the electronic edge, a sensitive measure of absorbed laser power can be obtained. With Nurmikko's technique, a thin sample can (and must) be used since electronic absorption by the probe beam is high. This factor is advantageous in that very high power heating lasers may not be required; however, thin sample preparation can often introduce strains so that stress dependence of the band edge must also be considered. Nurmikko's technique, which is discussed in detail in this volume, has been employed to measure absorption coefficients of ZnSe, CdTe, and some III-V compounds down to the 10^{-3} cm^{-1} range

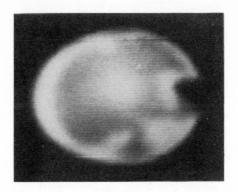

FIGURE 7.   Non-radial reradiation in a centrally laser irradiated
Si window.   The white areas are regions on intense 3-5μm reradiation.   The black area at 3 o'clock is an insulated sample
clamp.   (after Skolnik et al[28]).

at 10.6μm.   A detection limit of 5 x 10^{-5} cm^{-1} has been extrapolated
for this method.[27]

     Perhaps the obvious way of remotely sensing sample temperature
rise is to measure sample thermal emittance resulting from
laser irradiation.   This task, however, is experimentally difficult
in cases where temperature rises are small, and, in general, would
require cooling of the sample background and imaging optics to
low temperatures.   While spectral emittance as a direct measurement
loss technique will be pursued later, we briefly describe some
spatial emissivity measurements performed by Skolnik, Bendow,
Gianino and Cross.[28]   An infrared vidicon sensitive in the 2.5 -
4.5μm range was used to provide real time television pictures of
thermal reradiation from samples irradiated by a CO_2 laser.   The
vidicon, manufactured by General Electric Corporation, is liquid
nitrogen cooled and employs a copper II-doped Germanium sensing
layer.   It is capable of detecting an irradiance of 4.5x10^{-7} w/cm^2
incident on the sensing surface.   Signal irradiance from a sample
can be converted to temperature rise by measuring the ratio of
emittance at two frequencies.   Fig. 7 shows the spatial temperature
distribution developed in a Si window which was centrally irradiated
by a 50 watt CO_2 laser.   Note that the heat distribution is non-
radial because of incident laser beam scattering combined with subsequent absorption and reradiation.   Fig. 7 demonstrates the danger
of calculating time-temperature thermocouple response curves based
on radial heat flow modeling when scattering is significant.   Thermal maps obtainable with the high resolution vidicon system can
provide absorption homogeneity data on large samples.   This thermal
imaging technique is unaffected by directly scattered CO_2 laser
radiation since vidicon spectral response does not extend below

4.5μm. The method has also been used to measure absorption in coatings on transparent materials.[28]

We summarize some advantages of optical laser calorimetry: (1) direct scattering of irradiating laser energy does not influence the measurement; (2) measurements can be performed in times short compared to a thermal diffusion time so that thermal losses are minimized, pulsed laser measurement losses are possible, and calculations for β are simplified; (3) in the time short regime (Tαt), temperature increase depends on laser power density rather than total laser power; thus by focussing the heating laser beam, lower power lasers can be utilized for measuring small β; (4) interferometric scanning or thermal imaging techniques allows measurement of absorption in homogeneity or scattering effects; (5) interferometric calorimetry sensitivity is increased by using longer samples (Eq. 2), whereas to first order, neglecting heat loss effects, thermocouple calorimetric sensitivity is independent of sample length since sample mass increased proportionally with L (Eq. 1); (6) by the same token, interferometric methods are much less sensitive to surface absorption than are thermocouple techniques since, in the former, an overall change in total optical path is sensed and effective surface thicknesses are many orders of magnitude smaller than bulk sample lengths. Thus, interferometric calorimetry measurements are more indicative of bulk losses than of surface absorption.

Some disadvantages of most optical calorimetric methods are: (1) difficulty in implementation; i.e. more sophisticated (and expensive) equipment is required than is necessary for thermocouple calorimetry; (2) difficulty in easily obtaining equivalent thermocouple calorimetry sensitivities; (3) for interferometric methods, the need to account for thermally induced and/or residual stress.

Acoustic Calorimetry

The photoacoustic effect was first investigated by Tyndall, Roentgen, and Bell[29] in 1881. This phenomena occurs when a chopped beam of light is incident on an absorbing gas enclosed in a cell; light energy absorbed by the gas is converted to kinetic energy giving rise to pressure differentials in the cell. The rise in pressure, which can be detected by using transponders or sensitive microphones, is proportional to the temperature increase of the gas and therefore is a measure of absorption loss at the frequency of the irradiating source. Although the photo-acoustic effect has been used extensively for gas analysis,[30] it is only recently that this technique has been pursued for measuring optical losses in solids. Rosencwaig[31,32] has utilized optoacoustic spectroscopy in the visible region to study absorption and de-excitation processes in Cr_2O_3, Rhodamine-B dye, and in some biological

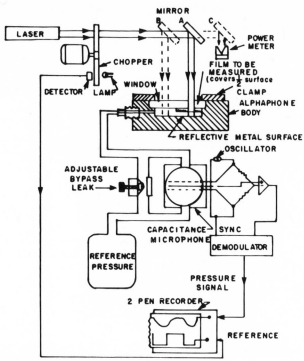

FIGURE 8. Alphaphone system for acoustically measuring absorption losses. (after Kerr[33]).

materials (Cytochrome c, whole blood, RBC, and Hb). Kerr[33] has employed "irsonics" for measuring optical losses in thin films deposited on transparent substrates. A diagram of his instrumentation, called the Alphaphone, is shown in Fig. 8. The sample which is partially coated on one side forms the windows of a chamber filled with a gas that does not absorb at the irradiating laser frequency. (Kerr used about 1.2 atmospheres of nitrogen, a CO_2 laser for the irradiating source, and measured CaF_2, ThF, and ZnS coatings on KRS-5 substrates.[33]) The chopped laser beam irradiates the sample and heat absorbed in the window is partially transferred to the gas causing a pressure rise within the cell. The pressure change is detected by the capacitance microphone and bridge electronics (a Diametrics Barocel) which, in this case, is capable of recording pressure differentials down to 10^{-8} atmospheres. (Rosencwaig,[32] on the other hand, placed the sample inside a gas filled cell and used an electret microphone and lock-in amplifier. This arrangement allowed him to detect 0.1 μ watt of heat transferred to the gas). Assuming a constant cell volume, the acoustic signal will be proportional to the temperature rise in the gas which can then be related to the absorption coefficient of the sample. A complete discussion of thermal analysis, calibration, operating procedure, and errors (such as thermal expansion of the window) can be found

in Ref. 33. Alphaphone sensitivity calculations show that losses
of 10^{-4}/surface can be measured with a signal to noise ratio of
10:1 using a laser power of only 10W.[33] As mentioned by Kerr, this
technique also has the advantage of being insensitive to surface
scattering. Although laser optoacoustic techniques have not been
widely exploited for measuring infrared bulk absorption losses,
it would appear that further pursuit is warranted. Indeed, might
not surface and bulk absorption be separable by varying the chopping rate? That is, rapid chopping rates would allow only heat
absorbed near the surface to be transferred to the surrounding gas
whereas chopping rates corresponding to many diffusion times would
permit bulk heat to diffuse to the surface thereby also contributing to gas temperature rise.

An acoustic calorimetry technique which utilizes thermoelastic
properties of solids has recently been developed by Hordvik.[34] When
a chopped or pulsed laser beam is incident on a solid, an acoustic
wave is generated because of absorptive heating of the material.
By attaching strain gauges or piezoelectric transponders to the
sample, the acoustic wave can be readily detected as a voltage output from these sensors. The duration of the laser pulse is adjusted so as to be substantially longer than the inverse elastic
resonance frequency while the repetition rate or chopping frequency
is long enough to allow the acoustic wave to decay between pulses.
(Typically 50-500 Hz). In actual operation, Hordvik uses two
Piezoelectric transponders spaced symmetrically about the incident

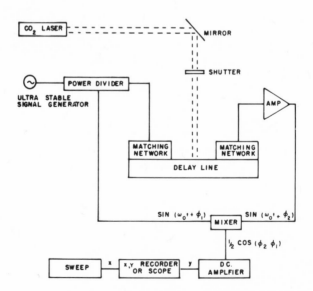

FIGURE 9. Surface Acoustic Wave (SAW) method for probing surface
absorption losses. (after Parks et al[35] and White et al[36]).

beam; one is bonded to the sample periphery while the other is placed in close proximity. By bucking the output voltages or by using a Wheatstone bridge, this two detector arrangement allows some compensation for scattering. Absolute calibration is accomplished by conventionally measuring the absorption coefficient at some wavelength where it is high, and at the same time, by using a laser at this same wavelength to measure the transponder voltage output per pulse input energy. Once voltage/mJ/cm^{-1} is determined at one wavelength, this conversion factor is used for absolute calculation of β at another laser wavelength where absorption is low. By using phase sensitive detection locked-in to chopping frequency, a sensitivity of nearly 10^{-5} cm^{-1} has been achieved in the visible region with only 200 mW Argon laser input power.[34] Sample scattering is the major effect limiting absorption loss detection.

Surface acoustic waves (SAW) have a velocity which is dependent on the temperature of the medium through which the waves are launched. Parks, et al[35] at Univ. of S. California, working with White, et al[36] at the Naval Weapons Center have taken advantage of the temperature dependent SAW velocity to measure optical absorption losses. The apparatus is diagrammed to Fig. 9. A surface acoustic wave is launched across a delay line by an interdigital transducer. The phase shift introduced in the SAW by the laser heated surface is detected by heterodyning part of the generator signal with the detected SAW frequency thereby measuring the rate of change of the phase shift across the delay line. If the laser beam intensity profile, coefficient of thermal expansion, and thermal coefficient of SAW velocity are known, then the optical absorption coefficient can be calculated from the measured phase change. The determination of β requires knowledge of acoustic wave interaction with heat transfer processes from surface to bulk of the delay line and vice-versa. By using surface acoustic waves of varying frequencies, the penetration depth of this interaction can be controlled, i.e. SAW energy is generally confined to a penetration depth of about one acoustic wavelength. Therefore, the possibility exists of separating surface and bulk absorption by varying the SAW input frequency; e.g., the SAW penetration depth is about 10µm at 300 MHz but increases to ~100µm at 30 MHz. In practice it may be difficult to directly launch acoustic waves on transparent non-piezoelectric materials such as KCl. For these cases Parks, et al[35] have developed a liquid coupling technique whereby the KCl substrate is coupled to a piezoelectric driver (e.g., quartz or LiNbO₃) via a thin fluid layer such as xylene or ethylene glycol. The SAW method can be made sensitive enough to detect millidegree changes, and therefore, with low laser powers, absorption coefficients in the 10^{-4} to 10^{-5} cm^{-1} range can be measured. Of course this method is also a sensitive one for detecting infrared radiation if a delay line is used that is strongly absorbing rather than transparent at the frequency of interest. A further discussion of the SAW technique is presented by Parks, et al in this volume.

In summary, acoustic calorimetric techniques, while still new, show great promise for determining both bulk and surface absorption losses as well as for measuring absorption coefficients of thin films. They offer particular advantage over conventional calorimetry because in acoustic measurements, a.c. rather than d.c. detection techniques are used. Narrow electrical bandwidths and phase sensitive detection can be employed so that millidegree temperature rises can be accurately measured. Correspondingly, absorption losses in the 10^{-5} cm^{-1} range are detectable using only a few tenths of a watt of laser power.

Thermal Lens Techniques

The thermal lens technique has been exploited primarily for measuring absorption in low loss liquids[37-40] and we mention its application to solids here. When a liquid or solid absorbs energy from a spatially non-uniform laser beam, thermal lensing can result. In solids, this effect is due to temperature gradients which arise in the material resulting in spatially varying changes in refractive index, thermal expansion, and stress-optic effects which cause the material to act as an aberrating lens. The focal length which can be induced in a transmitted laser beam by a liquid or solid sample will vary inversely with the optical absorption coefficient at the irradiating laser frequency. For a liquid, the induced steady state focal length, F, for a gaussian beam is given by:[38]

$$F = \frac{\pi k w^2}{0.24 \beta P L \frac{dn}{dT}} , \tag{3}$$

where K is the thermal conductivity, w is the beam radius, β is the absorption coefficient, P is the total laser power, L is the sample thickness, and dn/dT is the temperature change of the refractive index. Gordon, et al[38] and Solimini[39] have applied this technique to measuring absorption coefficients in a variety of low loss liquids at He-Ne laser frequency by measuring variations in the spot size of the output beam. Kohanzadeh and Auston[40] calculated the induced focal length by using the change in beat frequency between a strong fundamental TEM$_{00}$ mode and a weaker transverse TEM$_{01}$ probe mode. Typically, beat frequency changes on the order of 2 MHz were observed and an absorption coefficient as low as 5 x 10^{-5} cm^{-1} was measured for CCl$_4$ at .6328μm using 0.2 watts of laser power.[40]

Hordvik[41] has investigated the thermal lens method for measuring absorption losses in solids at 10.6μm. Several problems can be expected. First, the longest induced focal length that can be accurately determined by measuring spot size is three to four times the confocal parameter of the laser[41] ($Z_o = \dfrac{\pi W_o^2}{\lambda}$ where W_o is the waist radius). For longer focal lengths, the smallest beam size will not change much; attempts to induce shorter focal lengths by focussing the beam onto the sample results in no net improvement since F and Z both decrease in the same proportion as the intensity is increased.[41] Since confocal parameters are typically longer for IR than for visible lasers, a slight decrease in inherent sensitivity is expected for long wavelength absorption measurements. A more serious problem that is encountered is that the thermal conductivity for transparent solids of interest is generally two to three orders of magnitude greater than are those for organic liquids which have been measured using this technique. For example, K of KCl is 220×10^{-4} cal/cm sec ^{o}K[42] while K for CCl_4 is 2.5×10^{-4} cal/cm sec ^{o}K[37]. Hence from Eq. 3 for a fixed induced focal length, the measurable sensitivity is decreased by a factor of 100. Furthermore, a greater thermal conductivity lessens the thermal lens effect itself. Finally, it must be remembered that Eq. 3 is valid only for liquids, while for solids it is not only dn/dT that must be considered, but the linear expansion coefficient and stress-optic effects as well. Therefore, in some alkali halides where linear expansion coefficient and dn/dT have opposite signs, the sensitivity of this technique is further degraded. With moderate laser powers it appears that in the infrared region the thermal lens method is not useful for measuring absorption coefficients much below the 10^{-3} cm^{-1} level.

DIRECT LOSS MEASUREMENT TECHNIQUES

Direct Laser Loss Measurements

We briefly review here some sensitive non-calorimetric techniques which have been developed for directly measuring optical attenuation by using available laser sources. Two-beam bridge-balance methods[43] and direct power attenuation measurements as a function of length in which the sample is immersed in an index matching liquid[44] have been used for determining losses in fiber optics materials. These techniques have been applied both at discrete laser wavelengths and by using broad band W-I or Xe lamp sources to obtain loss spectra in the 0.5 to 1.1μm range; they have been reviewed by Miller, et al[45], and will not be included here.

Direct laser loss measurements are usually differential in nature involving two "identical" samples of different thickness or one sample of variable thickness. Several methods have been developed by Rehn et al[46] at NWC and one is diagrammed in Fig. 10. The sample is wedge or prism shaped. With the return mirror in position 1, combined bulk loss (absorption + scattering) is measured; concurrently bulk and surface losses are separated since the chopper mirror alternatively directs the laser beam through thin and thick portions of the sample. A lock-in amplifier and ratio technique are used to directly measure the difference in return signal received from the thick and thin portions of the sample; by recording the ratio of the derivatives of these return signals, surface scattering, reflectivity, and absorption parameters will cancel out.[44] Surface reflectivity and absorption are measured with the mirror in positions 2 and 3, and surface scattering is determined independently by using a Coblenz sphere. A complete analysis of this and other phase sensitive detection techniques for differential loss, reflectivity, and scattering measurements have been given by Rehn et al.[46]

A Q-meter bridge null technique for measuring β has been suggested by Birnbaum et al.[47] The apparatus which was originally designed for determining stress birefringence is shown in Figure 11. Collimated radiation from a stabilized laser passes through a polarizer and a quarter waveplate isolator. Circularlly polarized radiation excites two resonance modes in the Fabry-Perot interferometer (FPI) with $\|$ and $\perp$ polarizations when the resonance frequency of each mode is equal to the laser frequency. The resonance frequency of the FPI can be tuned by moving an interferometer mirror with a piezoelectric element. The FPI resonance modes, separated by a polarizing beam splitter are displayed concurrently on a dual-beam oscilloscope whose horizontal sweep is synced to the mirror drive. By measuring the change in width of the resonance curves (i.e. the change in Q) with and without the sample in the FPI, the absorption loss at the probe laser frequency can be

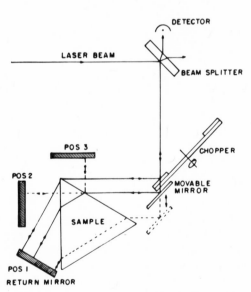

FIGURE 10. Laser differential loss method for separating bulk and surface effects. See text (after Rehn et al[46]).

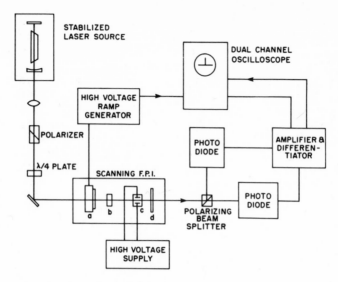

FIGURE 11.   Optical bridge Q-meter technique for measuring small β
(after Burnbaum et al[47]).

determined. However, accuracy will be limited by measurement of resonance pulse width and shape. Therefore, a mode-differentiation and
null technique is used wherein a compensating electro-optic cell
allows the measurement of a difference in zero crossing frequencies
between the two separated polarizations in terms of applied cell
voltage.

Spectroscopic Loss Measurements

The most severe limitation on laser calorimetric or direct
attenuation techniques is that they can only supply the absorption
coefficient at frequencies where suitable laser sources exist.
Often the frequency dependence of β over a broad spectral range is
required in order to unambiguously determine the mechanism(s)
responsible for residual losses in the transparent region.   The
transmission of a material for near normal incidence and neglecting
interference effects is given by

$$T = \frac{(1-R)^2 \, e^{-\beta L}}{1-R^2 \, e^{-2\beta L}} \; ; \quad R \simeq \frac{(n-1)^2}{(n+1)^2} \quad (\beta < 10^{+3} \, cm^{-1}), \tag{4}$$

where R is the reflectivity, β is the absorption coefficient at the
wavelength of incident light and L is the sample thickness.   For
small β such that βL<<1, Eq. 4 can be approximated by:

$$T \simeq \frac{2n}{n^2+1} (1 - \beta L) \tag{5}$$

Therefore, unless long samples are used, transmission measurements are insensitive to the value of β. Further, long samples can often cause spectrometer beam defocussing especially if the refractive index is high, and care must be taken to insure that defocussing effects do not appear as a false transmission loss. Also, the refractive index as a function of wavelength must be accurately determined. Deutsch[48] has successfully employed a differential spectrophotometer technique for measuring absorption coefficients as low as 2×10^{-3} cm^{-1} in the 2-40μm region. A dual beam Perkin-Elmer 457 Spectrophotometer was used with a thin sample placed in the reference beam and a thicker sample in the measurement beam. Since the spectrometer records the ratio of the two transmitted beams, the measured transmission is

$$T_{meas} = \frac{T_{sample}}{T_{ref}} = e^{-\beta \Delta L} \left[\frac{1-R^2 \, e^{-2\beta L_{sample}}}{1-R^2 \, e^{-2\beta L_{ref}}} \right], \quad \Delta L = L_{sample} - L_{ref} \tag{6}$$

For materials with low reflectivity such as KCl or CaF$_2$, Eq. 6 becomes $T_{meas} \simeq e^{-\beta \Delta L}$ and the absorption coefficient can be determined almost independently of reflectivity losses. For small β such that $\beta L \ll 1$, but for samples with large R, the absorption coefficient can be approximated from

$$\beta \simeq \frac{1-T_{meas}}{\Delta L} \left(\frac{1-R}{1+R} \right) \tag{7}$$

Deutsch has generated differential absorption spectra for a variety of fluorides, alkali halides, Al$_2$O$_3$, Si, Ge, GaAs, CdTe and ZnSe over broad frequency range in the infrared.[45] Sample path length differences of 2 to 5 cm were used. Since the accuracy of the transmission measurement was 1 percent, the uncertainty in β was 10% near T = .9 and ~3% for T = .5.

Another spectroscopic technique which has been used extensively by Deutsch[49] in the infrared is that of internal reflection spectroscopy (IRS) wherein the incident light is caused to make several internal reflections in the sample before exciting the material. IRS techniques are described in detail by Harrick in Ref. 50. Since the internally reflected light samples the surface of the material many times, this technique is sensitive to surface contamination and has been used to study effects of various surface preparation and cleaning procedures on fluorides and halides.[49]

Emittance spectroscopy is an inherently sensitive and direct technique for obtaining the frequency dependence of the absorption coefficient in a region of low optical loss. It has been used by Stierwalt and Potter[51], Stierwalt[52], and others[53-56] for investigating loss mechanisms in highly transparent solids. When a material is heated above or cooled below ist ambient temperature, it will emit or absorb radiation from its environment. Under thermal equilibrium conditions, emittance, E, and absorbance, A, will be identical at every wavelength:

$$E(\lambda,T) \equiv A(\lambda,T) = \frac{(1-R)(1-e^{-\beta L})}{1-R\,e^{-\beta L}} \quad , \tag{8}$$

where R, β, and L are as previously defined. From Eq. 8 for $\beta L \ll 1$, the emittance becomes

$$E \approx \beta L. \tag{9}$$

Therefore, in the highly transparent region, the emittance is directly proportional to the absorption coefficient and is nearly independent of reflection loss, in contrast to the transmission (Eqs. 4,5) which is strongly sensitive to reflectivity and nearly independent of β under highly transparent conditions. In a highly absorbing region, on the other hand, the transmission is nearly zero and the emittance approaches $1-r(\lambda,T)$ where r is the sample reflectance. Hence emittance spectroscopy can also be useful for providing reststrahl data[51] and for evaluating perfection of anti-reflection surface coatings.

Emittance measurements are performed by comparing the spectral irradiance of the sample to that of a reference blackbody at the same temperature. If low absorption coefficients are to be measured at room temperature, it becomes necessary to reduce background radiation to a low level by cooling the sample environment and spectrometer optics to cryogenic temperature. In addition, the spectrometer optics must be adequately baffled from stray radiation and the sample (and chopper) slightly angled so that stray radiation is not reflected by the sample towards the detector thus appearing as a false emitted signal. Sample temperature must also be accurately controlled during a spectral run (usually to within $\pm$ .05°K); temperature control can be accomplished with electrical heaters and with cryogenic and hot fluids. Solid state emittance techniques have been described by Stierwalt and Potter[51], by Stierwalt[52], and by Skolnik et al[54].

Absorption spectra obtained by emissivity measurements in the 2-20 micron region have been reported by Stierwalt[52] for NaF, KCl, and ZnSe. The latter two spectra are reproduced in Fig. 12. From this frequency and temperature dependence data it is evident that multiphonon absorption does not dominate the behavior of either of these materials in the mid- and near IR region. Emittance spectra

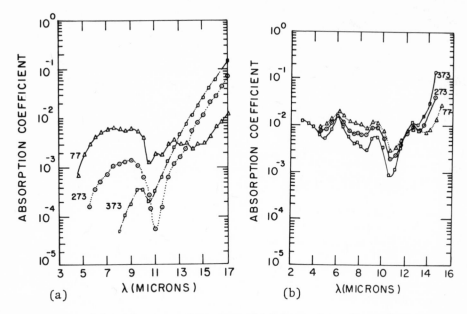

FIGURE 12.   (a) Emittance Spectra of high purity (reactive atmosphere processed) KCl.   (b) Emittance Spectra of chemical vapor
deposited ZnSe. (after Stierwalt[52]).

of CaF_2, BaF_2 and SrF_2 may be found in the work of Harrington et
al[55], while Skolnik et al[54] have also reported emittance spectra
of KCl, ZnSe and GaAs.   A comparison of absorption coefficient
values obtained by laser calorimetry and emittance spectroscopy
have been made by Lipson, Skolnik and Stierwalt[8,22] as a function
of surface finishing for KCl.   As predicted by Eq. 9, it was found
that emittance results were less affected by surface roughness
than were calorimetric absorption values; in general, lower values
for β were obtained by emissivity than by calorimetry for rough or
poorly cleaved surfaces.   For well polished and cleaned surfaces,
data obtained by both techniques were in good agreement.

In principle, given existing high D* infrared detectors, spectral emissivity should provide a sensitive way of obtaining frequency and temperature dependence data on low absorbing materials.
Indeed, calculations show that noise equivalent emissivities in the
10^{-7} to 10^{-9} cm^{-1} range[54,56] should be attainable at 10.6μm (depending on spectrometer throughput) for a sample held at 300°K against
a 77°K background.   Presently, however, light scattering effects
(e.g. from sample heater elements, etc.) have limited emittance
detection levels to 10^{-5} cm^{-1} around the 10μm region.

TABLE II

"READILY" MEASURABLE OPTICAL LOSS LEVELS

METHOD	ABSORPTION LOSS (cm^{-1})
Thermal Methods	
Thermocouple Laser Calorimetry	1×10^{-3} - 1×10^{-4} Many[1-7]
Laser Doppler Calorimetry	5×10^{-5} Skolnik[24]
Band Edge Shift Calorimetry	1×10^{-4} - 5×10^{-5} Nurmikko[27]
Photoacoustic Calorimetry	1×10^{-6} Kerr[33], Rosencwaig[31,32]
Thermoacoustic Calorimetry	5×10^{-6} Hordvik[34]
Surface Acoustic Wave Calorimetry	1×10^{-5} - 5×10^{-6} Parks[35]
Thermal Lens (IR-Solids)	1×10^{-2} - 5×10^{-3} Hordvik[41]
Direct Loss Methods	
Laser Differential Loss	1×10^{-4} - 1×10^{-5} Rehn[46]
Optical Bridge Q-Meter	1×10^{-5} Birnbaum[47]
Differential Spectrophotometry	5×10^{-3} - 2×10^{-3} Deutsch[48,49]
Emittance Spectroscopy	1×10^{-5} Stierwalt[51,52], Skolnik[54]

SUMMARY

We have presented a review of some recent developments in infrared measurement techniques for determining optical losses in highly
transmitting solids. Most of these methods have grown out of the
need to characterize improved low loss laser materials; it was the
purpose of this presentation to compare some of the more common
laboratory approaches on the basis of sensitivity and relative
merits of each. Although laser themocouple calorimetry is still the
most widely employed technique, it appears that this method is
rapidly approaching its sensitivity limit and it is probably not
fruitful to push these low level d.c. measurements much further.
Rather, forms of a.c. acoustic calorimetry discussed here show great
promise for accurately measuring absorption levels to near or below
10^{-6} cm^{-1} range even with low power lasers. These newer acoustic
methods are especially promising for measuring optical losses in
surfaces or in coatings. The major problem appears to be one of
absolute calibration. Emissivity measurements are the best way of
obtaining broad band spectral data on the frequency dependence of
β at least until high power, broadly tunable laser sources are
available. Further refinement in emittance instrumentation is
necessary before ultimate sensitivities can be realized; problems
of scattering, especially in the 2-5μm range and surface effects
remain to be dealt with. Finally, in Table II we summarize measurement sensitivities which are presently "readily" attainable by some
of the methods which we have discussed.   If optical materials such
as KCl or CdTe can be produced which approach their presently predicted intrinsic limits of 10^{-7} cm^{-1}, further refinements in
established techniques or development of new methods will be necessary for accurate optical loss characterization.

REFERENCES

Thermocouple Calorimetry Related Papers (Refs. 1-23).

1. F. Horrigan, C. Kline, R. Rudko, and D. Wilson, Microwaves
    8, 68 (1969).
2. R. Weil, J. Appl. Phys. **41**, 3012 (1971).
3. T. Deutsch and R. Rudko, Raytheon Research Div., Final Tech.
    Rpt. on Contract No. DAAH01-72-C-0194, 1973; also F. Horrigan
    and T. Deutsch, Raytheon Rsch Div. 2nd Quart. Rpt. on Contract
    No. DAAH01-70-C-1251, 1970.
4. B. Bendow, A. Hordvik, H. Lipson and L. Skolnik, Some Aspects
    of Optical Evaluation of Laser Window Materials, AFCRL Rpt.
    No. AFCRL-72-0404 (1972).
5. See, for example, First Conf. on High Power Infrared Laser
    Window Materials, AFCRL-71-0592 (ed. by C. Sahagian and C. A.
    Pitha) 1971.

6.   See, for example, Second Conf. on High Power Infrared Laser
     Window Materials, AFCRL-TR-73-0372 (I)&(II), (ed. by C. A.
     Pitha) 1972.
7.   A. Kahan, H. Lipson and L. Skolnik in Third Conf. on High
     Power Infrared Laser Window Materials, AFCRL-TR-74-0085 (I)
     (ed by C. A. Pitha and B. Bendow) 1974.
8.   H. G. Lipson, L. Skolnik and D. Stierwalt, in Ref. 7.
9.   H. Winsor in Ref. 5.
10. L. Shiozawa, D. Roberts and J. Jost in Ref. 6.
11. J. Davisson in Ref. 5, Vol. 1. Also, J. Davisson, J. Mat.
     Sci. 9, 1701 (1974).
12. D. Stierwalt and M. Hass, "Elimination of Surface Absorption
     in KCl" in Fourth Confer. on Infrared Laser Window Materials,
     (C.L. Andrews and C.L. Strucker, ed), AFML(LPL), WPAFB,Jan 75).
13. T. F. Deutsch and R. I. Rudko in Ref. 6. Also, T. F. Deutsch,
     J. Phys. Chem. Solids 34, 2091 (1973).
14. M. Hass, T. Davisson and J. Babiskin, "Improved Laser Calori-
     metric Techniques", in Ref. 11. Also, see Hass et al in this
     publication.
15. D. A. Pinnow and T. C. Rich, Appl. Optics 12, 984 (1973).
16. R. Weil, J. Appl. Phys. 40, 2857 (1969).
17. L. Skolnik, B. Bendow, P. Gianino, A. Hordvik and E. F. Cross,
     in Ref. 6, V. 1.
18. J. A. Harrington and M. Hass in Ref. 7. Also, J. A. Harrington
     and M. Hass, Phys. Rev. Lett. 31, 710 (1973).
19. C. P. Christensen, R. Joiner, S.T.K. Nick, W. H. Stieir in
     Ref. 7. W. Steier, C. Christensen, S. Nick, and R. Joiner in
     IR Window Studies (ed by F. Kroger and J. Marburger), Quart.
     Tech Rpt No. 5 on Contract F19628-72-C-0275 (Univ S/California),
     1973.
20. L. Skolnik, H. Lipson, B. Bendow, J. Schott, Appl. Phys. Lett.
     25, 442 (1974).
21. J. W. Davisson, M. Hass, P. H. Klein and M. Kulfedd in Ref. 7.
22. H. Lipson, L. Skolnik and D. Stierwalt, Appl. Optics 13, 1741
     (1974).

     Optical Calorimetry Related papers (Refs. 23-28).

23. A. Hordvik and L. Skolnik in Proc. of the AFSC 1972 Science
     and Engineering Symposium, San Antonio, Texas. AFSC Pub No.
     AFSC-TR-72-005 (U.S. GPO, Washington, D.C. 1972) Vol. II.
24. L. Skolnik, A. Hordvik and A. Kahan, Appl. Phys. Lett. 23,
     477 (1973). Also, L. Skolnik, A. Hordvik, A. Kahan in Ref. 7.
25. J. N. Dukes and G. B. Gordon, Hewlett-Packard J. 21, No. 12,
     2 (1970).
26. R. Bernal, "Preparation and Characterization of Polycrystal-
     line Halides for use in High Power Laser Windows", Quart. Tech.
     Rpt No. 4 on Contract No. DAHL-15-72-C-0227, Honeywell Corporate
     Res. Labs. 1973.
27. A. Nurmikko, Presentation at AFCRL, 13 Nov 1974. Also, see
     A. Nurmikko, Appl. Phys. Letters 26, 175 (1975).

28. L. Skolnik, B. Bendow, P. Gianino and E. Cross in Ref. 7. Also
     L. Skolnik, B. Bendow, P. Gianino and E. Cross, "Mid & Far Infra-
     red Vidicon Investigations of Thermal Lensing, Interference,
     and Thermal Radiation from Laser Windows, AFCRL Rpt., AFCRL-
     TR-74-0031 (1973).

Acoustic Calorimetry Related Papers (Ref. 29-36)

29. Early Investigations of the Photoacoustic Effect were reported
     by J. Tyndall, Proc. Roy. Soc. London 31, 307 (1881); W. C.
     Röntgen, Phil. Mag. II, 308 (1881), - and A. G. Bell, ibid
     p. 510.
30. See, for example, M. E. Delany, Sci. Progr. 47, 459 (1959).
31. A. Rosencwaig, Science 81, 657 (1973).
32. A. Rosencwaig, Optics Comm. 7, 305 (1973).
33. E. L. Kerr, Appl. Optics 12, 2520 (1973).
34. A. Hordvik, private communication, to be published.
35. J. Parks and T. Colbert in "IR Window Studies", Quart. Tech.
     Rpt No. 4 on contract F19628-72-C-0275 (F. A. Kroger and
     J. H. Marburger, ed) Univ S/California, USCEE Rpt 457, 15 Jun
     73. J. H. Parks, D. A. Rockwell, T.S. Colbert, K. M. Lakin
     and D. Nih, Appl. Phys. Lett. 25, 537 (1974). Also, J. Parks,
     D. Rockwell, T. Colbert, T. Georges in this volume.
36. D. White, R. King, P. Archibald and R. Sharp, in "High Energy
     Laser Mirrors and Windows", Semi Annual Rpt No. 5 on ARPA
     Order 2175, Michelson Laboratory, Naval Weapons Center, China
     Lake, Calif., Sept 1974.

Thermal Lens Papers (Ref. 37-42)

37. R. C. Leite, R. S. Moore and J. R. Whinnery, Appl. Phys. Lett.
     5, 141 (1964).
38. J. P. Gordon, R. C. Leite, R. S. Moore, S. P. Porto, J. R.
     Whinnery, J. Appl. Phys. 36, 3 (1965).
39. D. Solimini, "Loss Measurement of Materials at Light Frequencies
     using Interaction of a Laser Mode," M.S. Thesis, University of
     Calif., Berkeley, 1965. Also, D. Solimini, Appl. Optics 5,
     1931 (1966).
40. Y. Kohanzadeh and D. Auston, IEEE J. of Q. Elec., p.475-476,
     Jul 1970.
41. A. Hordvik in Ref. 5.
42. See, for example, "Compendium on High Power Infrared Laser
     Window Materials", (C. Sahagian and C. Pitha, eds.) AFCRL
     Rpt No. AFCRL-72-0171, 9 Mar 72.

Direct Loss Measurement Techniques Papers (Ref. 43-47)

43. See, for example, M. Jones and K. Kav, J. Sci. Instrum. (J.
     Phys. E) 2, 331 (1969); P. Laybourn, W. Gambling and D. Jones,
     Opto-Electron 3, 137 (1971); A. Tynes and D. Bisbee, Proc. of
     the Spring Meeting of the Opt. Soc., Denver Colo., Mar 13-16,
     1973.
44. P. Kaiser, Appl. Phys. Lett. 21, 1973.

45. S. Miller, E. Marcatili and T. Li, Proc. of IEEE 61, 1703 (1973).
46. V. Rehn, O. Kyser, V. Jones in Ref. 6. Also, D. Burdick, V. Rehn, D. Kyser and V. Jones in Ref. 36.
47. G. Birnbaum, E. Cory and K. Gow, Appl. Optics 13, 1660 (1974).

Spectroscopic Techniques Papers (Ref. 48-50)

48. T. F. Deutsch, J. Phys. Chem. Solids 34, 2091 (1973).
49. T. F. Deutsch and R. I. Rudko, "Research in Optical Materials and Structures for High Power Lasers", Final Tech Rpt on Contract No. DAAH01-72-C-0194, Raytheon Rsch Div., Waltham, MA, Jan 1973.
50. M. J. Harrick, Internal Reflection Spectroscopy, Interscience, Wiley, NY (1967).

Emittance Spectroscopy Papers (Ref. 51-56)

51. D. Stierwalt and R. Potter, J. Phys. Chem. Solids 23, 99 (1962). D. Stierwalt and R. Potter in Semiconductors and Semimetals 3, (ed by R. Willardson and A. Beer), Academic Press, NY (1967).
52. D. L. Stierwalt in Refs. 6 and 7.
53. J. K. Barr, Infrared Phys. 9, 97 (1969).
54. L. Skolnik, M. Clark, R. Koch, W. McCann and W. Shields in Ref. 12. Also, L. Skolnik, "Spectral Emittance Measurements on Some Laser Window Materials", AFCRL-TR-74-0590, Dec 1974.
55. J. Harrington, B. Bendow, K. Namjoshi, S. S. Mitra and D. L. Stierwalt, "Low Loss Window Materials for Chemical and CO Lasers", in Ref. 12.
56. G. Wijntjes, N. Johnson and M. Weinberg in "Laser Induced Damage in Optical Materials 1972", (ed by A. Glass and A. Guenther), NBS Special Pub. 372, USGPO, Wash DC (1972).

IMPROVED LASER CALORIMETRIC TECHNIQUES

M. Hass, J. W. Davisson, H. B. Rosenstock, J. A. Slinkman,

and J. Babiskin

U.S. Naval Research Laboratory

Washington, D.C. 20375

A number of improvements in laser calorimetric techniques are
described for the measurement of very low absorption coefficients
in materials. Of these the most important is the method for
separating surface and bulk absorption involving a long rod
geometry. It is shown that the initial slope of the thermal rise
curve for long rods is associated with bulk absorption alone. The
method is illustrated with KCl at 10.6 μm and with NaCl at 1.06 μm.
In the latter case an absorption coefficient at least as low as
7×10^{-6} cm^{-1} has been extracted and this is one of the lowest, if
not the lowest, absorption coefficient reported for a crystalline
material so far.

I. INTRODUCTION

A number of specialized techniques have been developed for the
determination of optical absorption coefficients of highly transparent materials. Among these are calorimetry,[1-5] spectral
emittance,[5-7] and dual beam transmittance.[8] Of these, the calorimetric or thermal rise method has so far been capable of obtaining
the lowest reported absorption coefficients. While this method is
limited to wavelengths at which sufficiently powerful laser
sources are available, it is less sensitive to scattered radiation
which is difficult or impossible to eliminate using other techniques.
At very low absorption levels, scattering and surface absorption
can in fact become much larger contributors to the total

attenuation then the bulk absorption itself. In this contribution,
emphasis will be placed on the development of a new method for the
determination of bulk absorption in the presence of even overwhelming surface absorption. A number of useful techniques
developed at this Laboratory over the past few years are also
presented.

II. PRINCIPLES

In most calorimetric or thermal rise experiments, a laser
source is employed to heat a sample contained in a blackened
thermally massive enclosure as illustrated in Fig. 1. A small
temperature sensor, most frequently a differential thermocouple,
is employed to measure the sample temperature relative to the
surroundings as a function of time. If the sample is allowed to
come to thermal equilibrium and a laser beam is turned on, the
sample will absorb energy from the beam and increase temperature.
From analysis of the heating curve, the absorption coefficient β
can be deduced.

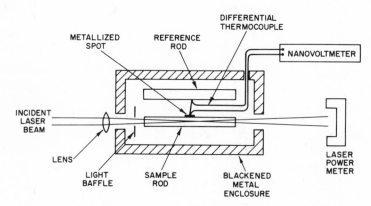

Fig. 1 Apparatus for improved laser calorimetry.

The shape of the heating curve depends upon the specific
heat c_p, mass m, length ℓ, transmitted power P_t, index of refraction n, heat loss parameters, and thermal diffusivity κ. In the
special case of a long narrow rod, it can be shown that the
initial slope of the thermal rise curve will not depend upon the
thermal diffusivity or heat loss parameters[2] and that the initial
slope dT/dt of the thermal rise curve will be linear and can be
related to the absorption coefficient by the expression

$$\beta \ell P_t = m c_p \left(\frac{dT}{dt}\right)\left(\frac{2n}{n^2 + 1}\right) \qquad (1)$$

This follows from very simple heat absorption considerations in a material which is sufficiently transparent so that the intensity is nearly uniform across the specimen. The term $2n/(n^2 + 1)$ is a correction for multiple reflections.[9] It will be argued shortly that this initial slope expression holds for the bulk absorption even in the presence of overwhelming surface absorption for long narrow rods.

For the long rod geometry, the principles are readily illustrated by a simple calculation. In the case of long rods, the transit time for the heat to be transported from the point of absorption to the temperature sensor can be estimated in the following way. Consider an infinite medium with an instantaneously generated heat pulse at time $t = 0$. The temperature at a point r distant from the source has its maximum value at a time $t = r^2/6\kappa$, where κ is the thermal diffusivity of the material.[10] To a first approximation, bulk heating can be assumed to arise along the axis of the rod and the corresponding transit time for a heat pulse to travel from the axis to the periphery will be $r^2/6\kappa$ where r is now the radius of the rod. Surface absorption can be assumed to arise at an end surface and the corresponding transit time from the end to the middle will be $(\ell/2)^2/6\kappa$, where ℓ is now the length of the rod. For a practical case such as a cylinder of KCl ($\kappa = 0.05$ cm^2/sec) one cm in diameter and 10 cm in length, the bulk transit time is about 1 sec and the corresponding surface transit time is about 83 sec. This simple calculation suggests that for slope measurement time less than 80 sec, only bulk absorption will contribute to the results. For longer times, both surface and bulk effects will be present and so the slope will be greater (assuming little or no heat losses). The exact thermal rise curve will depend upon the thermal diffusivity.

The fundamental foundations of laser calorimetry and especially the method of separation of surface and bulk absorption by the initial slope method can be demonstrated in a more satisfactory manner by direct calculations of the surface temperature of the sample at the location of the sensor. Such calculations have been carried out for the case of the long rod geometry where both surface and bulk absorption are included. The results are shown in Fig. 2 where the bulk absorption is kept fixed and the amount of surface absorption has been varied. It is immediately evident that the initial slope is independent of the amount of the surface absorption. This provides a more quantitative justification of the initial slope method for determination of bulk absorption. It is evident from Fig. 2 that after about 30 sec, the slope increases due to contributions from surface heating.

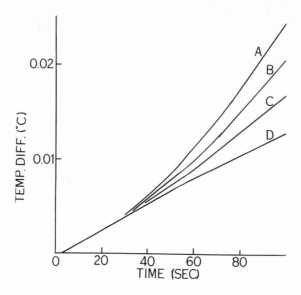

Fig. 2.   Calculated thermal rise curve for axial heating of a
cylindrical rod in which there is both surface and bulk absorption.
The various curves A, B, C, and D correspond to ratios of surface
to bulk power absorption of 3:1, 2:1; 1:1, and 0:1.   The initial
slope of the thermal rise curve is seen to be independent of surface
contributions.   The dimensions of the rod are assumed to be 1 cm
diameter and 10 cm in length.   The bulk absorption coefficient is
assumed to be 0.00008 cm^{-1} and other rod parameters assumed to be
same as for KCl.   The temperature difference between the middle
of the rod and the surroundings is plotted.   A laser power of 2
watts has been chosen.

By carrying out detailed measurements of the temperature profile along the rod, both the surface and bulk absorption coefficients could be extracted accurately.   An illustration of this is
shown in Fig. 3 where the temperature at different portions of the
rod have been calculated for a situation in which both surface and
bulk absorption occur.   Here a temperature sensor near the end
surface will be affected primarily by surface absorption while a
sensor at the middle will be affected by bulk absorption primarily.
The temperature profile after laser turnoff involves a thermalization to a nearly uniform rod temperature.   Experimental examples
corresponding to this have been observed.

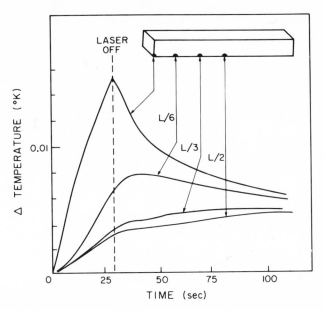

Fig. 3. Calculated thermal rise curve for square rod as a function of distance along the rod. The power absorption associated with surface and bulk effects are assumed to be equal. At the middle of the rod, the initial slope is associated with bulk effects. Near the end of the rod, surface contributions enter. The dimensions of the rod are 1 cm square and 10 cm long with other parameters as in Fig. 2. For long times, the rod approaches a uniform temperature as the heat losses are small for the small temperature rises employed and low emittance of KCl.

III. EXPERIMENTAL TECHNIQUES

The following techniques have been developed for obtaining the absorption coefficients of materials. In general, it is desirable to polish both the end faces and side faces of the material. Chemical polishing is used for alkali halides.[11] Differential thermocouples are generally used as temperature sensors for simplicity, but other small sensors may be employed. The technique of attachment is particularly important for low losses in the visible and near infrared. Here a small mirror is employed with a liquid coupling for crystals. For glasses, an evaporated mirror is employed. Visual inspection with a visible laser beam is helpful in identifying sources of scattered light on the thermocouples. Insertion of light baffles and focusing lenses is helpful in reducing scattered light.

In going down to extremely low values of the absorption, it is important to reduce noise and achieve stability in the system. Here, the noise limit in thermocouples can be reduced to about 3 nV by use of low-thermal solder and by twisting the thermocouple leads. The stability of commercial instruments such as the Keithley Model 148 Nanovoltmeter has been found to be quite adequate for much of this work. In air calorimeters and some vacuum calorimeters, air currents in the room can cause some fluctuations and limit the stability. With proper technique, temperature differences range of hundredths of a degree have been measured in our laboratory and elsewhere.

IV.   EXPERIMENTAL RESULTS

A.   NaCl at 1.06 μm

A crystal at 1.06 μm was studied in order to gain some idea of the lower limit of sensitivity of the system. Here a Harshaw NaCl crystal appear to show the lowest amount of visible light scattering and was employed in these investigations. The results are shown in Fig. 4. An absorption coefficient of 7×10^{-6} cm^{-1} can be extracted from the initial slope. This is comparable to a value of 5×10^{-6} cm^{-1} which has been reported for a water-free sample of fused silica.[3]

This is the lowest absorption coefficient for a crystalline material ever reported, to the best of our knowledge. While water-free fused silica and other fiber optic materials have lower absorption coefficients, there has not been a correspondingly large effort devoted towards crystals. In principle, there is no reason why crystals should not have equally low if not lower absorption coefficients than glasses, especially as bulk scattering effects should be lower.

B.   KCl at 10.6 μm

A KCl crystal at 10.6 μm was studied in order to demonstrate separation of surface and bulk absorption in a single long rod. The existence of both surface and bulk absorption in KCl at 10.6 μm has been shown using absorption measurements as a function of length. The results are shown in Fig. 5, which can be compared to the theoretical calculations for a sample of similar shape and absorption levels. The experimental curve closely resembles the theoretical curves for simultaneous surface and bulk absorption of approximately equal magnitude.

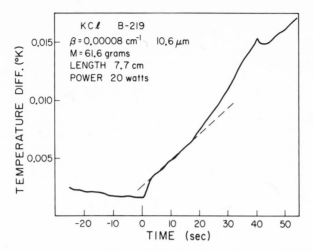

Fig. 4.   Thermal rise curve of NaCl at 1.06 μm.   The jump at t = 0 is associated with scattered light.   The initial slope (apart from the jump) corresponds to an absorption coefficient of 7 x 10⁻⁶ cm⁻¹.)

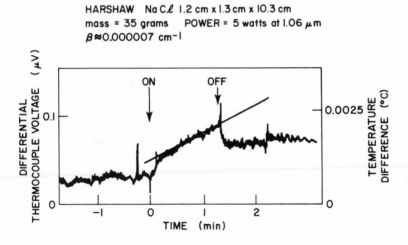

Fig. 5.   Thermal rise curve of KCl at 10.6 μm.   The initial slope corresponds to an absorption of about 8 x 10⁻⁵ cm⁻¹.   The small jumps corresponds are associated with scattered light and also indicated laser on-off times.   Note that temperature at the mid-point of the sample continues to rise after the laser is turned off and this is associated with thermalization of a rod with hot ends arising from end surface absorption.

The method described should be particularly advantageous for KCl near 10.6 μm as there is a surface absorption band near 9.5 μm whose intensity is much larger than the corresponding bulk absorption[1],[12]. As a result, only a specialized technique such as that described would be able to establish the bulk absorption.

V. CONCLUSION

It has been shown by experiment and by theoretical thermal analysis that the long rod geometry in laser calorimetric absorption measurements can be used to extract the bulk absorption independent of surface absorption by using a simple initial slope method. By a somewhat more extensive analysis, the surface contribution can also be obtained, although the thermal diffusivity and heat loss corrections are involved. This aspect will be considered more fully in subsequent work. A number of improvements in techniques to study low loss absorptions in crystals are described and the results are illustrated for KCl at 10.6 μm and with NaCl at 1.06 μm. In the latter case, an absorption coefficient of 7×10^{-6} cm^{-1} has been extracted. It is believed that use of these methods will allow determinations of much lower bulk absorption coefficients in materials than has been reported previously.

REFERENCES

1. M. Hass, J. W. Davisson, P. H. Klein, and L. L. Boyer, J. Appl. Phys. 45, 3939 (1974).
2. K. I. White and J. E. Midwinter, Opto-Electron. 5, 323 (1973).
3. D. A. Pinnow and T. C. Rich, Appl. Opt. 12, 984 (1973).
4. P. L. Cohen, K. W. West, P. D. Lazay, and J. Simpson, Appl. Opt. 13, 2522 (1974).
5. H. G. Lipson, L. H. Skolnick, and D. L. Stierwalt, Appl. Opt. 13, 1741 (1974).
6. D. L. Stierwalt and R. F. Potter, in Semiconductors and Semimetals (Academic Press, Inc., New York, 1967) Vol. 3.
7. J. K. Barr, Infrared Physics 9, 97 (1969).
8. J. P. Dakin and W. A. Gambling, Opto-Electron 5, 335 (1973).
9. R. Weil, J. Appl. Phys. 41, 3012 (1970).
10. H. S. Carslaw and J. C. Jaeger, Conduction of Heat in Solids, 2nd ed. (Oxford Univ. Press, London, 1959) p. 256.
11. J. W. Davisson, J. Mat. Sci. 9, 1701 (1974).
12. T. F. Deutsch, Appl. Phys. Lett. 25, 109 (1974).

MEASUREMENT OF SMALL ABSORPTION COEFFICIENTS FROM

THERMALLY INDUCED SHIFTS AT THE FUNDAMENTAL EDGE[†]

A. V. Nurmikko, D. J. Epstein, and A. Linz

Center for Materials Science and Engineering
and Department of Electrical Engineering
Massachusetts Institute of Technology

The measurement of small absorption coefficients in transparent infrared materials is currently a subject of considerable interest. The practical motivation for the work arises from the need for evaluating materials for high power laser windows but there is also a more fundamental interest in developing an understanding of intrinsic and extrinsic optical absorption mechanisms which limit material performance. A considerable number of measurement techniques [1] have been developed which utilize a variety of physical indicators for measuring low optical absorption.

In this article we describe a novel method for measuring small absorption coefficients in the infrared that utilizes the temperature dependent shift in the optical absorption edge as an indicator of the slight temperature increase produced in a sample by infrared illumination. [2] The method has been found to be experimentally expedient, sensitive and to offer advantages not available, for instance, in conventional calorimetric measurements. Experimental results are presented for CdTe, a material which has recently showed promise as a high-power infrared window candidate.

It is well known [3] that the bulk absorption coefficient β and surface absorption σ of a relatively transparent window sample can be obtained from measurement of the absorbed and transmitted infrared powers (P_{abs} and P_{trans}, respectively) by use of the relation:

[†]Research supported by the Advanced Research Projects Agency through the Air Force Materials Laboratory.

$$\beta \ell + 2\sigma = \frac{P_{abs}}{P_{trans}} \cdot \frac{2n}{n^2 + 1} \quad \text{when} \quad \beta \ell, \sigma \ll 1 \tag{1}$$

Here ℓ is the optical pathlength and n is the index of refraction. The measurement of P_{trans} is usually trivial and the real task of an experimenter is to measure P_{abs}. We will assume that in addition to the incident infrared laser beam a weak monochromatic probe beam also illuminates the sample. Unlike the long wavelength infrared laser, the wavelength of the weak probe beam is chosen in such a way that it lies within the fundamental electronic absorption edge of the sample material. As the sample heats up from absorption of the infrared laser energy changes occur in the transmitted probe beam intensity because of the temperature induced shift of the fundamental edge. It can be straightforwardly be shown that the measured fractional voltage change $\Delta V/V$ in the steady state response of the photodetector signal monitoring the transmitted probe beam gives the absorbed power via the relation:[2]

$$P_{abs} = \frac{(\Delta V/V)}{\tau} \cdot \left(\frac{d\alpha}{dE}\right)_{E_0}^{-1} \cdot \frac{dE_{gap}}{dT}^{-1} \cdot \frac{c_s m_s}{\ell'} \tag{2}$$

In this expression $(d\alpha/dE)_{E_0}$ is the slope of the electronic absorption coefficient α vs. energy curve evaluated at the probe beam photon energy E_0, (dE_{gap}/dT) is the temperature dependence of the energy gap, ℓ' is the probe beam pathlength in the sample and c_s and m_s are the sample specific heat and mass, respectively. The thermal time constant τ has a value determined usually by heat conduction losses arising from the physical mounting of the sample. Under the experimental conditions discussed below τ is much longer than the time τ_{th} required for thermal equilibrium to be achieved within the sample and consequently the assumption of an approximately uniform temperature distribution is justified.* For many semiconductors the energy gap dependence on temperature is well known so that in an actual experiment the only quantities needed for Eq. (2) are $\Delta V/V$, τ and $(d\alpha/dE)_{E_0}$.

Figure 1 shows a typical experimental arrangement used by us in measuring small infrared absorption coefficients in CdTe by the edge shift technique. Sample wafers of typical areas ~ 1 cm^2 and

---

*For samples of mass $m \lesssim 1$g typical values are $\tau \sim 30$ sec, $\tau_{th} \sim 1$ sec.

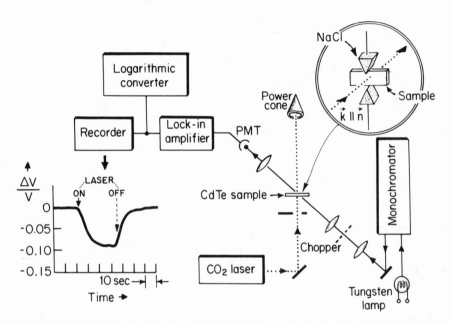

Figure 1:

Experimental arrangement for edge shift measurements.

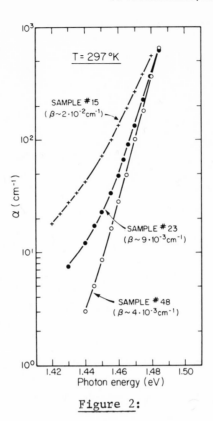

<u>Figure 2:</u>

Fundamental absorption edge in CdTe for materials
of varying IR transparency.

thicknesses varying from 100μ to 2000μ are cut from polycrystalline
bulk material and polished both mechanically and chemically.   The
weak probe beam is provided by a tungsten lamp monochromator combination producing a power of ∿ $10^{-7}-10^{-8}$ W at the sample. By passing
the synchronously detected transmitted intensity through a logarithmic converter a direct recording of α vs. E can be obtained from
which (dα/dE) can be determined. Figure 2 shows some characteristic
data of ℓnα vs. E for samples of CdTe obtained from several different suppliers. In parentheses are shown the approproximate average
infrared absorption coefficients β measured by us for each material.[*]

---

[*]For each curve at least two samples of different thicknesses
were used.

A linear slope of $\ln \alpha$ vs E is commonly said to be indicative of "Urbach's law" over some temperature and energy range:

$$\alpha = \alpha_0^{\gamma(E_0-E)/kT} \qquad (3)$$

where E_0 is the highest photon energy and γ is a coefficient on the order of unity. It is interesting to note that for the data of Figure 2 γ decreases ($\gamma \sim 3-2$) and deviation from linear slope increases monotonically with increasing infrared absorption coefficient β. Inasmuch as one mechanism for infrared absorption in CdTe is thought to be provided by local deviations from stoichiometry (precipitates; vacancies) which also affect the electronic band structure, the study of the fundamental absorption edge might be usable as a diagnostic tool for characterizing the overall optical quality of semiconductor materials such as CdTe for laser window applications.

The inset of Figure 1 shows the method used by us to support the semiconductor sample wafers in our experiments. The samples are gently held between the edges of triangular pieces of NaCl to minimize heat conduction losses and erroneous signals resulting from heating of the holder by scattered laser radiation. The incident infrared laser radiation is incident at the test sample at two different geometries. In the $\vec{k}||\vec{n}$ geometry ($\vec{k}$ is the direction of propagation of the infrared laser beam and $\vec{n}$ is the sample wafer face normal) the laser beam pathlength through the sample is quite short and surface absorption ordinarily dominates. In the $\vec{k}\perp\vec{n}$ geometry the bulk absorption makes a major contribution. The two independent measurements (by simple 90° rotation of a sample) can then be used in Eq. (1) to calculate both the bulk and surface absorption in one sample, a considerable advantage compared with conventional calorimetry where surface absorption is usually identified by extrapolating absorption data from samples of various thicknesses to zero thickness. The edge illumination in the $\vec{k}\perp\vec{n}$ geometry is performed by careful aperture control (and slight focussing if required) of the incident infrared laser beam for samples of thicknesses in excess of 500μ. The probe beam wavelength is chosen to optimize the expected signal consistent with noise considerations. (Larger ($d\alpha/dE$) implies a larger signal but at the expense of a reduced intensity in the transmitted probe beam.)

Our edge shift measurements have been compared with measurements of β, by standard calorimetry, on large bulk samples from which small thin wafers were later cut. The best state-of-the-art large windows of CdTe available to us have yielded values of $\beta \sim 1.10^{-3} - 2.10^{-3}$ cm^{-1} by laser calorimetry. In performing the edge shift experiments on samples derived from these materials we obtained general agreement with the calorimetric measurements. However we find considerable variation in measured values of β (as much as a factor of five) from sample wafer to the next one. Even the better quality laser window CdTe of today is somewhat inhomogeneous and this will reflect strongly in the optical probing of small volumes. The edge shift technique then has the advantage of rather unequivocally indicating the presence of material defects in those small volumes which can then be further probed by complementary studies, e.g. in an electron microscope.

We have found that very careful surface preparation of the samples is necessary to minimize effects of surface absorption. Commonly measured values by us indicate $\sigma \sim 1.10^{-4} - 2.10^{-4}$. The edge shift method itself is quite sensitive for measurement of small values of β and σ. For instance very modest infrared powers ($P_{trans} \sim 1$ W) are required to produce signals ($\Delta V/V \sim -.15$, $\tau \sim 30$ sec) with excellent signal-to-noise ratio for $\beta \sim 2.10^{-3}$ cm^{-1}.

Numerical estimates based on experimental results indicate that measurement of bulk absorption coefficients as small as $\beta \sim 10^{-5}$ cm^{-1} (for samples of unit length) are possible for incident infrared laser intensities less than 100 W/cm^{-2} ($P_T \sim 10$ W). This estimate does not imply an ultimate limit in sensitivity (determined in part by photomultiplier noise) but rather a conveniently practical and measurable value. Accordingly, surface absorption $\sigma \sim 10^{-5}$ should also be measurable.

The newly developed edge shift technique has several advantages which make it attractive, particularly when applied to measurement of small values of β and σ in semiconductor infrared materials. First, as a purely optical technique it is particularly convenient in those ambient conditions where electrical measurements (leads or samples) are cumbersome. Next, the small sample sizes should allow comparison between optical data and that obtained by microprobe techniques and electron microscopy thereby making it possible to better understand the role of material defects and imperfections in determining optical losses in IR transparent semiconductors. The possibility of measuring of small amounts of surface absorption makes this technique potentially attractive for evaluating multi- and single-layer dielectric coatings in the infrared region.

An additional and interesting application of the edge shift technique might be in the detection of fast infrared radiation. Consider the case of an infrared opaque thin film (t $\sim$ 1μ) deposited on a transparent substrate. A short burst of infrared radiation (at a wavelength in the region of fundamental lattice absorption) will be strongly absorbed in the thin film, the resulting transient temperature rise can then be measured by observing the thermally induced shift at the fundamental edge using a fast photomultiplier or a photodiode. If the infrared pulse is short enough (typically $t_p < 1$ μsec) heat conduction losses to the substrate will not be significant during the transient. As a crude estimate of the sensitivity of such a detection scheme we assume $(d\alpha/dE)_{E_0} \simeq 10^4$ cm^{-1}/ eV, t = 1μ, area of 0.1 cm^2, heat capacity of 0.2 cal-gm °K and a density of 3 g/cm^3. Neglecting thermal losses during the transient gives for a "conveniently" measurable minimum signal ($\Delta V/V \sim$ few %) a sensitivity of $\sim$ 1μ J of transient infrared energy. This order-of-magnitude sensitivity compares very well with that reported by Parks et al. [4] who have used a surface acoustic wave thermal probe for measurements of short optical transients. The method suggested here might be particularly useful in the measurement of fast single transients at the CO_2 laser frequencies (some material candidates are SiC, ZnO and BN). It would facilitate room temperature detection of infrared radiation at visible and near UV regions of the spectrum where sensitive and fast detectors are more readily available.

REFERENCES

1.   L. H. Skolnick, article in this volume.

2.   A. V. Nurmikko, Appl. Phys. Letters 26, 175 (1975).

3.   T. F. Deutsch and R. I. Rudko, in Conference on High Power Infrared Laser Window Materials, edited by C. Pitha, (AFCRL, Bedford, Mass, 1972), pp. 201-223.

4.   J. H. Parks, D. A. Rockwell, T. S. Colbert, K. M. Lakin, and D. Mih, Appl. Phys. Letters 25, 537 (1974).

AN ALTERNATIVE WAY TO DETERMINE ABSORPTION COEFFICIENTS IN HIGHLY TRANSPARENT SOLIDS: THEORY

Herbert S. Bennett and Richard A. Forman

National Bureau of Standards

Washington, D. C. 20234

ABSTRACT

The development of highly transparent solids requires improved methods to measure very low absorption coefficients at laser wavelengths. We have calculated the temperature profiles in the solid and the temperature and pressure profiles in the gas when a laser beam passes through a weakly absorbing solid which is surrounded by a confined and non-absorbing gas. Our calculations suggest that sufficient heat transfers from the solid to the gas to produce a detectable pressure rise in the gas.

I. INTRODUCTION

The developing technologies of optical communications, integrated optics, and high-power lasers all depend in part upon highly transparent solids with absorption coefficients less than 10^{-3} cm^{-1}.[1]

We may group the methods for measuring small absorption coefficients into four general techniques: calorimetric,[2] spectrophotometric,[3,4] emissive,[5] (i.e., the absorption coefficient is obtained after measuring the emissivity), and photoacoustic-barothermal. In the last method, the energy absorbed from a beam of radiation passing through the sample produces heat. The heat then diffuses from the solid to a non-absorbing and confined gas which is adjacent to the sample. The transfer of heat to the gas leads to a pressure rise in the gas. The last method is known as the photoacoustic method whenever the beam of

451

radiation is modulated at frequencies greater than the characteristic frequencies of the solid-gas system.[6,7] When the beam of
radiation is modulated at frequencies very small compared to the
characteristic frequencies of the solid-gas system, we refer to
the method as the barothermal method.

In this paper, we analyze theoretically the heat transfer
process in the barothermal method. We assume for this analysis
cylindrical configurations with no surface absorption of radiation
at the solid-gas interface. We calculate the temperature profile
in the cylindrical solid and the temperature profile and pressure
in the gas when a laser beam passes through the weakly absorbing
solid which is surrounded by the confined and non-absorbing gas.
Our calculations suggest that sufficient heat transfers from the
solid to the gas to produce a detectable pressure rise in the gas
and that thereby an absorbing layer at the solid-gas interface is
not essential for the production of detectable pressure rises in
the gas.

II. BAROTHERMAL DEVICE

Figure 1 shows the device which we have in mind for measuring
absorption coefficients. The insulating annuli support coaxially
the cylindrical transparent sample of radius r_s and length ℓ
inside a larger cylinder of radius r_i. A confined, non-
absorbing gas at an ambient pressure p_0 fills the space between
the two cylinders. The walls of the outer cylinder are at a
constant temperature T_0, which may be the room temperature. The
circular ends of the solid cylinder are also at temperature T_0.
A collimated laser beam, the energy source, propagates in the
z direction coaxially through the solid cylinder. The beam has
a power W_ℓ and an effective radius r_ℓ. We turn the laser beam on
at time $t = 0$, and by means of a pressure transducer located at
the heat sink-gas interface ($r = r_i$), monitor the pressure rise.
We do not show the pressure transducer in Figure 1.

III. THEORETICAL DESCRIPTION

We assume that the incident laser beam is cylindrically
symmetric and has a radial intensity profile

$$I_0(r) = \begin{cases} (W_\ell/\pi r_\ell^2) & \text{for} \quad 0 \leqslant r \leqslant r_\ell < r_s \\ 0 & \text{for} \quad r > r_\ell \end{cases} \tag{1}$$

where W_ℓ is the power of the laser beam and where r_ℓ is the radius
of the laser beam.

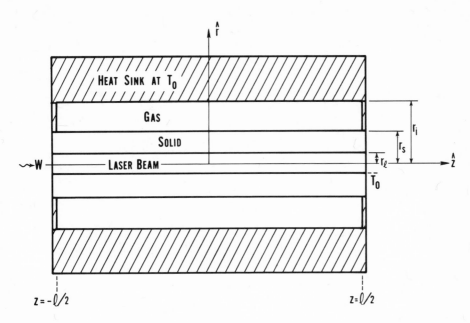

Fig. 1. Schematic of device for measuring absorption
coefficients in very transparent solids. Insulating annuli support
coaxially the cylindrical and transparent solid inside a larger
cylinder. A confined and non-absorbing gas fills the space between
the two cylinders. A collimated laser beam propagates coaxially
inside the cylindrical solid.

The power absorbed per unit volume of the sample as the beam
propagates has the form

$$Q_s(r,z) = \alpha_s \, I_0(r) \, \exp\{-\alpha_s[z + (\ell/2)]\}, \qquad (2)$$

where α_s is the absorption coefficient of the solid at the laser
wavelength. We define a highly transparent solid to be one for
which $\alpha_s\ell \ll 1$.

A non-absorbing gas is one for which the absorption coefficient of the gas is zero at the laser wavelength.

Because we choose $\ell \gg r_s$, we may neglect the z dependence
of the temperature when $\alpha_s\ell \ll 1$. We thereby neglect any end
effects and consider properties per unit length in the z direction.

The heat diffusion equations for the solid and the gas describe
the time and space dependences of the solid and gas temperatures;

namely,

$$k \; \nabla^2 \; T(r,t) \; + \; \frac{Q}{\rho C} \; = \; \frac{\partial T(r,t)}{\partial t} \quad , \tag{3}$$

where $T(r,t)$ is the temperature; $k = K/\rho C$ is the thermal diffusivity, K is the thermal conductivity, ρ is the density, and C is the specific heat at constant volume. A subscript s will denote quantities for the solid and a subscript g will denote quantities for the gas. We assume that no mass flow develops in the gas and thereby gas pressure gradients are zero everywhere. We also assume that only small temperature differences and temperature gradients occur, and thereby neglect, respectively, radiation of energy and convection currents in the gas.

The diffusion equations require statements of the boundary conditions before a particular solution can be determined. We write the temperature $T(r,t)$ in the form $T(r,t) = u(r,t) + v(r)$, where $v(r)$ is the steady state temperature and where $u(r,t)$ is the transient temperature behavior; namely,

$$\mathrm{Lim} \; u(r,t) = 0. \tag{4}$$
$$t \to \infty$$

We list the boundary conditions. At time $t = 0$, the temperature is everywhere equal to the ambient temperature T_0.

$$T(r,o) = T_0. \tag{5}$$

At any time t, the temperature at the heat sink surface $r = r_i$ is T_0;

$$T_g(r_i,t) = T_0 \quad \text{or} \quad u_g(r_i,t) = 0 \quad \text{and} \quad v_g(r_i) = T_0. \tag{6}$$

The solid and gas are in intimate thermal contact,

$$T_s(r_s,t) = T_g(r_s,t); \tag{7}$$

the heat flow is conserved across the solid-gas interface,

$$K_s \; (\partial T_s/\partial r) \Big|_{r=r_s} = K_g \; (\partial T_g/\partial r) \Big|_{r=r_s} ; \tag{8}$$

and the temperature is finite for all $r \geqslant 0$ and for all times.

Equations (1) through (8) represent an idealized description of the barothermal device given in Figure 1.

We do not give in this paper the details for solving the heat diffusion equations and for computing the pressure rise from the

equation of state of an ideal gas. We will report these in a
future paper. We quote here only those results required for computing the absorption coefficient from the pressure rise in the gas.

The steady state solutions yield the result that the relative
pressure rise (p_∞/p_0) is given by

$$(p_\infty/p_0) = [(r_i^2 - r_s^2)/2T_0 \langle v_g^{-1} \rangle]$$

where

$$\langle v_g^{-1} \rangle \equiv \int_{r_s}^{r_i} \{1/v_g(r)\} r dr,$$

and

$$v_g(r) = T_0 - (\alpha_s W_\ell/2\pi K_g) \ln(r/r_i).$$

The quantity p_∞ is the steady state pressure in the gas for large
times t such that $(u_g/T_0) \ll 1$. From this we see that the temperature rise, $v_g(r) - T_0$, is directly proportional to the product
$\alpha_s W_\ell$.

Whenever the quantity $|v_g(r) - T_0| \ll 1$ for $r_s \leqslant r \leqslant r_i$, we
express the absorption coefficient of the solid α_s in the approximate form

$$\alpha_s \sim \frac{4\pi T_0 K_g}{W_\ell} \{(p_\infty/p_0) - 1\} \{1 - \frac{2}{[(r_i/rs)^2-1]} \ln (\frac{r_i}{r_s})\}^{-1}. \quad (9)$$

Observe that the relative pressure rise $\{(p_\infty/p_0)-1\}$ is directly
proportional to the product $\alpha_s W_\ell$.

The transient solutions give the relative pressure rise in
the form,

$$[p(t)/p_0] = [(r_i^2 - r_s^2)/2T_0 \langle T_g^{-1} \rangle_t] , \qquad (10)$$

where

$$\langle T_g^{-1} \rangle_t = \int_{r_s}^{r_i} \{1/T_g(r,t)\} r dr;$$

and

$$T_g(r,t) = \sum_{j=1}^{\infty} A_j F_0(\xi_j,r_i,r) \exp (-\xi_j^2 k_g t) + v_g(r).$$

We have calculated the eigenvalues ξ_j for which the determinant of the coefficients of the simultaneous equations from the boundary conditions vanishes. The expansion parameters A_j and the function $F_o(\xi_j, r_i, r)$ are lengthy expressions involving zero and first order Bessel functions of the first kind, $J_n(\xi r)$, and of the second kind, $Y_n(\xi r)$. In addition, the coefficients A_j are all directly proportional to $\alpha_s W_\ell$ and are functions of r_ℓ, r_s, r_i, k_s, k_g, K_s, K_g, and ξ_j. The eigenvalues ξ_j also determine the characteristic times t_j for the solid-gas system; namely,

$$t_j = (\xi_j^2 k_g)^{-1}. \tag{11}$$

IV. NUMERICAL RESULTS

The several properties of the solid-gas system which enter our calculations may be divided into two groups. The first group consists of the physical properties such as the absorption coefficient, density, specific heat, and thermal conductivity. The second group contains external and geometric quantities such as power of the laser beam, ambient temperature, ambient pressure, and the dimensions of the apparatus.

In order to determine some of the quantitative predictions of the preceding model, we use the properties of a representative neodymium doped laser glass for the solid and of air at $T_o = 300$ K and $p_o = 1.013 \times 10^5$ N/m^2 (1 atmosphere) for the gas. Of course, unlike air, our ideal-fictitious gas does not absorb the laser beam. We cite these values in Table I.

Typical undoped laser glasses have absorption coefficients in the 1×10^{-2} cm^{-1} range. We therefore use $\alpha_s = 10^{-2}$ cm^{-1}. We also take the power of the laser beam to be $W_\ell = 10.0$ watts and the radius of the solid sample to be $r_s = 0.5$ cm. All numerical examples discussed in this section are based upon these values for

Table I. Values of physical properties. The quantities ρ, C, and K are, respectively, the density, specific heat at constant volume, and the thermal conductivity.

	ρ (g/cm^3)	C (J/g °C)	K (W/cm °C)
glass	2.6	0.630	0.0084
gas	0.00129	0.718	0.000261

Table II. Steady state solutions. The relative
pressure is (p_∞/p_0); the relative steady state temperature at $r = 0$ is $\{v_s(0)/T_0\}$; the radius of the laser
beam is $r_\ell = 0.45$ cm, the radius of the solid is $r_s = 0.50$ cm;
and the radius of the outer cylindrical surface is r_i in cm.
The relative quantities are dimensionless. The ambient
pressure and temperature are, respectively, $p_0 = 1.013 \times 10^5$ N/m^2 and $T_0 = 300$ K.

r_i (cm)	p_∞/p_0	$v_s(0)/T_0$
0.55	1.01	1.02
0.78	1.04	1.09
1.00	1.06	1.14

α_s, W, and r_s for illustrative purposes. Because α_s and W_ℓ occur
only as the product $\alpha_s W_\ell$, the illustrative numerical examples
remain unchanged for all α_s and W_ℓ such that $\alpha_s W_\ell = 0.1$(W/cm).

Using the above values for the input parameters, we have
written a computer program to evaluate the expressions in Section
III. After computing the steady state solutions, this computer
program then finds the first n eigenvalues and uses these to
compute the transient solution.

Referring to approximate Eq. (9), we observe that the quantity
$[(p_\infty/p_0)-1]$ is one measure of the performance or figure of merit
for the device. Also, approximate Eq. (9) suggests that to
maximize the relative pressure rise in the steady state we want to
choose the product $\alpha_s W_\ell$ and the ratio (r_i/r_s) as large as possible.
But, as the ratio (r_i/r_s) increases, the time which is required
to reach the steady state condition also increases.

Table II summarizes the steady state predictions of the model.
The 1% to 6% relative pressure rises should be readily detected
with present techniques.

Table III outlines the transient solutions. Observe that
the $j = 1$ term dominates the summations appearing in Section III.
The fact that the expansion coefficient (A_2/T_0) is negative for
some values of r_i is of no consequence because $|A_2/T_0|$ is so much
smaller than (A_1/T_0). We therefore consider the inverse of the
$j = 1$ eigentime t_1 to represent the characteristic frequency of
the solid-gas system; namely, $\omega_1 = (1/t_1)$. After times $2t_1$, the
relative pressure rises vary from 1% to 5% as the values for r_i
vary from 0.55 cm to 1.00 cm. But the eigentime for $r_i = 0.55$ cm

Table III. Transient solutions. The radius of the solid sample is $r_s = 0.5$ cm and the radius of the laser beam is $r_\ell = 0.45$ cm. The radius of the outer cylindrical surface is r_i; the eigentimes t_j in s are given by Eq. (11); the relative pressure at time $2t_1$ is $\{p(2t_1)/p_o\}$ and is given by Eq. (10); and the relative temperature at time $2t_1$ and at $r = 0$ is $\{T_s(0,2t_1)/T_o\}$. The jth eigenvalue in cm^{-1} is the jth solution for the determinant of coefficients to vanish and the ratios (A_j/T_o) are the expansion coefficients.

r_i (cm)	$\{p(2t_1)/p_o\}$	$\{T_s(0,2t_1)/T_o\}$	j	ξ_j (cm^{-1})	t_j (s)	(A_j/T_o)
0.55	1.008	1.02	1	0.209	81.0	0.466
			2	1.06	3.18	-0.807×10^{-3}
			3	1.91	0.978	0.106×10^{-3}
			4	2.75	0.468	0.149×10^{-3}
0.78	1.032	1.08	1	0.101	350.0	0.543
			2	1.04	3.29	-0.553×10^{-4}
			3	1.90	0.988	0.756×10^{-4}
			4	2.75	0.470	0.489×10^{-4}
1.00	1.046	1.13	1	0.080	551.	0.536
			2	1.04	3.30	3.30×10^{-4}
			3	1.89	0.989	0.989×10^{-4}
			4	2.75	0.471	0.471×10^{-4}

is 81 s and the eigentime for $r_i = 1.00$ cm is 551 s. The advantage of such readily detected pressure rises is offset in part by these rather long times before the pressure measurement is made. However, optical materials with larger thermal conductivities than glass are expected to have shorter times t_1. Hence, operating the proposed device in the barothermal mode with times comparable or greater than t_1 could be a promising alternative way to measure absorption coefficients for highly transparent solids. Because this technique tends to average over any absorption inhomogeneities in the sample, we expect that localized absorbing or scattering centers could be less of a problem for this technique than for the calorimetric techniques in which the placement of thermal couples may be important.

Acknowledgments

We thank R. D. Mountain, A. Feldman, and A. H. Kahn for many helpful discussions. We also gratefully acknowledge the assistance of S. T. Peavey and B. A. Peavey, Jr.

References

[1] K. I. White and J. E. Midwinter, Opto-electronics $\underline{5}$, 323 (1973).
[2] H. H. Witte, Appl. Opt. $\underline{11}$, 777 (1972); F. W. Patten, R. M. Garvey, and M. Hass, <u>Materials Research Bulletin</u> 6, 1321 (1971).
[3] K. C. Kao and T. W. Davies, J. Sci. Instrum. $\underline{1}$, 1063 (1968).
[4] M. W. Jones and K. C. Kao, J. Sci. Instrum. $\underline{2}$, 331 (1969).
[5] D. Stierwalt, Appl. Opt. $\underline{2}$, 1169 (1963); and Appl. Opt. $\underline{5}$, 1911 (1966).
[6] A. Rosencwaig, Opt. Commun. $\underline{7}$, 305 (1973).
[7] A. Rosencwaig, Science $\underline{181}$, 657 (1973).

A CALORIMETRIC TECHNIQUE FOR THE MEASUREMENT OF LOW OPTICAL ABSORPTION LOSSES IN BULK GLASS AND OPTICAL COMMUNICATION FIBRES

K I White and J E Midwinter

Post Office Research Centre, Martlesham Heath

Ipswich, Suffolk IP5 7RE, UK

In the bulk glass measurement, a pair of thin rod samples are rigidly mounted in a massive brass enclosure. As the rod ends are immersed in index matching liquid, no optical polishing is necessary. One sample is illuminated by a laser beam and the temperature rise due to the absorbed energy is measured by a thermocouple. For the fibre measurement, the solid rod is replaced by a fibre led through a silica tube filled with a low index liquid. A laser power of 100mW gives a sensitivity of 1dB/Km limited by thermocouple Johnson noise. Replacing the fibre by a fine resistance wire conducting a known current provides a calibration of the apparatus.

INTRODUCTION

It is desirable to be able to measure independently, at levels of a few dB/Km, the absorption and scattering losses as well as their sum, total attenuation of the bulk glasses and fibres used for optical communications. This is both an aid to understanding the power loss mechanisms and as a check on the accuracy of the experiments. Total attenuation can be measured in bulk glass [1] and fibres by insertion loss techniques. This is difficult for bulk glass since very large samples of good optical quality having highly polished and parallel end faces are required. Surface reflection losses are some 10^2-10^3 times the attenuation that is to be measured. Determinations of scatter loss in bulk glass [2] and fibres [3] have also been reported.

This paper describes a differential calorimetric method of
measuring bulk and fibre absorption losses. The temperature rise
of a thin rod of glass due to the power absorbed from an
incident laser beam is measured, relative to a reference sample,
by a thermocouple. The rod is thermally and mechanically
clamped to a massive hollow enclosure at two points along its
length. This decouples any end surface heating from the bulk
heating in the measurement region. The bulk samples studied
were from either ultrapure melts or experimental melts having
small (∿ 1 ppm) levels of doping of single impurities whose absorption was investigated as a function of glass making technique.
Fibre absorption is measured with the fibre running through a
liquid filled silica tube which replaces the solid rod of the
bulk sample.

THEORY

As a full presentation of the theory appropriate to the
present experiment is to be found elsewhere [4] only an outline
will be given here. The experimental geometry and definition
of co-ordinates used are shown in figure 1. The heat diffusion
equation in cylindrical polar co-ordinates is

$$\frac{\partial \theta\ (r,z,t)}{\partial t} = D \left[\frac{1}{r} \frac{\partial}{\partial r} \left(r\ \frac{\partial \theta\ (r,z,t)}{\partial r} \right) + \frac{\partial^2 \theta\ (r,z,t)}{\partial z^2} \right] \qquad 1$$

where $\theta\ (r,z,t)$ is the temperature relative to the environment,
k is the thermal conductivity, ρ is the density, C_p is the
specific heat at constant pressure and $D = k/\rho\ C_p$ is the
diffusivity. Assuming the variables to be separable,
equation 1 can be solved to give

$$\theta\ (r,z,t) = \sum_{m,n} \left[A_{mn} \ \text{Sin} \ (\gamma_m z) + B_{mn} \ \text{Cos} \ (\gamma_m z) \right] \ x$$

$$J_o\ (\alpha_n r)\ \exp\ (-t/T_{mn}) \qquad\qquad 2$$

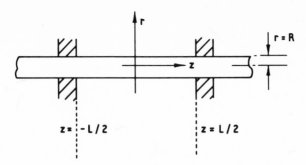

Figure 1. Definition of Co-ordinates.

with $\quad T_{mn} = \left[(\alpha_n{}^2 + \gamma_m{}^2) D \right]^{-1}$ $\qquad\qquad$ 3

$J_0 (\alpha_n r)$ is a Bessel function of the first kind in zero order.

The constants are determined by the following initial and boundary conditions:-
(1) The power absorption is assumed uniform along the sample. Thus $A_{mn} = 0$ to give a temperature distribution symmetrical about $z = 0$.
(2) The rod is maintained at the environment temperature at $z = \pm L/2$. Thus $\theta (r, \pm L/2, t) = 0$, which leads to $\gamma_m = (2m-1)\pi/L$.

(3) The rate of heat loss from the cylindrical surface of the sample is assumed proportional to the temperature difference between the surface and the surroundings.

$$\therefore \quad - k \left. \frac{\partial \theta (r,z,t)}{\partial r} \right|_{r=R} = \beta\, \theta\, (R,z,t) \qquad\qquad 4$$

where β is the heat loss rate/unit area/unit temperature difference. It thus follows that

$$(\alpha_n R)\ J_0' (\alpha_n R) + \frac{\beta R}{k}\ J_0 (\alpha_n R) = 0 \qquad\qquad 5$$

Tables are available giving the roots of this equation [5].

Under the conditions of the present experiment, $k \gg \beta$ and $R \sim 1\text{-}2$ mm.

$$\therefore \quad J_o (\alpha_1 R) = 1 - (\alpha_1 R/2)^2 \qquad\qquad 6$$

$$(\alpha_1 R)^2 = 2\beta R/k \qquad\qquad 7$$

For $n \geqslant 2$, $(\alpha_n R)$ and $J_o (\alpha_n R)$ are virtually independent of sample parameters.
(4) Since the power is absorbed from a focussed laser beam or in a fibre core guiding laser light, the heat source is a cylinder of radius $q \lll R$. If P Watts/cm is the power absorbed from the incident beam, an initial instantaneous source of energy, P dt, is considered.

$$\therefore \quad \theta (r,z,0) = \frac{P\ dt}{\pi\ q^2\ \rho\ C_p} \qquad \text{for } 0 \leqslant r \leqslant q$$

$$\qquad\qquad\qquad\qquad\qquad\qquad\qquad\qquad\qquad 8$$

$$= 0 \qquad\qquad \text{for } q < T \leqslant R$$

This is equated to $\theta (r,z,0)$ obtained from equation (2). B_{mn} is obtained on multiplying both sides by $r\ J_o (\alpha_p r)\ \mathrm{Cos}\ (\gamma_s z)$ and integrating r from 0 to R and z from $-L/2$ to $L/2$. Further integration over t gives the temperature distribution for a continuous source as

$$\theta (r,z,t) = \frac{4P}{\pi H} \sum_{m,n=1}^{\infty} \frac{(-1)^{m-1}\ \mathrm{Cos}\ (\gamma_m z)\ J_o (\alpha_n T)}{(2m-1) \left[1 + (\beta/\alpha_n k^2\right]\ J_o^2 (\alpha_n r)} \quad \times$$

$$T_{mn} \left[1 - \exp (-t/T_{mn})\right] \qquad\qquad 9$$

where $H = \pi R^2 \rho\ C_p$ is the heat capacity/unit length of sample. For parameter values appropriate to this experiment, T_{11} is the principal thermal time constant and the only one operative for times greater than approximately 30 seconds.

The surface temperature at the mid-section of the rod, i.e. the measured temperature, when the sample is in equilibrium with its surroundings is

$$\theta\ (R,0,\infty) = \frac{4P}{\pi H} \sum_{m,n=1}^{\infty} \frac{(-1)^{m-1}}{(2m-1)\left[1 + (\beta/\alpha_n k)^2\right]} \frac{T_{mn}}{J_o\ (\alpha_n R)} \qquad 10$$

On applying the approximations R << L and β << k, which are well satisfied for long thin glass rods in air subject to small temperature rises (< 1°C), equation (10) reduces to

$$\theta\ (R,0,\infty) = \frac{4P}{\pi H}\ T_{11}\ [1 - \delta] \qquad\qquad 11$$

where δ is a complicated function, only weakly dependent on sample parameters, having a typical value of ∿ 0.06. The final temperature is obtained directly from the heating curve - the trace of temperature against time on illuminating the sample. A semi-log plot of log [θ (R,0,∞) - θ (R,0,t)] against time has a straight region for t $\gtrsim$ 30 seconds whose gradient yields T_{11}. H is obtained from the measured mass/unit length and the specific heat calculated from a formula due to Sharp and Ginther [6]. A measurement of incident laser power completes the quantities required for a determination of the sample's absorption loss.

DESCRIPTION OF THE APPARATUS

The bulk glass absorption apparatus is shown schematically in figure 2. Two cylindrical samples are mounted in the apparatus and one illuminated by a focussed laser beam. The differential temperature rise, due to power absorption from the incident laser beam, is measured by a thermocouple having a junction on each rod.

The samples are obtained by pulling a rod from the glass melt and removing a 80 mm length by cleavage. There is no polishing or preparation of the ends. Brass collars of 6 mm diameter are fixed to the rod at 40 mm spacing by an epoxy resin glue to facilitate mounting in the apparatus. To obtain good sensitivity, small diameter rods (∿ 2 mm) and a collimated laser light source are obligatory. The sources used are a krypton ion laser and a Nd:Yag laser. In all, 11 wavelengths are available from 468 nm to 1060 nm with the most important being those at 799 nm and 1060 nm which monitor the IR absorption bands due to copper and iron impurities respectively. A liquid-tetrachloroethylene - having similar refractive index to the glass samples, facilitates launching of the laser beam by index matching the rod ends and, together with the brass collars,

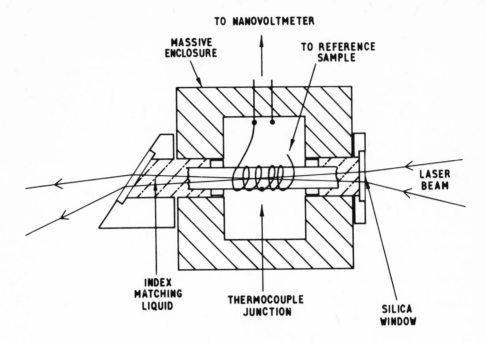

Figure 2.   Schematic Diagram of Bulk Absorption Apparatus.

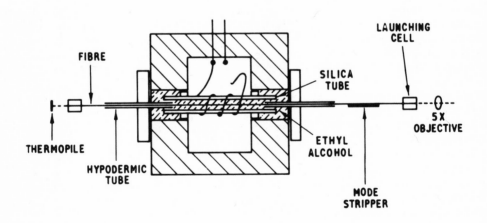

Figure 3.   Schematic Diagram of Fibre Absorption Apparatus.

provides a thermal short circuit for any end surface heating.
The latter has already been reduced by the use of a cleaved
surface. The liquid is filtered to minimise scatter which might
directly illuminate the thermocouple and the sample is placed
close to the silica windows to avoid beam divergence caused by
heating of the liquid. The exit window is at Brewster's angle
to reduce back scattered light. The massive hollow brass
enclosure shields the sample and reference from ambient
temperature fluctuations. Differential techniques give a good
signal-to-noise ratio when, as here, the two samples are strongly
coupled – all the rod ends are held at ambient temperature – and
thus external temperature fluctuations affect both rods
similarly. The thermocouple wire used is 50μm diameter
Pallador I [7]. It is wound onto the rod in a spiral with the
junction halfway along the rod. Spots of glue secure the wire
to the rod. Since energy conducted along the thermocouple wires
is extracted from a different part of the rod to that at which
the temperature is measured, the required temperature is not
perturbed by the act of measurement. We have found that this
noble metal thermocouple with soldered junctions gives a
reproducible thermoelectric coefficient. The thermal emf is
detected by a Keithley 148 nanovoltmeter [8] whose lead goes
directly into the apparatus and terminates on electrically
insulated brass blocks which act as heat sinks for temperature
fluctuations conducted down the lead wires. The thermocouple
wires are soldered directly to the nanovoltmeter leads to
minimise the number of junctions on the wire. The temperature
rise is displayed on a chart recorder. Sensitivity is Johnson
noise limited to 1 dB/Km for 100 mW laser power and a 2 mm
diameter sample.

A similar apparatus for measuring fibre absorption is shown
in figure 3. The "sample" in this case consists of a silica tube
having a 1 mm bore containing the fibre and ethyl alcohol. The
fibre is led in and out of the apparatus by hypodermic tubes.
Ethyl alcohol is chosen for its low refractive index (1.33)
which results in light travelling at up to 30° to the forward
direction being trapped in the fibre. Direct illumination of
the thermocouple by tunnelling modes which could give a
spuriously high indicated loss, is thus avoided. The fibre is
terminated in cells of index matching liquid. The laser beam
is launched by a 5X microscope objective and cladding modes
removed by a mode stripper of benzyl alcohol on a metal plate
placed immediately prior to the entrance hypodermic tube.

The use of larger bore silica tubes and entrance and exit
windows, enables bulk samples to be measured in this way. This
is particularly useful if there is substantial forward scattering
of the laser beam. This can arise through a radial strain

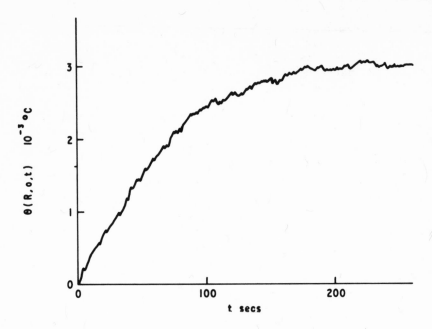

Figure 4.   The Surface Temperature at the Mid-Point of a 2mm
            Diameter Sprasil-W1 Rod as a Function of Time.

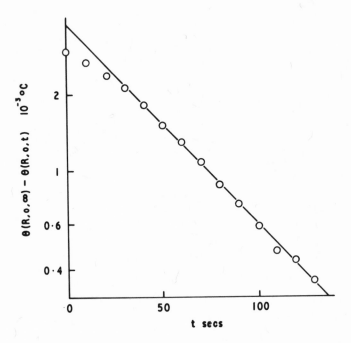

Figure 5.   A Logarithmic Representation of the Data of Fig. 4.

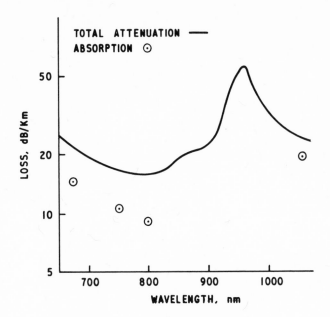

Figure 6.   Attenuation and Absorption Spectra of a Sodium
            Borosilicate Glass Fibre.

distribution, and hence a refractive index variation caused by the
rapid cooling of the sample on pulling from the melt.   The rod
acts as a weakly diverging lens.   Scattering may also arise in
the refractive index liquid and from microbubbles in the sample.
As distinct from the fibre case, the liquid must perform the
dual roles of both completing the sample thermally and assisting
launching of the laser beam.   Tetrachloroethylene is normally
used.   Light scattered at up to 15° is trapped within the
silica tube and direct illumination of the thermocouple is
substantially reduced.   The use of the silica tube increases the
thermal capacity over that of the rod itself, resulting in a loss
of sensitivity.   However, since the tube and thermocouple are
permanently mounted in the apparatus, sample loading is simple.

                              CALIBRATION

     As a check on the operation of the apparatus, the fibre
can be replaced by a fine resistance wire of the same diameter
as the fibre and a known quantity of electrical power applied.
All the quantities in equation (11) are thus known or are

determined by the experiment. The measured and calculated final temperatures agree to within 2%. This shows both that the approximations applied in the theory are valid for this experiment and also that the thermocouple arrangement does in fact measure the required temperature. Also verified is the shape of the heating curve for pure absorption from an axial heat source. Direct illumination of the thermocouple wires produces an additional heat source which alters the shape of the heating curve. Spuriously high results can thus be identified and rejected.

Since both the glass fibre and the calibration wire form less than 0.1% of the total sample heat capacity, the temperature response determined during calibration can be applied directly for any fibre. Treatment of the results is thus very simple compared with that for the bulk sample. For a silica tube of nominal bore 1 mm and wall thickness 0.5 mm having $L = 40$ mm, the loss is calculated from $\alpha = 10.8 \ \theta(\infty)/P_L$, where $\theta(\infty)$ is the final thermal emf in μV and P_L is the laser power in Watts. The time constant for this arrangement is 50 seconds.

Bulk samples immersed in the silica tube may be individually calibrated by running a resistance wire through the liquid between the rod and the tube. Alternatively, having determined the total heat capacity of the tube, liquid and glass rod, the loss may be obtained from equation (11) and the full heating curve. Agreement to better than 1% has been obtained between those two methods.

RESULTS

The heating curve for a sample of Suprasil W-1 measured at 2.9 dB/Km at 1060 nm with 400 mW laser power is shown in figure 4. The same information is presented in figure 5 as a semi-log plot where the straight line region in which T_{11} is the only operative time constant is clearly apparent. Figure 6 shows the results of a continuous insertion loss measurement of total attenuation and the point determination of absorption on a recent Post Office sodium borosilicate fibre. The absorption loss falls to ∿ 9 dB/Km at 799 nm. The difference between absorption and attenuation is partly due to Rayleigh Scattering, which falls from 5.3 dB/Km at 647 nm to 2.3 dB/Km at 799 nm, and partly to modes tunnelling out of the fibre.

CONCLUSIONS

This technique provides a quick, sensitive measurement of

optical absorption on small samples of bulk glass and optical fibre and requires the minimum of sample preparation. Over 300 samples have been measured at up to 11 wavelengths and many of the results have been correlated with other data on the loss mechanisms to provide a convincing check of the absolute accuracy of the technique.

ACKNOWLEDGEMENTS

Acknowledgement is made to the Director of Research of the Post Office for permission to publish this paper.

REFERENCES

[1]   K G Kao and T W Davies, J. Sci. Instr $\underline{1}$ (1968) 1063-1068
       C R Wright and K C Kao, ibid $\underline{2}$ (1969) 331-335
[2]   D A Pinnow, T C Rich, F W Ostermeyer, Jr., and M Di Domenico, Jr., Appl. Phys. Lett. $\underline{22}$ (1973) 527-529
[3]   F W Ostermeyer Jr., and W W Benson, Appl. Opt. $\underline{13}$ (1974) 1900-1902
[4]   K I White and J E Midwinter, Opto-electronics $\underline{5}$ (1973) 323-334
[5]   H S Carslaw and J C Jaeger, Conduction of Heat in Solids, p 493 Clarendon Press, Oxford (1959)
[6]   D E Sharp and L B Ginther, J. Am. Ceram. Soc. $\underline{34}$ (1951) 260-271
[7]   Johnson Matthey Metals Ltd, London UK
[8]   Keithley Instruments, Ohio, USA

RAMAN SCATTERING TECHNIQUE TO EVALUATE LOSSES IN GaAs DIELECTRIC

WAVEGUIDES

J. L. Merz, R. A. Logan and A. M. Sergent

Bell Laboratories

Murray Hill, N. J. 07974

ABSTRACT

Loss measurements in dielectric slab waveguides
of GaAs cladded by AℓGaAs epitaxial layers have been
made utilizing spontaneous Raman scattering. In
these experiments, incident radiation at 1.06 μm is
focused onto a cleaved edge of the waveguide layer
and Raman-scattered photons are detected at 90°. The
attenuation of the resulting signal is measured as a
function of the propagation distance along the guide.
Spatial resolution of 0.2 to 0.5 mm has been achieved
for samples 3 to 8 mm long. Losses from 0.25 to 8.8
cm^{-1} have been measured to ±0.1 cm^{-1}. The experimental technique is described in detail, and its use
illustrated by results obtained for a variety of GaAs
waveguide layers.

INTRODUCTION

The rapid growth of interest in the possibility of optical
communications in recent years has stimulated the investigation of
many different materials for this purpose. Of particular interest
is the III-V compound-semiconductor GaAs. With the development of
the continuously operating GaAs-$Al_xGa_{1-x}As$ double heterostructure
laser[1,2], light modulators[3], and distributed-feedback lasers[4-6],
GaAs shows future promise for its application to integrated optical communications systems. However for such applications, passive
waveguides should be fabricated from GaAs or AℓGaAs with losses of
the order of 1 cm-1 at the wavelengths of interest, ∼0.8 to 1.2 μm.

473

The measurement of low losses in GaAs waveguides is complicated by the difficulties of coupling light into micron-sized layers of this high refractive-index material, and by the relatively short path lengths of uniform material available. This paper reports the utilization of spontaneous Raman scattering to measure losses in dielectric slab waveguides of GaAs cladded by $Al_xGa_{1-x}As$ epitaxial layers. In these experiments, incident radiation at 1.064 μm from a cw Nd:YAG laser is focused onto a cleaved edge of the waveguide layer, coupling energy into a low-order mode of the guide. The photons in the guided wave interact with lattice vibrations of the GaAs layer, coupling with acoustic phonons (Brillouin scattering) or optical phonons (Raman scattering). Photons may either lose energy by the emission of a phonon (Stokes lines) or gain energy by phonon absorption (anti-Stokes). Scattered photons are observed on both sides of the laser line, shifted in energy by the energies of the phonons involved. In this work, anti-Stokes Raman-scattered photons are detected at 90° to the direction of propagation. The attenuation of the resulting signal, which is proportional to the laser power in the guide, is measured as a function of the propagation distance through the layer.

This technique has several advantages over more conventional loss measurements. When the scattered laser frequency is detected directly, the attenuation curve is dominated by numerous strong peaks due to imperfections, making a loss determination impossible. Standard transmission measurements of loss require accurate knowledge of the coupling efficiency of the optical input, or multiple measurements must be made under identical coupling conditions on identical guides of different lengths. The intensity of the spontaneous Raman scattering, on the other hand, depends only on the intensity of light within the guide; knowledge of the details of input coupling is not required. Furthermore, these scattering measurements yield a continuous measure of losses in the guide, and exhibit specific regions of guide inhomogeneity and gross layer imperfections. Its chief disadvantage is weak signal strength.

In this paper, the experimental technique for loss evaluation in GaAs waveguides employing Raman scattering will be described in detail. The usefulness of this technique is illustrated by results obtained on a variety of samples: lightly doped p-type layers, n-type layers varying between 10^{16} and 10^{18} cm^{-3}, and $Al_xGa_{1-x}As$ waveguide layers with $x \leq 0.1$. The results indicate that the losses in n-type guides are a sub-linear function of doping, dominated at low doping by layer inhomogeneities. Further discussion of these results, along with loss data at other energies below the GaAs bandgap, will be provided in a later paper.

EXPERIMENTAL

The samples studied were grown by liquid phase epitaxy (LPE) on (100) surfaces of GaAs substrates.[2] Layers of $Al_xGa_{1-x}As$ for optical confinement were grown before (x=0.2) and after (x=0.4) the LPE GaAs waveguide layer. For the series of Sn-doped wave-guide layers, most of the samples were grown consecutively to mini-mize variations in layer properties associated with growth parame-ters. Carrier concentrations of the guide layers were determined by Schottky barrier capacitance measurements, and layer thicknesses were determined on cleaved test pieces. Waveguide layer thicknesses ranged from 5 to 10 μm.

A block diagram of the apparatus used for the scattering experiments is given in Fig. 1. The following crystal geometry was employed: the incident beam was in the $\langle 110 \rangle$ crystallographic direc-tion, scattered light was detected at 90° in the $\langle 001 \rangle$ direction, and the incident beam was polarized in the $\langle \bar{1}10 \rangle$ crystal direction to excite the TE_0 mode of the guide. The resulting Raman-scattered beam is also polarized ⊥ scattering plane, allowing discrimination against unpolarized luminescence and stray light. A Sylvania Model 605A Nd:YAG laser was used as source (single mode output power ∿1/4 watt cw), with Glan prisms to polarize and attenuate the beam. The resulting beam of 100 to 200 mW power was focused astigmatically onto a cleaved edge of the waveguide layer by means of a cylindrical lens (L1) and an appropriate microscope objective (L2) to match the focused beam to the layer thickness, reducing diffraction spreading of the beam in the plane parallel to the layer. The sample was mounted on a stage with 6 degrees of translational and rotational motion to allow accurate waveguide alignment with the beam. Light scattered out of the waveguide in the direction normal to the layer was collected by a 58 mm Minolta lens (L3) of small f-number (1.4), and focused onto the entrance slit of the first of a pair of mono-chromators. The two monochromators were used in a "crossed" con-figuration; i.e., oriented so that the grating axis of each was perpendicular to the other. The first monochromator was a low reso-lution, high intensity Bausch and Lomb, used as a filter to reduce scattered laser light; it was oriented with entrance slit parallel to the direction of propagation of light in the waveguide. The second monochromator was a 0.3 meter McPherson with a dispersion of 106 Å/mm. Its entrance slit width determined both the spatial reso-lution at the sample surface and the spectral resolution of the detected radiation; a 500 μm slit width was usually used (effective sample resolution of 0.5 mm), although resolution to 0.20 mm has been achieved with adequate signal. This combination of monochroma-tors was chosen as a reasonable compromise between the conflicting requirements of high signal intensity and adequate spectral and spatial resolution. Because of the rapid decrease in sensitivity near 1.1 μm of the photomultiplier employed (RCA C70007A, S-1 response), anti-Stokes scattered photons were detected with the

photomultiplier cooled to -100°C. Photon counting techniques were
used with a SSR Model 1140 Quantum Photometer and preamplifier.

Care was taken to correct the resulting signal in the following way. A large area Si Schottky barrier detector was located a
few cm behind the waveguide output edge, centered at the peak of the
waveguide far-field intensity pattern. The analog signal from the
quantum photometer (Raman signal) was divided by the signal from the
Si detector (monitor) using a PAR Model 230 Analog Multiplier-divid-
er, with a time constant of 1-3 sec. The sample, monitor, and input
coupling objective L2 were rigidly attached to a motor-driven stage
which was translated in the direction of the incident beam at a rate
of 0.64 mm/min, so that the sample moved past the signal collection
lens L3 (cf. Fig. 1). During the course of such an experimental run
(attenuation scan), the monitor signal was maximized by translating
the sample in the direction normal to the waveguide layer with the
use of a piezo-electric micrometer capable of submicron transla-
tions. In this way, the monitor was used to correct the Raman signal not only for changes in the laser output power, but also for
changes in the light intensity coupled into the waveguide as a result
of any mechanical motion of the sample. The corrected intensity of
the Raman-scattered light was stored in a Nicolet Model 1074 signal
averager, using the time base as a measure of the propagation distance in the waveguide. At least three runs were made on each
sample to average out noise fluctuations. Photon counting rates
were usually of the order of a few hundred counts/sec for each run.

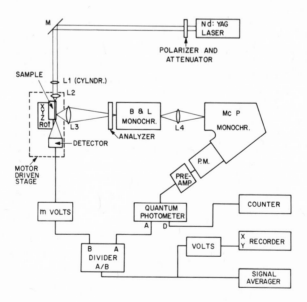

Fig. 1. Block diagram of the experimental apparatus used
for Raman scattering measurements of losses in GaAs waveguides.

For each sample studied, the spectrum of the anti-Stokes emission was measured with the sample stationary at several positions along the waveguide. These spectra were used to determine the appropriate grating settings for subsequent attenuation scans at some convenient wavelength for background scattering as well as the Raman scattering.

DATA ANALYSIS

The capability of the apparatus described above to resolve and detect Raman scattering is clearly shown in Fig. 2. In the spectrum of Fig. 2(a) the LO and TO phonon peaks are unambiguously resolved. The arrows indicate the anticipated shift of these phonon-scattered photons from the central laser line at 1.064 μm.[7] Figure 2(b) shows the appearance of the same spectrum under the conditions used for subsequent attenuation measurements (slit-widths of 500 μm). The LO and TO phonons are no longer resolved, but the Raman peak dominates this region of the spectrum. These spectra are characteristic of the GaAs waveguide layer, rather than the cladding layers or substrate material, since waveguide confinement of the optical energy is extremely high for the choice of layer parameters used here.

In addition to the Raman peak at 1.034 μm, additional light has been observed between the Raman peak and the Rayleigh-scattered laser light. This extraneous radiation appears to be dependent on the concentration of donors or acceptors in the waveguide layer, but its origin is not presently understood. This "background" emission is generally weak for low-doped samples, but often does not exhibit the same exponential attenuation as the Raman signal with penetration through the waveguide. For high-doped samples

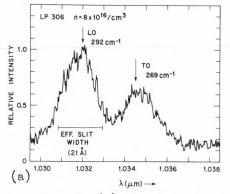
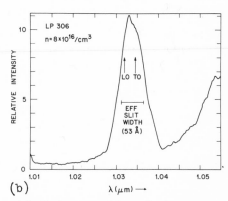

Fig. 2. (a) Raman spectrum of anti-Stokes scattered phonons with LO and TO phonons resolved. The phonon energies for GaAs are given. (b) Same spectrum as (a) under the conditions of lower resolution used for loss measurements.

$(n \geq 10^{18}/cm^3)$, a strong, broad band of radiation is observed from about 1.05 μm merging into the Raman peak. However, in this case radiation from all portions of the band shows strong exponential attenuation with the same decay constant, α. Due to the presence of this background radiation, attenuation scans were made at both the wavelength of the Raman peak (1.034 μm) and a "background" wavelength, usually corresponding to a minimum in the background spectrum on the long wavelength side of the Raman peak (~1.040 to 1.045 μm). The Raman data were then corrected for background effects as described below.

The advantage of detecting Raman-scattered light rather than Rayleigh-scattered laser light is clearly illustrated in Fig. 3, along with some of the background effects discussed above. In Fig. 3(a) attenuation scans are shown for light scattered at the laser wavelength (I_L), the Raman wavelength (I_R), and a background wavelength of 1.041 μm (I_B). Any attempt to derive an estimate of loss from the laser scattering would clearly be unproductive. Visual examination of the surface of this sample through an infra-red image converter showed bright spots at the positions of the peaks at 1.1, 2.0, and 3.2 mm past the input edge of the sample. These peaks result from gross imperfections in the layer, which scatter light out of the waveguide and into the detection system. However, the amount of energy scattered out of the guide is a small fraction of the energy which remains, as evidenced by the smooth attenuation of the Raman signal. The different decay constants for Raman and background scattering are also obvious in Fig. 3(a). The input and output edges of the sample are clearly seen for I_L, whereas for I_R and I_B the scattered light intensity is reduced to approximately half its maximum value at the sample edges, as expected for a finite effective slit width.

Since it is difficult to judge the extent to which the background radiation affects the attenuation of the Raman peak, the data were analyzed in the following fashion. The curves I_R and I_B were read out of the signal averager, and a background correction was made by forming the difference $I(x) = I_R(x) - fI_B(x)$, where x is the propagation distance through the waveguide, and f could be varied between f=0 and f=1. For a given value of f, a least-squares fit was made of the experimental values of $I(x)$ to the expression for the field intensity inside the waveguide, neglecting phase effects:[8]

$$\frac{I(x)}{I_o} = (1-R)e^{-\alpha x}\left\{\frac{1+Re^{-2\alpha(L-x)}}{1-R^2 e^{-2\alpha L}}\right\}, \tag{1}$$

where I_o is the power incident on the input cleaved edge of the sample, R is the power reflectivity at the waveguide-air cleaved interface (R≈0.31 for GaAs at 1.06 μm), L is the sample length,

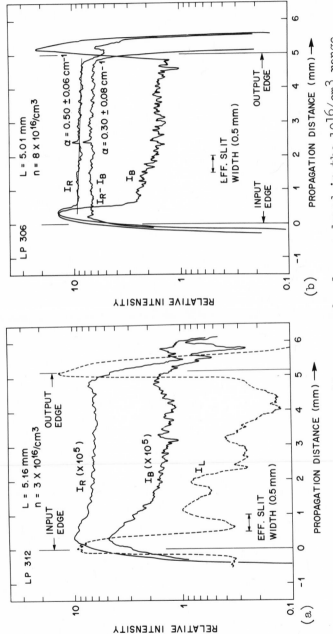

Fig. 3. Typical attenuation scans for layers doped in the $10^{16}/cm^3$ range. Scattered light intensity is plotted vs. propagation distance through the wave-guide at the wavelengths for Raman scattering, I_R ($\lambda = 1.034~\mu m$), background radiation, I_B ($\lambda = 1.041~\mu m$), and incident laser, I_L ($\lambda = 1.064~\mu m$). Effective slit width indicates spatial resolution on sample. (a) Illustrates difficulty of making loss measurements using laser (Rayleigh) scattering. (b) Indicates measurement of attenuation coefficient α by fitting data to Eq. (1), with and without background subtraction.

and α is the attenuation coefficient. The factor within the braces of Eq. (1) results from multiple reflections between the cleaved edges of the sample, and can be significant for small α. This fit of the data to Eq. (1) yields the best value of the attenuation coefficient α, with the experimental error ($\Delta\alpha$) given by one standard deviation from the least-squares fit. Usually this error was a minimum for values of f between 0 and 0.5, as expected, since the background intensity was usually decreasing between its measurement wavelength and the Raman peak at 1.034 μm. The value of $\alpha \pm \Delta\alpha$ reported from these measurements was then deduced from the results between f=0 and f=0.5; the details of this procedure will be reported elsewhere. For heavily-doped samples, the results were essentially independent of the value of f. An example of this procedure for a moderately-doped sample is given in Fig. 3(b) showing the extreme cases of no background correction and maximum background correction. The results are α = 0.50 ± 0.06 cm^{-1} and α = 0.30 ± 0.08 cm^{-1} for f=0 and f=1, respectively. The solid line through the upper trace represents the best fit of the data to Eq. (1) with f=0. Also evident for the two samples shown in Fig. 3 is the fact that the stray light seen near the input and output edges of the sample in the background radiation is significantly reduced at the Raman wavelength.

DISCUSSION OF RESULTS

The results of the loss measurements made in this work are presented graphically as a function of dopant concentration in Fig. 4. The lowest loss measured, α = 0.25 ± 0.15 cm^{-1}, was observed in a Sn-doped sample with n = 3×10^{16}/cm^3. For the majority of the samples studied, α varies between 0.5 and 1.0 cm^{-1}, with no consistent trend until the impurity concentration reaches the mid-10^{17}/cm^3 range, when a gradual increase in loss accompanies increasing concentration. The highest loss measured was α = 8.8 ± 0.1 cm^{-1} for n = 3×10^{18}/cm^3; this was a test measurement of the scattering technique in a typical substrate upon which the layers are grown. An attempt to connect the lower-lying points in this figure yields the solid straight line, which represents a sublinear relationship between α and doping. The expected linear relationship between α and n for free carrier absorption is represented by the broken line in the figure, drawn through the high-loss data so that all the experimental points are above it. The results indicate that for impurity concentrations below ~10^{17}/cm^3, the losses are independent of doping, and that another loss mechanism is operative in these waveguides. Since the Raman technique as applied here does not distinguish between absorption and losses due to scattering out of the waveguide, it seems probable that layer inhomogeneities are contributing to the measured loss. A few of the layers studied were found to be quite nonuniform. For example, the nominally undoped sample (n = 1×10^{16}/cm^3) had opaque regions in the

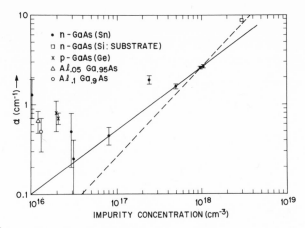

Fig. 4. Values of attenuation coefficient α, measured by Raman scattering, plotted as a function of donor or acceptor concentration. The broken line indicates the linear relation expected for free carrier absorption.

layer; Raman scattering measurements in regions of higher transmission through this layer were erratic, yielding a relatively high error, $\alpha = 1.3 + 0.6$ cm^{-1}. Published infrared absorption data of Spitzer and Whelan[9] for n-type bulk GaAs clearly showed that an unidentified loss mechanism was contributing to the absorption in their samples, in addition to free carriers. Their data give an approximately linear relation between absorption and doping at 1.06 μm: $\alpha = 0.5 \times 10^{-17}$ n cm^{-1}, where n is the donor concentration. However, this is approximately a factor of 15 higher than the expected free-carrier absorption extrapolated from his data at longer wavelengths, and a factor of 2 larger than the linear relation shown in Fig. 4.

The most significant result of this study is the fact that losses well within the requirements of integrated optics ($\alpha \lesssim 1$ cm^{-1}) have been routinely obtained in LPE waveguide layers of low concentrations. In particular, the addition of small amounts of Aℓ to form the higher band-gap ternary Aℓ_xGa$_{1-x}$As have not significantly added to the losses observed. This is important because the ternary compound layers, which are transparent to radiation obtained from GaAs double-heterostructure lasers ($\lambda \sim 0.9$ μm) are promising candidates for passive waveguide connections in future integrated optical structures.

The authors would like to thank C.H. Henry and F.K. Reinhart for many stimulating discussions during the course of this work, D.V. Lang for assistance with experimental problems, and H.G. White for growing the waveguide layers.

REFERENCES

1. I. Hayashi, M.B. Panish, P.W. Foy and S. Sumski, Appl. Phys. Letters $\underline{17}$, 109 (1970).

2. Zh. I. Alferov, V.M. Andreev, D.Z. Garbuzov, Yu. V. Zhilyaev, E.P. Morozov, E.L. Portnoi and V.G. Trofim, Sov. Phys. Semicond. $\underline{4}$, 1573 (1971). [Fiz. Tekh. Poluprovodn $\underline{4}$, 1826 (1970)].

3. F.K. Reinhart and B.I. Miller, Appl. Phys. Letters $\underline{20}$, 36 (1972).

4. M. Nakamura, A. Yariv, H.W. Yen, S. Somekh and H.L. Garvin, Appl. Phys. Letters $\underline{22}$, 515 (1973).

5. C.V. Shank, R.V. Schmidt and B.I. Miller, Appl. Phys. Letters $\underline{25}$, 200 (1974).

6. D.R. Scifres, R.D. Burnham and W. Streifer, Appl. Phys. Letters $\underline{25}$, 203 (1974).

7. A. Mooradian and G.B. Wright, Solid State Comm. $\underline{4}$, 431 (1966).

8. Effects due to phase coherence during multiple reflections, which could lead to Fabry-Perot resonances, introduce an additional constant factor (independent of x) multiplying this expression. This does not affect the curve-fitting procedure used, and will be discussed in a later publication.

9. W.G. Spitzer and J.M. Whelan, Phys. Rev. $\underline{114}$, 59 (1959).

ACCURATE SPECTROPHOTOMETER FOR THE ATTENUATION

MEASUREMENT OF LOW-LOSS OPTICAL MATERIALS

D. Krause

Res. Lab., Jenaer Glaswerk Schott and Gen.

D-65 Mainz, Hattenbergstrasse 10, Germany

A sensitive, unusual spectrophotometer
has been developed. The experimental setup is
discussed and main emphasis is on error analysis. Systematic errors are mainly due to
the surface properties of the sample. For
30 cm long samples the sensistivity is better
than 1 dB/km and the absolute accuracy is
$\pm$ 5 ^{o}/o $\pm$ 3 dB/km. The spectral distributions
of the total loss is Suprasil W 1 and flint
glass samples are shown.

1. INTRODUCTION

One way for the production of communication waveguides is by drawing fibres from low loss bulk glass.
In this case fibre losses are caused by the attenuation
of the core material as well as by the drawing process.
So methods are of considerable interest that are able
to measure very small optical losses in the relatively
short path length of bulk materials. The required accuracy is in the 10^{-5} range for total losses of about
1 to 10 dB/km (*) in samples in the 10 cm-range.

(*) A loss of dB/km is equivalent to an absorption
coefficient $\alpha = 2,3 \cdot 10^{-6}$ cm^{-1}.

The often used calorimetric measurements have their own problems: 1) They cannot use the large rods that we want to investigate prior to further treatment and fibre drawing, 2) They have the disadvantage to be relatively simple only at wavelengths of high power laser lines, and these are not available in the most interesting region around 850 nm.

On the other hand several sensitive one-beam and two-beam arrangements for optical transmission measurement have been described in the literature.

The fundamental advantage of two-beam arrangements, that time effects are compensated, is opposed by the disadvantage of the difficulty to find wide-band mirrors, detectors, etc. that are sufficiently homogeneous and spectrally identical. For spectral precision measurements the advantages are thus often canceled by the required time for adjustment and calibration.

In one-beam arrangements the fluctuations with time can be eliminated to a high degree by stabilisation measures and a sophisticated evaluation of the results. Their great advantage is the far-reaching identity of all optical elements in the beam which is used for measuring and reference as well. This turned out to be extremely important so that our setup worked the better the less components were used within the beam.

2. EXPERIMENTAL SETUP

The leading idea for the photometer was the elimination of the influence of the sample's reflectivity. This was found to be time dependent in complex glasses. An extreme (and not representative) behavior is shown in Fig. 1. The refractive index as calculated from the reflectivity of the sample falls down due to the water attack to the surface. These changes do already occur during the measuring time and differ on different samples.

We therefore used a principle first reported by Koppelmann and Krebs[1] (1959) for thin film measurements. The modified setup is shown in Fig. 2. The problem is to collect the primary beam as well as the higher order reflected beams. Therefore the light beam is made nearly parallel over a distance of several meters by the images of two diaphragms which limit the beam to a diameter of 3 to 10 mm. Such a beam has a strongly reduced

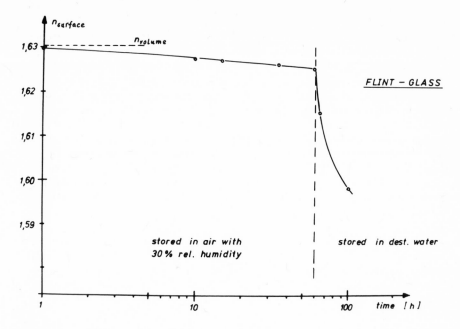

Fig. 1   Time dependence of refractive index of flint glass.

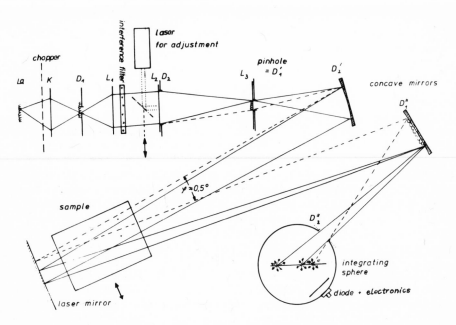

Fig. 2   Modified Koppelmann and Krebs optical setup.

intensity proportional to $(\text{diameter})^4/(\text{length})^2$. This
is, however, necessary to collect high order reflections.
After passing the sample this beam is reflected by a
high quality, dielectric, multilayer, broad band mirror
and passes the sample a second time. The various reflected beams are then collected by an extremely homogeneous, integrating sphere, It took a long time to
manufacture a sphere, homogeneous to about 10^{-4} over an
area of some cm^2. The electronics of the photometer is
shown in Fig. 3. A measurement is now performed by comparing the intensities with and without the specimen. Due
to the high sensitivity, dust must be kept away and the
system can only be used under clean room conditions and
is placed under a laminar flow bench.

Fig. 4, shows an original measurement curve under
intentionally bad conditions: drift, fluctuations and
a line voltage drop. Unaffected by these influences,
the result of the correlation analysis of the two upper curves gives a constant result of 16 ± 1 dB/km for
the Suprasil W 1 sample with air as reference. Normally the stability is at least one order of magnitude
higher, including all influences from optics and electronics, leading to a statistical relative error below
0.5 dB/km for samples of 30 cm length. However, this
is not the correct value for this specimen. The attenuation is only 9 to 10 dB/km what is proved by differential measurements.

3. ERROR ANALYSIS

To find the systematic contributions to the absolute error, the theory of the measurement is examined.
Fig. 5 gives the notation for the intensities and their
interrelations according to Fresnel's formulas. It is
assumed that the faces of the specimen are not identical
and not ideal, the laser mirror is not ideal. First we
assume ideal conditions: Fig. 6. The measured variable R is then nearly independent of the sample's reflectivity for high transmittance, as can easily be
shown by the examination of the second term, and the
sensitivity is doubled if compared with a conventional
transmission measurement. When taking real conditions,
Fig. 7, it is seen, however, that the measured variable
also includes terms caused by the surface losses, the
inhomogeneities of mirror and integrating sphere as well
as cut-off errors due possibly to the lost high-order
reflections as a consequence of a slightly diverging

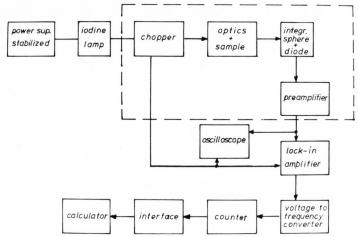

Fig. 3 Photometer electronics.

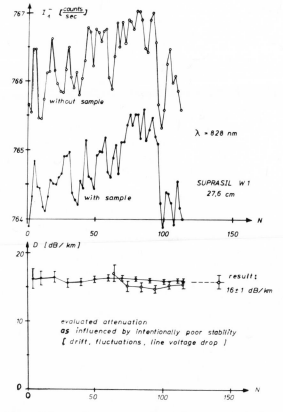

Fig. 4 Results of measurements under intentional bad conditions (see text).

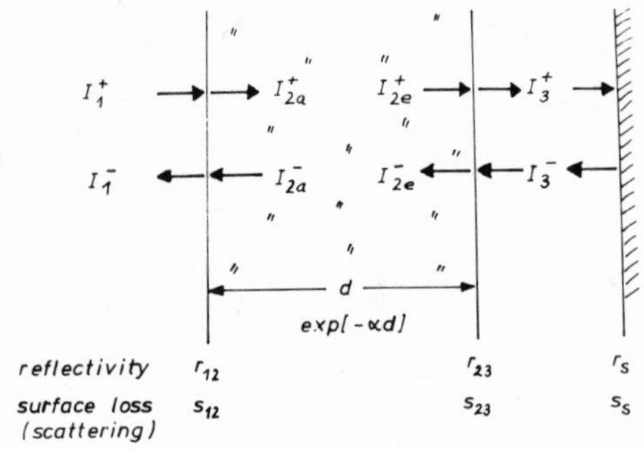

$$\begin{bmatrix} I_1^+ \\ I_1^- \end{bmatrix} = \frac{1}{1-r_{12}\cdot s_{12}} \begin{bmatrix} 1 & -r_{12} \\ r_{12} & (1-r_{12}-s_{12})^2-r_{12}^2 \end{bmatrix} \cdot \begin{bmatrix} I_{2a}^+ \\ I_{2a}^- \end{bmatrix}$$

$$\begin{bmatrix} I_{2a}^+ \\ I_{2a}^- \end{bmatrix} = \begin{bmatrix} exp[+\alpha d] & 0 \\ 0 & exp[-\alpha d] \end{bmatrix} \cdot \begin{bmatrix} I_{2e}^+ \\ I_{2e}^- \end{bmatrix}$$

$$\begin{bmatrix} I_{2e}^+ \\ I_{2e}^- \end{bmatrix} = \frac{1}{1-r_{23}-s_{23}} \begin{bmatrix} 1 & -r_{23} \\ r_{23} & (1-r_{23}-s_{23})^2-r_{23}^2 \end{bmatrix} \cdot \begin{bmatrix} I_3^+ \\ I_3^- \end{bmatrix}$$

$$I_3^- = (1-s_s)\, I_3^+ = r_s\, I_3^+$$

Fig. 5   Notation for the intensities and their interrelations according to Fresnel's formulas.

ideal conditions: detector sensitivity: E , homogeneous

$$s_{12} = s_{23} = 0$$
$$r_{12} = r_{23} = r = \left(\frac{n-1}{n+1}\right)^2$$

internal transmittance $exp\{-\alpha d\} = \vartheta$
mirror reflectivity $r_s = 1$, homogeneous

reflected intensity $R = \dfrac{I_1^-}{I_1^+} = \dfrac{r+(1-2r)\cdot\vartheta^2}{1-r\,\vartheta^2} = \vartheta^2 + \dfrac{r(1-\vartheta^2)^2}{1-r\,\vartheta^2}$

measured variable $\dfrac{r_s-R}{r_s} = 1-R \approx 2\alpha d$

conventional transmission measurement:

$$1 - \frac{I_3^+}{I_1^+} \approx \alpha d + 2r$$

Fig. 6   Reflected intensity and measured variable under
ideal conditions.

real conditions: E inhomogeneous: ΔE

$$s_{12} , s_{23} \ll 1 \neq 0 \qquad (\approx 10^{-5} - 5 \cdot 10^{-4})$$

$$r_{12} \neq r_{23}$$

$$r_s < 1 ; \qquad 1 - r_s = s_6 \ll 1 \quad \text{inhomogeneous: } \Delta s_s$$

beam divergence

measured variable $\dfrac{r_s - R}{r_s} = 2\alpha d + 2(s_{12} + s_{23}) + \Delta E + 2r\Delta s_s - 2rs_s^2 + \text{cutoff error}$

conv. transm. meas. $1 - T = \alpha d + r_{12} + r_{23} + s_{12} + s_{23} + \Delta E + \text{cutoff error}$

example: $D = 10\ dB/km$; $\Delta d = 30\ cm$ $\frown$ $\alpha \cdot \Delta d = 7 \cdot 10^{-4}$

for error $\leq \pm 3\ dB/km$ in a difference meas.

$$4\Delta s + 2r\Delta s_s + \Delta E + \Delta cutoff \leq 5 \cdot 10^{-4}$$

conv. trans. meas. $2\Delta r + 2\Delta s \qquad + \Delta E + \Delta cutoff \leq 2 \cdot 10^{-4}$

Fig. 7   Measured variable under realistic conditions.

ATTENUATION AT $\lambda = 850\ nm$ IN dB/km

			HVA 489 IV	HR 18/3 I
		mean value	$63,8 \pm 0,7$	$27,2 \pm 1,5$
conventional transmission measurement	difference $d_1 = 32\ cm$ $d_2 = 1\ cm$	fine polish	62,9	30
		rapid polish	63,7	26
		16 h in water	64,4	28
this method	difference $d_1 = 32\ cm$ $d_2 = 1\ cm$	rapid polish		28
		16 h in water		29
	$d = 32\ cm$ reference: air	rapid polish surf. scatter $\bar{s} = 1,5 \cdot 10^{-4}$ $\triangleq$		34 ± 2 / $4 \pm 0,5$
				$30 \pm 2,5$
		16 h in water surf. scatter $\bar{s} = 7,3 \cdot 10^{-4}$ $\triangleq$		49 ± 3 / 21 ± 2
				28 ± 5

Fig. 8   Comparison of results from conventional and present methods.

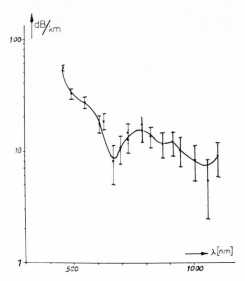

Fig. 9   Spectral distribution of the total losses in Suprasil W 1.

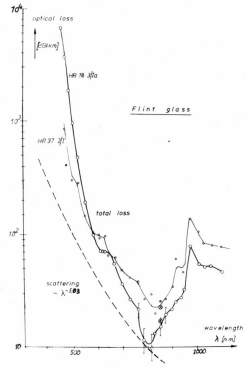

Fig. 10   Spectral distribution of the total losses in flint glass.

beam. Moreover a term is introduced which requires a
minimum reflectivity of the laser mirror. For the in-
tended sensitivity of 10^{-5} we require $r_s \geq 99^o/o$, what
can easily be fulfilled.

To prevent the disturbing terms from getting fully
effective there are two alternative ways. We can measure differences between two samples of different length,
or measure the scattering loss and inhomogeneity independently which turned out to be the most important quantities. A typical example underlines that this method
has some advantages over a conventional transmission measurement concerning stability and perfection of the elements and therefore saves time for preparation and measurement.

4. RESULTS

Some results of systematic investigations are given
in Fig. 8. The agreement of the results from different
methods is satisfactory and shows that the leached surface of the multi-component glass has mainly scattering
and not absorption loss (which would not influence the
above given analysis). This can be checked by testing
Lambert's law by plotting the measured loss over $1/d$.
We expect and find a straight line giving the bulk loss
as intersection with the ordinate and the surface loss
as slope. An independent scattering measurement corrects
for surface scattering. Within the error limits we find
consistent results from these and differential measurements.

The difficulties coming from the surface scatter of
flint glasses are somewhat reduced in silica, borosilicate or soda lime glasses. Nevertheless, it should be
emphasized that the problems are generally caused by the
preparation of the samples and not by the instrument.

Fig. 9 and 10 show the spectral distribution of the
total losses of Suprasil W 1 and lead alkali flint glass
specimens, respectively. The loss minimum of 10 dB/km
near 800 nm in flint glass shows that mainly scattering
is responsible in this region. This is the best melt we
achieved, but today the glasses can reproducibly be melt-
ed with a loss of approximately 20 dB/km in volumes of
about 5 litres.

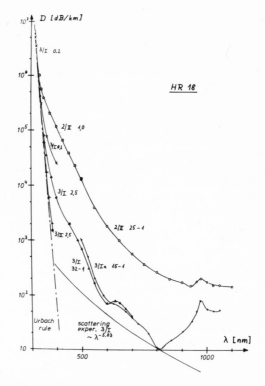

Fig. 11 Variation of the loss distribution due to the refinement of the glass.

Fig. 11 shows the variation of the loss distribution due to the refinement of the glass. Keeping the melt for long times at high temperatures generally reduces the loss. While first a deviation from the exponential edge is observed, the loss of the better refined glasses tends asymptotically to an Urbach rule behavior. Only at very low losses light scattering is another limiting intrinsic loss mechanism. In the long wavelength region the losses increase again (Fig. 10 and Fig. 11). One overtone of the OH stretching vibration can be observed at 975 nm.

REFERENCES

1.   G. Koppelmann, K. Krebs, Z. Phys., _156_, 38 (1959).

SURFACE STUDIES WITH ACOUSTIC PROBE TECHNIQUES*

D. A. Rockwell, T. S. Colbert, J. H. Parks

Departments of Physics and Electrical Engineering
University of Southern California
Los Angeles, California 90007

A new measurement technique is discussed which utilizes acoustic surface waves to detect the surface
depth dependence of radiative absorption. We present
a theoretical analysis of the detection process which
relates these measurements to a quantitative model of
surface absorption phenomena. This technique is generally applicable to study weak surface absorption at
wavelengths for which a material is essentially transparent. Surface wave phase variations induced by the
absorption of $10.6\mu m$ radiation have been measured on
KCl surfaces. The surface absorption σ of these samples was also measured by calorimetry to be $\sigma \sim 2 \times 10^{-4}$.
A radiative pulse of 100 msec and 5 joules incident en-
ergy has been observed to induce an 80 mrad phase
change. Shorter radiative pulses (~ 10 msec) are necessary to probe the temperature depth gradient arising
from a depth variation of the absorption coefficient.
Initial results of recent studies to measure the depth
dependence of surface radiative absorption on etched
alkali halide surfaces are presented.

*Research supported by Advanced Research Projects
Agency

1. INTRODUCTION

Infrared absorption measurement techniques have been developed [1] which avoid the effects of the surface contribution. However, the surface absorption represents a practical limitation of high power laser optical materials and an understanding of the details is needed. The principle effort of our research program has been to develop a new measurement process [2] which utilizes acoustic surface waves to detect the surface depth dependence of radiative absorption. In addition a detailed theoretical analysis of this detection process was derived to relate these measurements to a quantitative model of surface absorption phenomena.

These new experimental techniques and the related theoretical analysis involved advances beyond the current state of acoustic surface wave technology. For this reason, considerable research was necessary to develop a detailed understanding of surface wave propagation in the presence of a nonuniform temperature distribution. Reliable and well studied acoustic materials were used for this purpose. During this effort, infrared window materials were involved primarily to demonstrate applicability of these techniques rather than as controlled samples for analysis. The principle accomplishment of this program has been the successful development of the surface wave measurement process and its associated theoretical description as reliable research tools. As a result, specific studies have recently been initiated to measure the depth dependence of surface radiative absorption on etched alkali halide surfaces.

2. SURFACE WAVE DETECTION OF RADIATIVE ABSORPTION

A fundamental understanding of the physics involved in the surface wave detection process was obtained in quantitative studies on α-quartz presented in Ref. [2]. It was found that surface wave propagation in a region heated by radiative absorption could be completely described by a laser induced transient phase change given by

$$\Delta\phi(t) = \frac{4\pi\nu}{v_{so}} f_B \int_0^L [\epsilon_r(t) - \alpha_v \Delta T(t)] dr. \qquad (1)$$

The phase change arises from a change in propagation path length via thermoelastic strains, $\epsilon_r(t)$, and also a change in wave velocity through the temperature variation of the elastic coefficients represented by $\alpha_v \Delta T(t)$. In Eq. (1), ν is the frequency and v_{so} is the velocity of the surface wave, $2L$ is the acoustic path length, and r is the distance along the propagation path. The laser induced temperature change $\Delta T(t)$ is found by solving the heat equation, and has the spatial dependence of the gaussian laser profile.

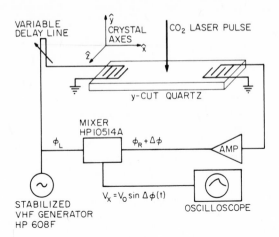

Fig. 1 Experimental apparatus for measuring laser
        induced phase changes.

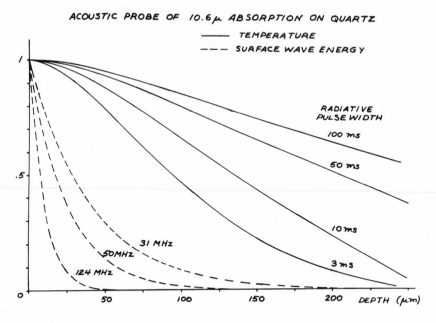

Fig. 2 Temperature depth profiles for various laser pulsewidths
        compared with surface wave energy profiles for
        several frequencies.

The factor f_B accounts for the incomplete overlap of the acoustic beamwidth and the temperature profile. For the experimental geometry used, f_B was calculated to be $\simeq .9$. Equation (1) is valid when the temperature distribution is essentially uniform over the surface wave penetration depth, or $\partial T/\partial z << T/\lambda_s$. Here λ_s is the acoustic wavelength and z is a distance normal to the surface.

Experimental measurements of $\Delta \phi$ on α-quartz were obtained as shown in Fig. 1, and then compared with theoretical calculations of Eq. (1) using no adjustable parameters. The quantitative agreement between theory and experiment was within 15%. In these experiments, the radiative pulse widths were long enough for heat diffusion to produce a uniform temperature within the surface wave penetration depth, ($\sim 15\mu m$ at 124.5 MHz). In this way, the need to account for propagation in the presence of an axial temperature gradient was avoided, and the resulting analysis provided a clear picture of the fundamental surface wave-thermal interaction leading to the induced phase changes.

3. THE EFFECT OF A TEMPERATURE GRADIENT

Consider in detail the depth dependence of the temperature for various radiative pulsewidths. Figure 2 shows that for lower acoustic frequencies it is not valid to assume that the temperature is uniform over the surface wave penetration depth. This spatial inhomogeniety results in dispersive surface wave propagation. Given a valid dispersion theory, measurements of the frequency dependence of the induced phase change yield information about the temperature distribution and thus the surface absorption properties.

A theory has been developed to calculate the velocity variation of the surface wave in the presence of an arbitrary temperature gradient [3]. The calculation is based on a perturbative solution of the wave equation for the case in which the elastic coefficients exhibit spatial variation through their known temperature dependence. We have solved this problem for the case of propagation in an isotropic material and are presently extending the theory to describe anisotropic materials, in particular those with cubic symmetry. The perturbation theory approach allows a precise model of surface absorption to be easily incorporated into the analysis. This calculation results in an explicit expression for the velocity variation, Δv_s, in terms of the parameters of the model and the acoustic wavelength λ_s. The term $\alpha_v \Delta T$ in Eq. (1) is then replaced by $\Delta v_s/v_{sO}$. To emphasize the dispersive character of $\Delta \phi$, the trivial frequency dependence is eliminated by defining a reduced phase $\Delta \phi_R = \Delta \phi/\nu$. The dispersive behavior predicted by this theory for fused quartz is shown in Fig. 3. The exceptionally high bulk absorption coefficient (241 cm^{-1}) in quartz

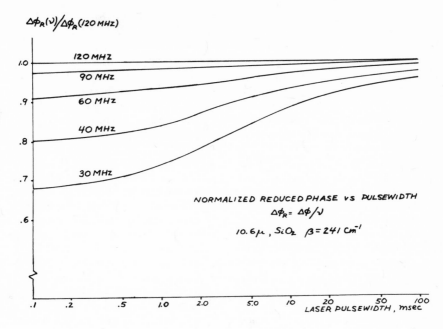

Fig. 3 Phase change dispersion predicted by the isotropic theory.

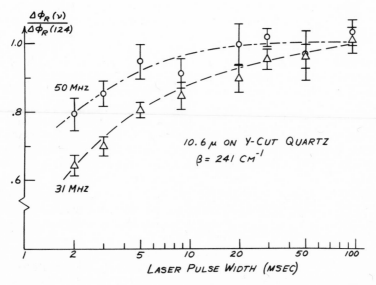

Fig. 4 Dispersive phase change measurements on Y-cut quartz.

is responsible for the significant temperature gradient over the
acoustic penetration depth.

Experimental measurements of the dispersive phase change
have been made on Y-cut α-quartz and are shown in Fig. 4. Although quantitative comparison with theory awaits a complete
anisotropic treatment, these data show a qualitative agreement
with predictions of the isotropic calculations.

4. KCl SURFACE ABSORPTION

In KCl the bulk absorption of good samples is known [4] to be
as low as $7 \times 10^{-5} \text{cm}^{-1}$. However, radiative absorption occurring
in a thin surface layer is significantly greater, and leads to
appreciable temperature gradients within the acoustic penetration
depth. The simplest absorption model is one in which the absorption coefficient varies stepwise as a function of depth. We
have studied a "two-β" model, where $\beta=\beta_s$, the surface value, for
depths less than a characteristic length d, and $\beta=\beta_B$, the bulk
value, for depths greater than d. For such a model in which
$\beta_s = 10 \text{cm}^{-1}$, $d = 2 \mu m$, and $\beta_B = 2 \times 10^{-4}$, a 1 msec laser pulse would
induce a temperature profile similar to that shown in Fig. 5.
Also indicated in the figure are the surface wave depth profiles
for four frequencies: 28MHz, 40 MHz, 62MHz, and 124 MHz.
All curves are normalized to their values at the surface. It is
clear that the 28 MHz wave samples a considerable temperature
variation, while the 124 MHz wave samples essentially a uniform
distribution. This will result in dispersive phase changes of the
surface wave on KCl similar to that observed on α-quartz in Fig.4.
Experimental measurements of this dispersion, combined with the
theory discussed above, will yield values of β_s and d for the KCl
sample under study.

The application of the surface wave measurement technique
to KCl requires the ability to propagate waves on the surface of a
non-piezoelectric material. This wave propagation must be
generated by a technique which does not affect the surface properties under analysis. This condition eliminates the deposition of
piezoelectric films and acoustic transducers directly on the
sample surface, a standard technique in current surface wave
technology. An alternative technique, shown in Fig. 6, relies on
the excitation of a wave on a piezoelectric material and the subsequent coupling of this wave through a fluid layer to the nonpiezo-
electric surface. Although this method has previously received
relatively little development, it is of a particular advantage for
our purpose. Our research has included an experimental study
[5] of a fluid coupling technique which has resulted in successfully coupling surface waves to KCl. Phase changes on piezo-

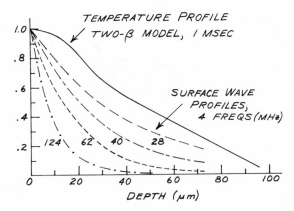

Fig. 5 Temperature depth profile predicted by the two-β model
for KCl compared with surface wave energy profiles.

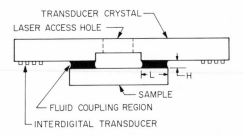

Fig. 6 Schematic of the fluid coupling technique. Fluid interfaces
couple the surface wave from (to) the piezoelectric
transducer crystal to (from) the KCl sample.

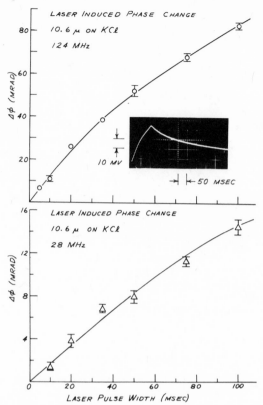

Fig. 7 Laser induced phase change measurements at two acoustic frequencies taken with 50 watt laser pulses. Insert
shows a transient phase change of 80 mrad for a 100 msec
pulse.

electric samples using the fluid coupling technique were compared with the results using direct surface wave excitation by
interdigital transducers. This comparison verified that the
laser induced phase change was reliably and reproducibly
coupled through the fluid regions. An experimental transducer
configuration has been designed which accepts arbitrary sample
sizes and provides ample access to the sample surface for the
infrared laser source.

Initial measurements of induced phase changes on KCl,
taken with 50 watt CO_2 laser pulses at surface wave frequencies
of 124 MHz and 28 MHz, are plotted in Fig. 7. The KCl was an
Optovac sample obtained from S. Allen, Hughes Research Laboratories, which had been polished and then etched for 90 sec in
an H Cl-water solution. The surface absorption σ was measured

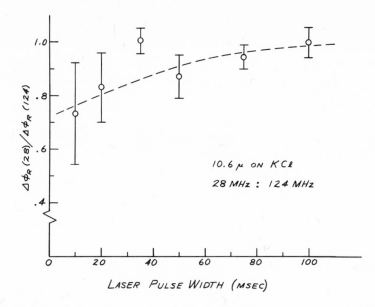

Fig. 8 Dispersive phase change measurements on KCl.

to be approximately 2×10^{-4}. The insert in the figure shows a
typical data trace obtained using a 100 msec pulse at 124 MHz.
A minimum measureable phase change of ~ 1.5 mrad, corresponding to an electrical signal of $\sim .6$ mv, is obtained at 28 MHz
with a 10 msec laser pulse. This limit was set by low frequency
60 cycle electrical noise and not by the high frequency phase
noise.

Reduced phase changes were calculated for each frequency,
and the ratio $\Delta\phi_R$ (28 MHz)$/\Delta\phi_R$(124MHz) was taken at each
pulsewidth. These ratios, shown in Fig. 8 display the characteristic behavior expected for a thin absorbing surface layer.
In future studies, additional data at 40 and 62 MHz will help establish a more complete knowledge of the details of this surface
layer.

The present results indicate the sensitivity of these surface
wave techniques. If it is only required to obtain a relative measurement of surface absorption for different samples, the measurement can be taken at a single frequency and a single pulsewidth. We have found a phase change of 80 mrad at 124 MHz,
for a 100 msec, 50 watt pulse on a surface with $\sigma\sim 2\times10^{-4}$. The
minimum phase change (1.5 mrad) for the same laser pulse
would be observed for a surface having $\sigma\sim 5\times10^{-6}$. This lower

limit could be reduced further by operating at higher laser power or higher acoustic frequency. However, dispersive phase change measurements require data at lower frequencies and shorter pulsewidths. Since the phase change is directly proportional to the frequency, the sensitivity will be reduced if depth dependence of the absorptive properties is of interest. For example, the minimum detectable phase change at 28 MHz was measured for a 10 msec, 50 watt pulse on a surface with $\sigma \sim 2 \times 10^{-4}$. This represents a lower limit at this power for measuring depth dependence parameters. This limit is comparable to the lowest surface absorption obtained on etched samples [4].

5. CONCLUSION

We have shown that an acoustic surface wave technique can be applied to the measurement of radiative absorption at material surfaces. The details of the technique, including a theoretical understanding, have been firmly established. Dispersive measurements at different acoustic wavelengths were shown to be related to the depth dependence of the absorption. This technique has been applied to studies of KCl after a method was developed to couple surface waves to non-piezoelectric samples. Preliminary results on KCl indicate a high sensitivity to the details of low level surface absorption. Current research plans include multi-frequency experiments on KCl, and an analysis of the observed dispersion to obtain the surface absorption coefficient β_s and the characteristic depth of variation d for polished and etched samples.

The authors wish to thank S. Allen, Hughes Research Laboratories, for the KCl samples and calorimetric measurements. They also thank D. White and R. King, China Lake NWC, for transducer fabrication and for the use of an RF power amplifier.

REFERENCES

1. M. Haas, J. W. Davisson, H. B. Rosenstock, and J. Babiskin, Proc. Int. Conf. Opt. Prop. Highly Transparent Sol., paper VI-B (Waterville, N. H. 1975).

2. J. H. Parks, D. A. Rockwell, T. S. Colbert, K. M. Lakin, and D. Mih, Appl. Phys. Lett. 25, 537 (1974).

3. D. A. Rockwell and J. H. Parks (to be published).

4. T. F. Deutsch, Appl. Phys. Lett. 25, 109 (1974).

5. D. A. Rockwell and J. H. Parks, (to be published).

A 10.6 MICRON MODULATED LIGHT ELLIPSOMETER[*]

S.D. Allen, A.I. Braunstein, M. Braunstein,
J.C. Cheng[**] and L.A. Nafie[**]
Hughes Research Laboratories
3011 Malibu Canyon Road
Malibu, California 90265

ABSTRACT

A modulated light ellipsometer capable of measurements from the ultraviolet to the infrared with particular emphasis on 10.6 μm has been developed for measuring
the refractive index and absorption coefficient of bulk
materials and thin films. This paper describes in detail the operating principles of the Hughes Research
Laboratories ellipsometer and the specialized optical
components required to permit operation of the instrument at 10.6 μm: a waveguide CO_2 laser, ZnSe wiregrid
polarizers, ZnSe acousto-optic modulator, and PbSnTe
detector. In contrast to manual ellipsometers, phase-
sensitive amplifiers are employed to process the signal
received at the detector, and the standard ellipsometric
parameters are derived from the signals at the fundamental and second harmonic of the modulator and the DC
signal. An analysis of sources of error and the limits
of sensitivity of this instrument will be presented.

[*]Work partially supported by Defense Advanced Research Projects
Agency under the technical cognizance of Air Force Cambridge
Research Laboratories.

[**]University of Southern California.

INTRODUCTION

The impetus for the development of the polarization modulated light ellipsometer at Hughes Research Laboratories (HRL) was the unavailability of an easy method of measuring the optical constants of materials at 10.6 μm. A knowledge of the absorption coefficient and refractive index, at the laser wavelength of interest, of bulk and thin film materials is necessary for the design of optical components for systems utilizing that laser. Optical constant measurements in the infrared have been made previously by making specialized optical components, e.g., measuring the dispersion of prisms.[1] These methods are time and material consuming and could suffer from unwanted surface finishing effects. The ellipsometer that we have developed is based on one that was designed by Jasperson[2] for operation in the visible. There are two of these in the United States — one at the Naval Weapons Center at China Lake and one at Princeton University. Our design differs slightly in experimental detail from Jasperson's in that our optical alignment is maximized for transparent substrates, while the Jasperson instruments are utilized for reflectors. The unique characteristic of the HRL ellipsometer is its ability to measure ellipsometric parameters at the CO_2 laser wavelengths. Optical components which function in the midinfrared were developed at HRL and in cooperation with the University of Southern California. The end result of this development will be an ellipsometer which can be used from the uv to the midinfrared with simple changes in optical components.

OPTICAL COMPONENTS

The schematic in Fig. 1 illustrates the ellipsometer with visible and near uv components and the polarization vectors of the light after each component. The source is a tungsten iodide lamp with a collimator and appropriate lenses yielding unpolarized light. An optically contacted quartz Rochon prism set at +45° to the plane of incidence gives plane-polarized light at that angle. The photoelastic modulator (PEM)[3] is essentially a dynamic variable quarter wave plate in which the maximum retardation (Fig. 2) can be set to correspond to λ/4 at the wavelength of interest. This component is the crux of the instrument, as the variation in light flux received by the photodetector due to the changes in polarization from the PEM can be Fourier analyzed to yield parameters proportional to the standard ellipsometric parameters. In our optical design, the PEM is set at 0° to the plane of incidence and for visible-uv operation is a commercially available quartz device.* The next components are the sample and adjustable position sample

*Morvue Electronics, Tigard, Oregon.

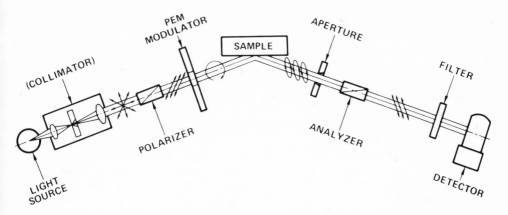

Fig. 1. Schematic of polarization modulated ellipsometer with visible components.

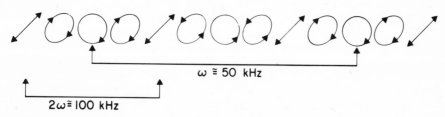

Fig. 2. Polarization vector representation of output of PEM with polarizer at 45° and PEM at 0°.

stage followed by a variable aperture. If the polarization of the light from the PEM is right circular polarized, as illustrated in the diagram, after reflection from a sample it will be elliptically polarized. An analyzer (also an optically contacted quartz Rochon) at -45° to the plane of incidence follows the aperture, yielding linearly polarized light of a constant polarization but varying amplitude as seen by the detector. In the experimental apparatus for visible measurements the detector is a 1P21 photomultiplier tube[*] which is protected from stray light by an interference filter. The basic components for visible operation exclusive of the PEM were purchased from Rudolph Research Corporation, Caldwell, N.J.

---

[*]RCA, Inc., Electronic Components and Devices, Harrison, N.J.

The 10.6 μm optical components were all developed at HRL and in some cases are unique. Because of this uniqueness, they will be described in some detail.

The light source is a waveguide CO_2 laser which is tunable and amplitude stabilized.[4] It is quite compact (Fig. 3), which is convenient for mounting on the optical bench. The output is up to one watt TEM_{00}, but for our purpose a lower intensity (several hundred milliwatts), achieved by lowering the gas pressure, is sufficient.

The polarizers are gold wire gratings on antireflection-coated ZnSe.[5] The spacing of the grid is 1000 lines/mm and the extinction ratios are typically 300, although some of them have extinction ratios as high as 500. One advantage of having ZnSe as a substrate for the polarizers is that they are transparent in the visible, allowing the use of a He-Ne alignment laser. The "wires" are relatively rugged and can be cleaned with standard techniques developed for ZnSe surfaces.[6]

The photoelastic modulator, illustrated in Fig. 4 was developed in cooperation with the co-authors at the University of Southern California. It is a double driven, resonantly operated ZnSe stress optic device which is capable of λ/4 maximum retardation to 12 μm for a 1/4 in. thickness. The characteristics of this device will be discussed in detail in a later section.

The detector for 10 μm operation is a PbSnTe photovoltaic device with a 0.5 mm by 0.5 mm surface area and a ZnSe window.[7] D*'s for these devices are normally approximately 10^9, but this particular example exhibited a $D* \simeq 10^8$. With the amount of power available from the CO_2 laser, extreme sensitivity in the detector was not necessary.

SIGNAL PROCESSING

The equations for the amplitude of the signal detected at the photomultiplier for the optical train shown in the previous figure are given below. (Constant factors common to all equations have been neglected.)

$$I = 1/8(r_\parallel^2 + r_\perp^2) - \frac{r_\parallel r_\perp}{4} \cos(\Delta + \phi) \tag{1}$$

$$= 1/8 [r_\parallel^2 + r_\perp^2 - 2r_\parallel r_\perp J_0 (\phi_m) \cos \Delta]$$

$$+ \frac{r_\parallel r_\perp}{2} J_1(\phi_m) \sin \Delta \sin \omega t - \frac{r_\parallel r_\perp}{2} J_2(\phi_m) \cos \Delta \cos 2\omega t$$

$$+ \ldots , \tag{2}$$

$$= I_{DC} + I_{\omega} + I_{2\omega} \tag{3}$$

$r_{\|}, r_{\perp}$ = amplitudes of reflection coefficients for $\|$ and $\perp$ polarization

Δ = phase difference between $\|$ and $\perp$ polarizations on reflection

ϕ (retardation angle) = $\phi_m \sin \omega t$

ω = frequency of modulator.

The expression $\cos(\sin \omega t)$ in eq. (1), when Fourier analyzed, produces a series with Bessel functions as coefficients. Expanding and combining terms yields eq. (2) which can be broken down into a constant or DC term, I_{DC}, a term at the fundamental frequency of the modulator, I_{ω}, a term at the second harmonic of the modulator frequency, $I_{2\omega}$, and higher order terms which contain no new information and are not generally measured. The extraction of the ellipsometric parameters from eq. (2) is shown in eqs. (4) through (6). For $J_0(2.405) = 0$,

$$\frac{I_{2\omega}}{I_{DC}} = -4J_2(2.405) \frac{r_{\|}/r_{\perp}}{1 + (r_{\|}/r_{\perp})^2} \cos \Delta \cos 2\omega t \tag{4}$$

$$\frac{I_{\omega}}{I_{DC}} = 4J_1(2.405) \frac{r_{\|}/r_{\perp}}{1 + (r_{\|}/r_{\perp})^2} \sin \Delta \sin \omega t \tag{5}$$

$$\frac{I_{\omega}}{I_{2\omega}} = -\frac{J_1(2.405) \sin \omega t}{J_2(2.405) \cos 2\omega t} \tan \Delta \tag{6}$$

$\tan \psi = r_{\|}/r_{\perp}$.

At a retardation amplitude chosen so that $J_0 = 0$, the ratio of the signals at ω, and 2ω (eq. (6)) yields a constant times $\tan (\Delta)$. Knowing Δ, the second ellipsometric parameter, ψ, can be calculated from either eq. (4) or (5). If the absorption coefficient is small,

Fig. 3.
Waveguide CO_2 laser.

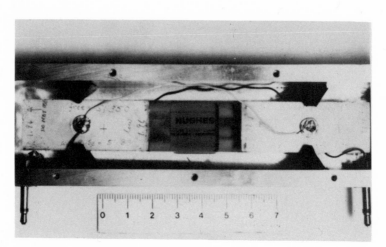

Fig. 4.   Zinc selenide double-driven photoelastic modulator.

i.e., Δ is near $0°$ or $180°$, eq. (4) can be used to calculate the real part of the refractive index directly for bulk samples, as $\cos(\Delta) \approx 1$.

Synchronous detection at three different frequencies is used to extract the three signals of interest in eq. (3). Besides the signals at the first and second harmonic of the modulator frequency, it is convenient for stabilization purposes to mechanically chop the CO_2 laser at a low frequency, 43 Hz in our case. This also allows synchronous detection of the DC signal.

An NBS computer program[8] is available for data processing. Computer handling of the raw data is particularly advantageous in cases where absorption must be considered or where thin films are being analyzed. In the latter case, data obtained at different angles are used to provide enough information to allow the unambiguous determination of the optical constants of the thin film/substrate complex.

ZINC SELENIDE PHOTOELASTIC MODULATOR

The calibration of the ZnSe photoelastic modulator involves determining the functional dependence of ϕ_m, the amplitude of the phase, on the applied voltage.

$$\phi_m = P \frac{2\pi d}{\lambda} V Q, \tag{7}$$

where

d = thickness of the modulator

V = voltage applied to quartz transducers

Q = quality factor of the whole modulator

λ = wavelength of the measurement

P = proportionality factor containing the stress optic coefficient of ZnSe and the piezoelectric constant of quartz.

It has been shown[9] that for the drive voltages of interest the Q of the system remains constant with changes in V, so that $\phi_m = P'V$, where P' is a new proportionality constant containing all those factors in eq. (7) which are constant for a particular modulator. For a fixed wavelength, P' can be determined by plotting I_{DC} or $I_{2\omega}$ versus applied voltage with no sample. From eq. (2), $I_{DC} = 1/4(1 - J_0(\phi_m))$ with no sample. At short wavelengths (He-Ne laser at 0.6328 µm) the condition $J_0(\phi_m) = 0$ is reached several times before the maximum attainable V. Calibration is thus accomplished relatively easily by curve fitting. For 10.6 µm, however, ϕ_m is relatively small even at the maximum voltage, and direct curve fitting is very difficult for either I_{DC} or $I_{2\omega}$. For the optical system shown in Fig. 1, $I_{DC} = 1/4(1 - J_0(\phi_m))$. If the polarizer is set at -45° or parallel to the analyzer, $I'_{DC} = 1/4(1 + J_0(\phi_m))$. We can then measure the quantity

$$A(V) = \frac{I_{DC}}{I'_{DC}} = \frac{1 - J_0(\phi_m)}{1 + J_0(\phi_m)}$$

and solve for $J_0(\phi_m)$. The 10.6 µm calibration curve for the double driven ZnSe modulator is given in Fig. 5.

The maximum applied voltage is determined by the elastic limit of the modulator material, and for ZnSe the stress necessary to obtain λ/4 modulation at 10.6 µm falls just slightly below the range of breaking stresses for this material. In one case we succeeded in breaking a modulator bar at a calculated maximum stress of

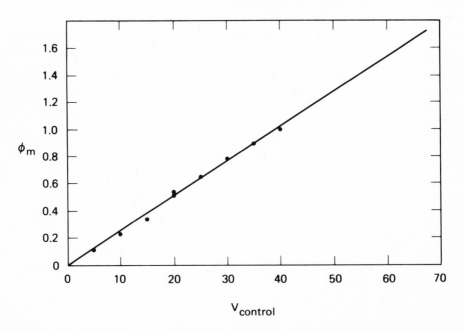

Fig. 5.   Calibration curve for double driven ZnSe modulator.
Control is a DC proportional to the peak-to-peak
voltage applied to the quartz transducers.

4400 psi which is a low but reasonable value of the breaking stress.*
This particular bar exhibited some residual strain which could explain the relatively low value.

If the ZnSe modulator bar is not antireflection coated, its
use and calibration become very difficult. For 10.6 µm the faces
of the modulator bar are parallel enough to act as an etalon; i.e.,
for the coherent radiation of the laser, the modulator can be a
resonant cavity if the thickness is an integral number of half
wavelengths. Experimentally, the modulator draws a small amount
of power and as it heats up and expands, it tunes in and out of
resonance. Antireflection coating the ZnSe serves as a viable alternative to temperature controlling the environment of the PEM.

PERFORMANCE AND MEASUREMENTS

The performance of the modulated ellipsometers can be compared with the best of the manual ellipsometers on the basis of
inherent sensitivity and the advantages of PEM ellipsometry —
speed, precision, easy computer interfacing and, for the HRL ellipsometer, the wide wavelength range — are readily apparent.
There are, however, a number of systematic errors which must be
eliminated before the accuracy of the optical constants obtained
by PEM ellipsometry can compare with those obtained by well-estab-
lished methods. The sources of these errors (some of which are
shared with conventional ellipsometers) include component imperfections, misalignment of components, and stability of detector, PEM,
light source, and electronics. In this section, we will discuss
imperfections in the polarizer and the PEM which have been found
to contribute most significantly to the systematic errors. If the
polarizer is rotated through an angle θ_p, the error signals at the
different frequencies I_ω^{err}, $I_{2\omega}^{err}$, and I_{DC}^{err} will evidence different angular dependences, depending on which component is responsible
for the error signal. In this way, the specific source of various
error signals can be pinpointed as shown in Table 1.

The effect of the finite extinction ratio of the polarizers
can be corrected for explicitly in eq. (2), ignoring any birefringence effects. The dominant error seems to be inherent strain
birefringence in the modulator as shown by the angular dependence
of the error signal at the first harmonic of the modulator.
Further work is in progress to verify this result and to attempt
to minimize the error and/or develop correction factors.

Measurements taken in the visible on quartz and sapphire yield
refractive indices which are comparable with those in the literature.

---

*A range from 4000 to 9000 psi has been reported.

Table 1. Error Signals for Optical System of Fig. 1 with no Sample
as a Function of Polarizer Angle θ_p with IR components and a
Light Source Polarized $\perp$ to the Plane of Incidence

Errors	Signals Caused by Polarizer (First Order)	Signals Caused by PE Modulator (First Order)
No errors	$I_{DC} = 1/4\ (1 - J_o(\phi_m))\ \sin 2\theta_p\ \sin^2\theta_p$ $I_\omega = 0$ $I_{2\omega} = 1/2\ J_2(\phi_m)\ \sin 2\theta_p\ \sin^2\theta_p$	
Angular alignment	No errors for $\theta_p = \pm 45$ $\theta_{mod} = 0^o$ $\theta_a = \mp 45^o$	$I_{2\omega}^{err} \propto \sin^2\theta_p\ \cos 2\theta_p$ $I_{DC}^{err} \propto \sin^2\theta_p\ \cos 2\theta_p$ $I_\omega^{err} = 0$
Birefring-ence	$I_{DC}^{err} \propto \sin 4\theta_p$ $I_\omega^{err} \propto \sin^2\theta_p$ $I_{2\omega}^{err} \propto \sin 4\theta_p$	$I_\omega^{err} \propto \sin^2\theta_p\ \sin 2\theta_p$ $I_{2\omega}^{err} \propto \sin^2\theta_p\ \sin 2\theta_p$ $I_{DC}^{err} = 0$

The precision of these measurements was dependent on the rather
elementary electronics used in the initial measurements and has
been improved considerably. At 10.6 µm the refractive index of
ZnSe was measured as a function of angle of incidence. For angles
below the Brewster angle (67.45^o) values have been measured which
agree within a few percent of published values.[10] It is obvious
from the angle dependence of the data, however, that systematic
errors are present which must be eliminated to obtain greater
accuracy.

As a corollary to the calibration of the ZnSe PEM at two different wavelengths (0.6328 and 10.6 μm) the relative stress optic coefficient is determined. Using NBS data[10] for the stress optic coefficient at 0.6328 μm ($C_\lambda^{0.6328}$ = -13.7 Brewsters), we have calculated ($C_\lambda^{10.6}$ = -14.8 Brewsters. Because of a lapse of several weeks between the measurements at the two wavelengths, this value of the stress optic coefficient should be taken as approximate only. With careful measurements, however, this experiment should provide a relatively easy method of determining the stress optic coefficient.

Future plans for improving the performance of the modulated 10.6 μm ellipsometer include replacement of the 1/4 in. thick ZnSe modulator with one 1/2 in. thick. The thicker bar should enable us to reach λ/4 (ϕ_m = π/2) with ease and probably λ/2 (ϕ_m = π). Interpretation of the data becomes much easier if the J_0 (2.405) = 0 condition can be reached. We also plan to replace the initial polarizer with a polarized laser at the appropriate angle, eliminating any possible birefringence effects of a polarizer. The stability of the modulator with time is being investigated and a feedback system to ensure constant retardation has been designed. In addition, a new amplitude stabilization procedure for the laser designed to keep the I_{DC} level constant will be tried.

Measurements of interest in the near future at 10.6 μm include the optical constants of KCl and ZnSe as a function of surface finish. As discussed elsewhere,[11] there is a difference in the absorption coefficient of both substrates, depending on the surface finish, i.e., etched surfaces versus polished surfaces. We also plan to measure the optical constants of thorium fluoride, both in bulk and as thin films on KCl and ZnSe.

ACKNOWLEDGMENTS

The authors wish to acknowledge the enthusiastic support of Dr. Harold Posen of AFCRL and Dr. C. Martin Stickley of DARPA. Our sincere thanks also to Dr. James Pappis and Dr. Perry Miles of Raytheon Research Division for their generosity and cooperation in supplying ZnSe for use in the PEM modulators.

REFERENCES

1.   A. Feldman, I. Malitson, D. Horowitz, R.M. Waxler, and M. Dodge, Laser Induced Damage in Optical Materials, NBS Special Publication (1974).

2.   S.N. Jasperson, D.K. Burge, and R.C. O'Handley, Surface Science 37, 548-558 (1973); S.N. Jasperson and S. Schnatterly, Rev. Sci. Instrum. 40, 761 (1969).

3.   J.C. Kemp, J. Opt. Soc. Am. 59, 950 (1969).

4.   R.L. Abrams and W.B. Bridges, IEEE J. Quantum Electron. QE9, 940 (1973); R.L. Abrams, Appl. Phys. Lett. 25, 364 (1974).

5.   H.L. Garvin, J.E. Kiefer, and S. Somekh, "Wire-Grid Polarizers for 10.6 μm Radiation," Proc. of the 1973 IEEE/OSA Conf. on Laser Engineering and Applications, Washington, D.C., May 30-June 1, 1973, p. 100.

6.   M. Braunstein and J.E. Rudisill, Final Report, AFML, Contract F33615-73-C-5044, Wright-Patterson AFB, Ohio (1974).

7.   L.H. DeVaux and H. Kimura, Final Technical Report, Contract DAAK02-72-C-0348, Night Vision Laboratory, U.S. Army Electronics Command, Fort Belvoir, Virginia, December 1973.

8.   F.L. McCrackin, NBS Technical Note No. 479 (1969).

9.   J.C. Cheng, L.A. Nafie, S.D. Allen, and A.I. Braunstein, "Infrared Photo-Elastic Modulators," submitted to Appl. Opt.

10. C.A. Feldman, I.H. Malitson, D. Horowitz, R.M. Waxler, and M.J. Dodge, Proc. Fourth Annual Conf. on Infrared Laser Window Materials, Tuscon, Arizona (1974). Published in the Proceedings.

11. S.D. Allen, M. Braunstein, C. Giuliano, and V. Wang, "Laser Induced Damage in Optical Materials," NBS Special Publication No. 414 (1974); M. Braunstein, Proc. of the Third Laser Window Conf., Hyannis, Mass. (1973).

MEASUREMENTS OF STRESS-OPTIC COEFFICIENTS IN THE TRANSPARENT REGIME OF SOLIDS[*]

C. A. Pitha and J. D. Friedman

Air Force Cambridge Res. Labs.

Hanscom AFB, Mass, 01731

ABSTRACT

The lack of data on infrared stress-optic coefficients is due primarily to experimental difficulties in adapting existing methods, which were devised for measurements in the visible. This report describes a method readily adaptable to laser frequencies in both the visible and infrared, and sensitive enough for measurement of materials of low elastic limit, such as the alkali halides. We treat the specimen as a linear retarder and determine the phase change between two linearly polarized rays oriented parellel and perpendicular to the applied stress direction, from which the birefringence, $q_{11}-q_{12}$, and the constant q_{44} may be calculated. We report the results of measurements at $0.63\mu m$ and $10.6\mu m$ on a variety of solids, including alkali halides, alkaline earth fluorides, and chalcogenides.

---

[*]Text not available.

MEASURING PHOTOELASTIC AND ELASTIC CONSTANTS OF TRANSPARENT

MATERIALS BY APPLICATION OF STATIC STRESS*

A. FELDMAN, R. M. WAXLER AND D. HOROWITZ

NATIONAL BUREAU OF STANDARDS

WASHINGTON, D. C.   20234

ABSTRACT

The shift of Twyman-Green and Fizeau fringes as a function of
applied uniaxial and hydrostatic stress have been measured on transparent solids. These data permit us to calculate all the photoelastic and elastic constants of a material. At the wavelength
10.6 µm, where fringe shifts are small, we have measured photoelastic constants using a modified Twyman-Green interferometer,
which is capable of detecting fringe shifts $\sim$0.01 λ by electronic
means. Data on polycrystalline ZnSe grown by chemical vapor deposition are presented.

INTRODUCTION

Photoelastic and elastic constants are parameters that enter
the theories of Brillouin scattering [1], Raman scattering [2],
electrostriction [3] and electrostrictive self-focusing [4,5].
Measured values of these constants provide data to test
theories [6-8] that predict the photoelastic [9-12] and elastic [13]
properties of transparent materials. In addition, the data are
needed to evaluate the performance of acousto-optic modulators [14]
and to calculate the effects of clamping stresses and thermally
induced stresses on the optical transmission quality of optical
components [15].

*This work was supported in part by the Advanced Research Projects
Agency of the Department of Defense.

In this paper we present techniques for measuring the stress-optical constants of transparent materials using Twyman-Green and Fizeau interferometry. From the shift of fringes as a function of applied uniaxial and hydrostatic compression, the necessary and sufficient data for determining all stress-optical and elastic constants are obtained. In certain cases, a material either possesses small stress-optical constants or cannot sustain large stress without plastic deformation; consequently, the shifts obtained by the above methods are less than one fringe. Because this problem is particularly troublesome in the infrared, a modified Twyman-Green interferometer has been developed which is capable of measuring fringe shift ∿0.01 λ at 10.6 μm. The method, however, requires a knowledge of the elastic constants for calculating the stress-optical constants.

In this study, we report on measurements of the stress-optical constants and elastic constants of polycrystalline ZnSe grown by the chemical vapor deposition method (CVD ZnSe). The data suggest that the photoelastic and elastic constants are altered if we exceed the elastic limit of the material, but that over an undetermined period of time the constants return to their original values. Because the material is polycrystalline it was treated as being isotropic. This assumption is not strictly correct, since there is evidence that the crystallites in the material are preferentially oriented [16].

EXPERIMENTAL

The stress-optical constants, q, and elasto-optical constants, p, of a material are defined by the relationships

$$\Delta(\kappa^{-1})_{ij} = q_{ijkl}P_{kl} = p_{ijkl}\,\varepsilon_{kl} \tag{1}$$

where $(\kappa^{-1})_{ij}$ is the reciprocal dielectric tensor, P_{ij} is the stress tensor and ε_{ij} is the strain tensor. Our objective is to measure either all the q coefficients or all the p coefficients where q and p are related to each other via the elastic constants by $p = qc$ or $q = pS$, where c is the elastic stiffness and S is the elastic compliance. In cubic materials of crystal classes $\bar{4}3m$, 432, m3m, there are three independent q (or p) components which in contracted notation are q_{11}, q_{12} and q_{44}. Cubic crystals are the most complicated materials considered throughout this study.

Several methods have been employed to measure photoelastic constants. In recent years, the elasto-optical constants have been measured successfully by acousto-optic scattering [17,18]; however, this method depends on the possession of a reference specimen whose photoelastic constants are already known. Also, the stress-optical constants have been obtained by directly measuring the refractive indices of prismatic specimens as a function of uniaxial compression [19], but the data obtained with this method have large errors.

Many measurements of stress-induced birefringence have been made to obtain the relative stress-optical constants. These are straight-forward measurements and may even be made on specimens of poor optical quality. Unfortunately stress-induced birefringence, which depends on $q_{11}-q_{12}$ and q_{44}, does not determine the absolute stress-optical constants, q_{11} and q_{12} individually. In this study the measurement of stress-induced birefringence has been combined with interferometric technique [20] in order to measure all the photoelastic constants and the elastic constants as well. This approach makes use of specimens in the form of rectangular prisms subjected to uniaxial and hydrostatic stress.

Twyman-Green Measurements

Standard Twyman-Green interferometers were assembled from optical components with lasers operating at 0.6328 μm, 1.15 μm, and 10.6 μm as the radiation sources. The specimen, which must have sufficient optical quality so that fringes are observed at the interferometer output, is mounted in a stressing frame located in one arm of the interferometer. The fringes in the visible and near infrared are typically viewed with a Si matrix tube and TV monitor, while at 10.6 μm the fringes are observed with either a liquid crystal sheet or a thermal image plate. The requirements for optical quality at 10.6 μm are less than in the near infrared and the visible.

The shift of fringes as a function of applied uniaxial stress is measured in the visible and near infrared with photomultiplier tubes and at 10.6 μm with a pyroelectric detector. The fringe shift per unit applied stress for stresses parallel and perpendicular to the polarization of the radiation are respectively

$$\frac{\Delta N_1}{\Delta P} = \frac{2t}{\lambda}\left[\frac{n^3}{2}\, q'_{11} - (n-1)\, s'_{12}\right] \tag{2}$$

$$\frac{\Delta N_2}{\Delta P} = \frac{2t}{\lambda}\left[\frac{n^3}{2}\, q'_{12} - (n-1)\, s'_{12}\right] \tag{3}$$

where n is the refractive index at the radiation wavelength λ, and t is the specimen thickness. Compressive stress is positive in the convention used. The primed quantities are linear combinations of the standard quantities. In isotropic materials or cubic material stressed along [111], the primes can be eliminated.

Fizeau Measurements

The sample itself can function as a Fizeau interferometer if the faces are polished sufficiently parallel. Fringes are then obtained from the interference between reflections from the front and back surfaces and the shift of these fringes is measured in the same manner as the shift of the Twyman-Green fringes. The fringe shifts per unit applied stress for stresses parallel and perpendicular to the polarization of the radiation are, respectively,

$$\frac{dN_1'}{dP} = \frac{2t}{\lambda} \left(\frac{n^3}{2} q_{11}' - n\, s_{12}' \right);$$

(4)

$$\frac{dN_2'}{dP} = \frac{2t}{\lambda} \left(\frac{n^3}{2} q_{12}' - n\, s_{12}' \right).$$

(5)

Birefringence Measurements

Additional experimental data are obtained by measuring the stress-induced birefringence. In this experiment the axis of the applied stress is at 45° with respect to plane of the incident polarization. The output beam then passes through an analyzer set at 90° with respect to the incident polarization and is photometrically detected. The fringe count per unit applied stress is

$$\frac{\Delta N_B}{\Delta P} = \frac{m}{2} \frac{n^3}{\lambda} qt \begin{cases} q = q_{11} - q_{12} & \text{for [110] stress} \\ q = q_{44} & \text{for [111] stress} \end{cases}$$

(6)

where m is the number of passes the radiation makes through the specimen. Increased sensitivity is obtained by allowing multiple passes.

Hydrostatic Stress Measurements

By performing the above measurements on a sufficient number of specimens of the proper crystallographic orientations, it is possible to calculate, using equations (2)-(6), all the stress-optic constants and all but one of the elastic constants, s_{11}. This constant can be obtained if, the shift of Fizeau fringes as a function of hydrostatic pressure is also measured. The fringe shift per unit applied stress is

$$\frac{\Delta N_H}{\Delta P} = \frac{2t}{\lambda} \left[\frac{n^3}{2} (q_{11} + 2\, q_{12}) - n(s_{11} + 2s_{12}) \right]$$

(7)

Modified Twyman-Green Interferometer

In certain cases we are unable to produce a minimum shift of one
fringe using the procedures of the earlier sections either because
the material cannot withstand the stresses necessary or because the
photoelastic constants are small so that sufficient stress cannot
be applied. This is a problem especially in the infrared at 10.6 μm.
For this case a modified Twyman-Green interferometer has been constructed that is capable of measuring ∿0.01 fringe shift at 10.6 μm.
A schematic diagram of the interferometer is shown in Fig. 1. The
two arms of the interferometer are in close proximity in order to
minimize instabilities due to air currents and vibrations. The
effects of vibrations are also minimized by mounting the diagonal
mirror on the same base as the germanium beam splitter and by
mounting the two end mirrors on a common base. The specimen arm
end mirror, which is mounted on a piezoelectric translator, undergoes a sinusoidal translation along the axis of the interferometer
thus modulating the output intensity of the interferometer according

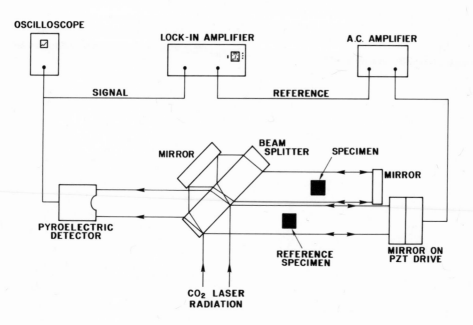

Figure 1.   Modified Twyman-Green interferometer for measuring
            fractional wavelength changes of optical path at 10.6 μm.

to the equation

$$I = \frac{I_o}{2} \left[1 - \cos\left(\frac{4\pi A \sin\omega t}{\lambda} + \Phi\right) \right] \tag{8}$$

where I_o is the intensity at the fringe maximum, A is the maximum excursion of the mirror about its average position and Φ is a phase related to the mean optic path difference of the two arms of the interferometer. The reference specimen is a crystal of Ge in a compression apparatus. It is self calibrating because a shift of many fringes can be produced as a function of the stress in the Ge. Fractional fringes are then obtained by linear interpolation of the data. In operation, the Ge is stressed until the interferometer is at a null, which occurs when the fundamental harmonic of the output intensity is zero. A given stress applied to the unknown specimen will shift the interferometer away from null, but this shift is compensated by applying an incremental stress to the Ge. From these data we then calculate the stress-optical constant using equations (2) or (3) provided the elastic constants are known.

RESULTS

In Table I we list the photoelastic constants obtained at 0.6328 μm and 10.6 μm for CVD ZnSe. Two sets of values are given for 0.6328 μm. The first set [21] was calculated from measurements made on a specimen shortly after it had been stressed beyond the elastic limit. The second set was calculated from measurements on the same specimen 10 months later. The latter set agrees with

Table I.   Photoelastic Properties of CVD ZnSe

λ (μm)	0.6328	10.6
q_{11} $(10^{-12}$ m^2/N)	-1.32^a, $-1.44 \pm .04$	-1.46
q_{12} $(10^{-12}$ m^2/N)	0.28^a, $0.17 \pm .05$	0.51
$q_{11}-q_{12}$ $(10^{-12}$ m^2/N)	-1.60^a, $-1.60 \pm .01$	-1.97
p_{11}	-0.10, -0.13	-0.10
p_{12}	-0.01, -0.04	0.007

[a] See Ref. 21.

average values obtained from three other specimens. A comparison of
the two sets of data suggests that the absolute photoelastic
constants q_{11} and q_{12} are altered when the material is stressed
beyond the elastic limit, but after an undetermined period of time
these constants return to their original values. However, it is
interesting to observe that the stress-induced birefringence
$(q_{11}-q_{12})$ appears to be unaffected by stressing the material beyond
the elastic limit.

The photoelastic data given for 10.6 μm were obtained from
birefringence data, modified Twyman-Green data, and elastic constant
data obtained at 0.6328 μm (first column Table II). A calculation
of dn/dP yields -3×10^{-12} m^2/N, a number which differs significantly from a theoretically calculated value [12]. In fact, there is a
sign difference between theory and experiment.

A summary of the elastic moduli we have measured for CVD ZnSe
are presented in Table II. Also shown are the single crystal data
of Berlincourt et al. [22]. It is possible to compare the single
crystal data to the polycrystalline data by averaging the single
crystal moduli. Two different approximations may be made which
yield extremal values for the averaged moduli, between which the
polycrystalline moduli should fall. The first approximation, due
to Voigt [23], who assumes strain continuity in the specimen, yields
upper limits to the rigidity modulus G and bulk modulus K which we
denote by G_V and K_V. The second approximation, due to Reuss [24],
who assumes stress continuity in the specimen, yields a lower limit
to G and K which we denote G_R and K_R. Values for K_V, G_V, K_R and G_R
are listed in Table II. Our value of G falls between G_V and G_R as
predicted by theory [25], but our value of K is larger than K_V,

Table II. Elastic Properties of ZnSe

	CVD	Single Crystal[a]	Polycrystalline[b]
$s_{11}(10^{-12} m^2/N)$	13.9 $\pm$ 0.6	22.6	
$s_{12}(10^{-12} m^2/N)$	-4.4 $\pm$ 0.2	-8.49	
$s_{44}(10^{-12} m^2/N)$		22.7	
$G(10^{10} N/m^2)$	2.74 $\pm$ 0.12		2.60 (G_R), 3.29 (G_V)
$K(10^{10} N/m^2)$	6.6 $\pm$ 0.6		5.95 $(K_V=K_R)$

[a] Berlincourt et.al. See Ref. 22.

[b] Calculated from single crystal data. See Ref. 25.

where theory predicts that for cubic crystals $K_R = K_V$. Because of the large experimental uncertainty in K, one must be cautious in drawing conclusions from the latter result. However, Martin [13] suggests that the single crystal data may be in error. He found that of all fifteen ZnS structure semiconductors that he had studied, only the ZnSe and CuCl data deviated significantly from his theory of elastic properties.

In obtaining interferometric data on CVD ZnSe it is imperative that temperature fluctuations be minimized. This is because of the large optic path changes that occur as a function of temperature due to a large positive value of the thermal coefficient of refractive index which adds to the large expansion effect.

SUMMARY

The measurement of photoelastic and elastic constants provides input data and tests for theories of the electronic and lattice properties of transparent materials. Methods for measuring the photoelastic and elastic constants of transparent materials by application of static stress were presented. The shift of Twyman-Green and Fizeau interferometer fringes as a function of uniaxial stress and the shift of Fizeau interferometer fringes as a function of hydrostatic pressure were measured. For materials in which the fringe shifts are small, a modified Twyman-Green interferometer is used, but in this case, the elastic constants must be obtained by other methods.

We have presented data for CVD ZnSe and found that dn/dP does not fit the theory. The data, however, must be used with caution because the material has been treated as being isotropic whereas X-ray evidence suggest preferential orientation of the crystallites [16]. Further measurements currently are underway to ascertain whether the preferential orientation of the crystallites produces an anisotropic stress-optical effect and whether the properties of CVD ZnSe vary from batch to batch. In addition, we plan to examine the effects of plastic deformation on the elastic and photoelastic constants.

ACKNOWLEDGEMENTS

We gratefully acknowledge the receiving specimens from Dr. Carl Pitha of the Air Force Cambridge Research Laboratory, Dr. Perry Miles of the Raytheon Corporation and John Fenter of Wright Patterson Air Force Base.

REFERENCES

1. I. L. Fabelinskii, Molecular Scattering of Light, Plenum Press,
    New York (1968).
2. A. A. Maradudin and E. Burstein, Phys. Rev. 164, 1081 (1967).
3. H. Osterberg and J. W. Cookson, Phys. Rev. 51, 1096 (1937).
4. E. L. Kerr, Phys. Rev. A 4, 1195 (1971).
5. A. Feldman, D. Horowitz and R. M. Waxler, IEEE J. Quant. Elec.
    QE-9, 1054 (1973).
6. A. Feldman, Phys. Rev. 150, 748 (1966).
7. A. Gavini and M. Cardona, Phys. Rev. 177, 1351 (1969).
8. D. K. Biegelsen, Phys. Rev. Letters 32, 1196 (1974).
9. J. A. Van Vechten, Phys. Rev. 182, 891 (1969).
10. S. H. Wemple and M. DiDomenico, Jr., Phys. Rev. B 1, 193 (1970).
11. D. L. Camphausen, G. A. N. Connell, and W. Paul, Phys. Rev.
    Letters 26, 184 (1971).
12. Y. F. Tsay, S. S. Mitra and B. Bendow, Phys. Rev. B 10, 1476
    (1974).
13. R. M. Martin, Phys. Rev. B 1, 4005 (1970).
14. R. W. Dixon, J. Appl. Phys. 38, 5149 (1967).
15. B. Bendow, P. D. Gianino, A. Hordvik and L. H. Skolnik, Optics
    Comm. 7, 219 (1973).
16. S. A. Kulin and H. Posin, in Third Conference on High Power
    Infrared Laser Windows, Nov. 1973, ARCRL Special Reports,
    No. 174, p. 463, Ed. by C. A. Pitha, A. Armington, and H. Posin.
17. R. W. Dixon and M. G. Cohen, Appl. Phys. Letters 8, 205 (1966).
18. D. K. Biegelsen, Appl. Phys. Letters 22, 221 (1973).
19. A. A. Giardini, J. Opt. Soc. Amer. 47, 726 (1957).
20. F. Twyman and J. W. Perry, Proc. Phys. Soc. (London) 34, 151
    (1922).
21. A. Feldman, I. Malitson, D. Horowitz, R. M. Waxler and M. Dodge
    in Laser Induced Damage in Optical Materials: 1974, NBS Special
    Publication 414, Ed. by A. J. Glass and A. H. Guenther (U.S.
    GPO SD Catalogue No. C13.10:414) p. 141.
22. D. Berlincourt, H. Jaffe and L. R. Shiozawa, Phys. Rev. 129,
    1009 (1963).
23. W. Voigt, Lehrbuch der Kristallphysik, p. 962 (Leipzig:
    Teubner, 1928).
24. A. Reuss, Z. angew. Math. Mech. 9, 55 (1929).
25. R. W. Hill, Proc. Phys. Soc. (London) A 65, 349 (1952).

LIST OF CONTRIBUTORS

SUBJECT INDEX

Absorption (see also multi-
    phonon absorption,lat-
    tice absorption)
    anharmonic, 59-68
    by electronic transi-
        tions, 245-252
    defects and impurities,
        249-251
    edge, 132, 138, 174,
        246-248, 251, 323,
        353, (see also fun-
        damental edge, ener-
        gy band edge, elec-
        tronic absorption
        edge).
    extrinsic mechanisms,
        110-118
    free carrier absorp-
        tion, 251-252
    frequency dependence,
        3, 4, 48, 72, 80, 81,
        88, 109-118, 265,
        271, 277, 292, 414
    high frequency spectra,
        45-58
    induced, 323-324
    intrinsic, 246-249
    main peak, 290
    materials, 413-414
    of insulators, 3
    residual intrinsic,
        114, 221
    temperature dependence
        in IR transmitting
        materials, 413-414
    theory, 3-20, 60-63,
        110-111

transient in pure KCℓ,
    233-240
two photon, 307-314,
    358, 365
U centers, (see U-
    centers).
Absorption Bands
    F, 233-234, 236
    H, 234
    photochromic in SrTiO$_3$,
        231-232
    color center, 234
Absorption Coefficient, 23,
    36-38, 50-58, 90, 92,
    94, 136, 137, 147,
    148, 221-245, 262,
    276-284, 287-296,
    355-361, 406, 416,
    419, 421, 425-426,
    426-427, 435-442,
    504
    experimental arrange-
        ment for measurement
        in IR region, 444-
        446
    in highly transparent
        solids, 451-459
    measurement in IR re-
        gion, 443-449
    of highly transparent
        materials, 435-442
Acoustic Calorimetry, 418-
    422
Acoustoelectric Domain, 135
    probe (see probes)
Alkali Halides, 13, 191-220,
    234-244

529